International Series in Operations Research & Management Science

Volume 329

The book series **International Series in Operations Research and Management Science** encompasses the various areas of operations research and management science. Both theoretical and applied books are included. It describes current advances anywhere in the world that are at the cutting edge of the field. The series is aimed especially at researchers, advanced graduate students, and sophisticated practitioners.

The series features three types of books:

• Advanced expository books that extend and unify our understanding of particular areas.

• Research monographs that make substantial contributions to knowledge.

• Handbooks that define the new state of the art in particular areas. Each handbook will be edited by a leading authority in the area who will organize a team of experts on various aspects of the topic to write individual chapters. A handbook may emphasize expository surveys or completely new advances (either research or applications) or a combination of both.

The series emphasizes the following four areas:

Mathematical Programming: Including linear programming, integer programming, nonlinear programming, interior point methods, game theory, network optimization models, combinatorics, equilibrium programming, complementarity theory, multiobjective optimization, dynamic programming, stochastic programming, complexity theory, etc.

Applied Probability: Including queuing theory, simulation, renewal theory, Brownian motion and diffusion processes, decision analysis, Markov decision processes, reliability theory, forecasting, other stochastic processes motivated by applications, etc.

Production and Operations Management: Including inventory theory, production scheduling, capacity planning, facility location, supply chain management, distribution systems, materials requirements planning, just-in-time systems, flexible manufacturing systems, design of production lines, logistical planning, strategic issues, etc.

Applications of Operations Research and Management Science: Including telecommunications, health care, capital budgeting and finance, economics, marketing, public policy, military operations research, humanitarian relief and disaster mitigation, service operations, transportation systems, etc.

This book series is indexed in Scopus.

José Manuel García Sánchez

Building and Solving Mathematical Programming Models

50 Practical Examples

Springer

José Manuel García Sánchez
Higher Technical School of Engineering
University of Seville
Seville, Spain

ISSN 0884-8289 ISSN 2214-7934 (electronic)
International Series in Operations Research & Management Science
ISBN 978-3-030-97625-5 ISBN 978-3-030-97626-2 (eBook)
https://doi.org/10.1007/978-3-030-97626-2

This Springer imprint is published by the registered company Springer Nature Switzerland AG
The registered company address is: Gewerbestrasse 11, 6330 Cham, Switzerland

To my students

Preface

The book presents the construction and resolution of 50 optimization problems, divided into five levels of difficulty, although assigning a level of difficulty is subjective in some cases. The first levels are dedicated toward problems oriented to learning the model construction methodology. The difficulty increases with the levels, up to the complex optimization environments presented in the last level. Despite this, understanding all the problems, including the last levels, is feasible after knowing the modeling methodology and progressing with analysis of the problems from the beginning.

A wide range of problems are presented, from problems associated with games (Unblock Me, sudokus) to logistical problems, plant distribution, production, operations scheduling, management, and resource allocation. Many of them with an innovative nature, since they had not been modeled until now, and even others present scenarios with new characteristics.

Besides all this, another fundamental characteristic of the book is that all the models are implemented in an optimization library, LINGO, and are solved in order to analyze the correct construction of the model. The book incorporates a simple guide to build and solve models in LINGO, easily and in any format for the input and output data (text files, spreadsheets, or databases). I have preferred to include the LINGO code for the building of each model within the proper section to each problem, instead of in an annex, so that the reader can more easily visualize the way in which each component of the model is implemented in LINGO.

Something very important in the book is the first chapter. It contains a summary of the main characteristics of the modeling methodology. All the references used for the modeling of logical propositions are presented, adding new details and explanations. Obviously, not all of the details of the methodology are included in this chapter. Having access to the book *Modelling in Mathematical Programming: Methodology and Techniques* (Springer, 2021) is an important aid to understand the modeling techniques used.

At the beginning of each chapter, the characteristics of all the problems are presented in a structured way, with the use of tables, according to the modeling methodology. These characteristics are defined throughout Chap. 1. The level of difficulty of the problems has been defined on the basis of these characteristics.

The format for modeling each problem is identical, which makes the models easier to read and understand. For some problems, two versions are presented for modeling. The versions differ in the configuration of the elements or the decision activities. It is not the object of the book to carry out experimental analyses of the different modeling alternatives.

Seville, Spain José Manuel García Sánchez

Contents

Chapter 1
Introduction to Modeling Methodology

1.1 Introduction

The methodology used for the modeling of all the problems presented in the book is described in detail by García (2021). In this chapter, we are only going to present a summary of some relevant aspects of it, noting all the references described in the methodology for the modeling of specifications. This summary will help us to represent the ideas of the methodology in a more schematic way. We will also take advantage of the summary to point out a couple of corrections and reductions on what is presented in the methodology.

However, some methodological aspects are left out of this summary. If a reference is made to any of them in problems, the exact reference of its section in the methodology book will be noted.

The methodology is structured on the basis of its components:

- Elements
- Decision activities
- Calculations
- Specifications
- Objective criterion

1.2 Elements

Elements are individual actors or entities that participate in the system. They are all the inputs the system has, the resources it uses, and the outputs it produces. They can be of any nature: materials, tools, personnel, places, objects, etc. Also considered as elements are the typologies that could be defined on other elements and even on resources of space or time.

J. M. García Sánchez, *Building and Solving Mathematical Programming Models*,
International Series in Operations Research & Management Science 329,
https://doi.org/10.1007/978-3-030-97626-2_1

They usually have associated information that we will call attributes or data of the elements, and that has to be numerical information. They participate in the decisions that occur in the system and also support the specifications of the system.

1.2.1 Attributes of an Element

A data or attribute is a specific property of an element. The data must have a numerical value assigned.

Elements can also have properties whose value is not determined because it has to be defined in the optimization process.

Data are used in specifications and optimization criteria.

An attribute has two components:

- A definition scheme: The scheme will have the following information:
  A Name
  - A parameter with which it is identified in the compact model
  - The belonging of the data
  - The type of value
- A numerical value.

Belonging of the Data
Regarding the belonging of the data, they can be of two types:

- **Own data**: The data and its value are only properties of an element. The attribute does not refer to any other element.
- **Shared data**: The data relate elements. Its value is obtained from the association between several elements of the system. Logically, the shared data can be assigned to all the elements that share them. The value of the data is not distributed by a function between the elements that share it, but it is fully attributable to each of the elements that share it.

When a shared attribute refers to a value of an intrinsic property of one of the elements, we will say that this element is the primary element in the relationship with the shared data.

Type of Data Value
Every data has a numerical value associated with it. The type of associated value can be:

1. Continuous or integer quantity
   This refers to some continuous or integer magnitude. If it is continuous, we can specify the unit of measure of the value. If it is an integer, it is measured in number of units.
2. Ordinal

The data have an integer value that refers to an ordinal of a list of options reflected or not as elements. The use of an ordinal value requires a simple association between the owner element of the data and the list of elements that the ordinal identifies.

This type of value is always optional in a problem. We can incorporate each option as a binary attribute, and we can even incorporate the options into the list of system elements and convert the ordinal value to a binary value that expresses the link between the elements.

3. Binary

   Binary values are used to define properties of the elements that are answered with a yes or a no, and they are used in attributes that represent links of belonging or compatibility between elements.

   The binary value is used to represent an ordinal value of an attribute in a flexible way, allowing multiple associations to be represented. In this case, it may be convenient to have the options in the list identified as elements.

1.2.2 The Quantitative Nature of the Elements

An element has a typology referred to its quantitative character. Thus, an element can be of two types:

- **Individual element**

  Represents an individual entity whose discrete quantity in the system is the unit, the element itself.
- **Collective element**

  The element represents a concept that has or will have a number of discrete units in the system, whether that amount is determined (known quantity) or undetermined. The element represents a collective entity that is quantified discretely. We will name items for each of those units. The items of a collective element are identical, that is, they have the same functionality and the same data, represented in the collective element.

  We will say that a collective element has a container and a content. The container is the concept, and the content is its items. On very few occasions, the decisions about a collective element are made on all its items together. In that case, we say that decisions are about the container, rather than the content.

Example 1.1
A shirt model of which we have 100 units. It is a collective element, where the container is the shirt model and the content is 100 units.

An individual element can, depending on the decisions configured in the system, be of two types:

Table 1.1 Notation of typologies of elements with respect to the quantitative nature

Type of element	Notation
I. Individual	I
I.1. Unitary individual	I_U
I.2. Measurable individual	I_M
II. Collective	C
II.1 Determined Collective	C_D
II.2 Undetermined Collective	C_I

- Unitary: The element does not have any continuous property that is measured in the decisions of the system.
- Measurable: The element has some attribute or continuous property that will be measured necessarily using continuous variables in the decisions of the problem.

The notation used for the quantitative nature of the elements is shown in Table 1.1.

The most common continuous properties that can appear in a system on an element:

- Decide on its continuous amount (weight).
- Decide on its length.
- Decide on its area.
- Decide on its volume.
- Decide its position in space (one decision for each dimension). Decide its angle of position.
- Decide on its scheduling (without considering finite-time options).
- Decide on its duration.

1.2.3 Grouping of Elements in Sets

Certain elements of a system can be grouped into sets when they have the same functionality in the system, are of the same quantitative nature, and have the same definition of data. Grouping into sets introduces the use of subscripts to reference each element of the set.

The scheme we use to represent the elements in all problems is shown in Tablc 1.2.

1.3 Decision Activities

Decision activities are independent actions of an undetermined value that are carried out in the system. They are translated into the decision variables of the problem.

A decision activity is made up of the following components:

Table 1.2 Framework of table of elements with sets

Elements	Set	QN	Data				
			Name	Parameter	Type	Belonging	Value
Name of set	Subscripts = 1…number of elements	$\{I/I_U/I_M/C/C_D/C_I\}$	Attribute Name	Attribute Parameter	$\{C/I/O/B\}$	$\{W/S\}$	…
			…	…	…	…	…
…							

The action that determines the activity: this corresponds to a verb.
The elements that participate in the activity: elements of the system must participate in the action that corresponds to the decision activity.

Element participation can occur in two ways:

- As a direct object to the action.
- As an indirect object to the action.

The quantitative meaning of the action: the quantification of the action defines the type of value of the variables. The meaning of an action can be of two types:

- **Measure**: The result of the action is a value referring to any continuous measure (liters, kilos, time, etc.) or discrete measure (number of units) of an element. It corresponds to continuous or integer variables, respectively. If the measurement of units is bounded superiorly by one, the integer variable can be defined as binary. It is necessary that a measurable element (measurable individual or collective) participates in a measurement activity as a direct object of the action.
- **Logical**: the action corresponds to an activity of choice or selection for an element or group of elements where the response is evaluated with a logical value, True or False. The activity is determined with binary variables (1/0). The "1" corresponds to a positive or true evaluation and the "0" to a negative. All activities in which only unitary elements participate will be logical.

The factors mentioned above are summarized in the definition of a decision variable. The definition of a variable must contain the elements that participate, the action, and the unit of measurement if the action is quantified as measure.

In a compact model, with grouping of elements in sets, the identification of the decision activities is also carried out collectively or jointly. The decision activity represents a set of events where an event is a valid association of elements participating in the action. Each event corresponds to an individual decision activity with respect to the joint decision activity and will give rise to a decision variable of the problem.

Elements: Decisions Relationship

In a decision activity, an element is always defined as a direct object of the action, and the rest of the participating elements are always defined as indirect objects. Sometimes, in logical decisions, this configuration is not unique, so several elements can be the direct object of the action (Assign Object to Position = Assign Position to Object).

According to this, the relationship between the elements that act as a direct object of the decision, according to its quantitative nature, and the type of decision is Table 1.3.

Table 1.3 Elements: decision variables

	Direct object	Decision activity	Decision variable
Action	Individual element	On a continuous property or continuous attribute	Continuous
		Logical	Binary
	Collective element	On content	Integer
		On container	Binary

1.4 Calculations

The calculations represent the variables that can appear in an optimization problem and whose values can be obtained from the value of other variables, decision variables in general, or even previous calculations. They fall on elements of the system, and, like the activities, they have a type of value associated.

Calculations can sometimes be considered as activities due to their semantics or their importance in the system, but they will never be of primary decision since their value will be obtained from the value of other variables. This determines the most significant aspect of these variables:

They always carry constraints that define their meaning, that is, they define the calculation.

The calculations in a system can be of three types:

- Auxiliary calculations
- Logical calculations
- Lower/upper bounds calculations

1.4.1 Auxiliary Calculations

Its value is calculated directly through a direct linear function on the variables on which it depends. Its use is optional, except when we want to impose the requirement that the result of the function be an integer value. They help to work more comfortably with the restrictions of the system. They give rise to the auxiliary variables of the model.

1.4.2 Logical Calculations

They are formulated by means of a logical proposition. Its use is necessary. The value is binary in its great majority, except for the propositions that are defined to collect the value of a variable when a condition is met, where non-binary calculations are used to be able to express the function in a linear way. The logical proposition that defines the calculation can also be understood as another specification of the system.

Definition
Binary logical calculation:

$$\beta = \begin{cases} 1 & \text{if the condition is satisfied} \\ 0 & \text{otherwise} \end{cases}$$

Logical proposition:

$$\beta = 1 \text{ IF AND ONLY IF Condition}$$

Non-binary logical calculation:
$y =$ value of x when the condition is satisfied
Logical propositions:

$$\text{IF Condition} \quad \text{THEN } y = x$$
$$\text{IF NOT(Condition)} \quad \text{THEN } y = 0$$

Reduction of the Definition of a Logical Calculation
As the calculations have been defined, your modeling will always be correct. However, in many models, but not in all, it is possible to reduce its definition, not being necessary to define the two values of the calculation for the case of binary logic calculations, or the value 0 for the case of non-binary logic calculations. This is fundamentally motivated by the objective function of the problem that may allow one to not have to define one of its values, but also by the type of specification in which we use the calculation.

The Proposition of a Binary Logical Calculation:

$$\beta = 1 \text{ IF AND ONLY IF Condition}$$

It is equivalent to the following.

$$\text{IF Condition THEN } \beta = 1 \text{ (Definition of value 1)}$$

$$\text{IF} \beta = 1 \text{ THEN Condition} \Rightarrow \text{IF NOT(Condition) THEN } \beta = 0 \text{ (Definition of value 0)}$$

In general, for a binary logic calculation, if the variable assumes a cost, it will not be necessary to define the value 0, because the solving process will set the variable to 0 whenever possible. It will only be necessary to define value 1. And if it supposes a profit, it will be, on the contrary, necessary to define the value 0.

Something equivalent happens with non-binary calculations: The objective function can make it unnecessary to define the value 0 if the calculation involves a cost. If it is a profit, it will be enough to define the value 0 and set the value of x as the upper bound for the calculation y.

This reduction is applicable to any logical calculation that is defined in the problem, including binary logical calculations that are defined as auxiliary in the modeling of a compound proposition with several operators. This may not be completely clear in the methodology due to some problems in the final version edition. Next, in the specification summary, we will make this same idea clear in compound proposition modeling.

Therefore, whenever there are doubts about the feasibility of reducing the definition of values, the definition of the two values will always lead to correct modeling. In the methodology, the reduction of values is expressed with the reference f_{S_V} (simplification of values). In the broad set of problems, we will see many cases where values are reduced, and also cases where values cannot be reduced in the definition of the calculation.

1.4.3 Lower/Upper Bound Calculations

They calculate an upper or lower bound of the values of other variables. Its use is also necessary. The definition of the calculation is carried out in a simple way by imposition of bounds.

If there is a set of variables x_i $(i = 1 \ldots n)$ that can define decision activities or calculations made in the system, the standard definition of a calculation of the lower/upper bound would be:

Variables

$$x_{\mathrm{UP}} = \text{upper bound of } x_i (i = 1 \ldots n) / x_{\mathrm{LOW}} = \text{lower bound of } x_i (i = 1 \ldots n)$$

Constraints defining the calculation:

$$\forall i : x_i \leq x_{\mathrm{UP}} / \forall i : x_i \geq x_{\mathrm{LOW}}$$

Conditioned bounds calculations

When the lower/upper bound calculation is made on a set of variables where each variable depends on a condition to be part of the bound calculation, the constraints that define the calculation are established by a logical proposition.

There is a set of variables x_i $(i = 1 \ldots n)$ representing activities or calculations:

Variable:

$x_{\mathrm{UP}} =$ Upper bound of $x_i(i = 1 \ldots n)/x_{\mathrm{LOW}} =$ Lower bound of $x_i(i = 1 \ldots n)$

Constraints defining the calculation
Constraints resulting from the modeling of the following logical proposition:

$$\forall i : \text{IF condition} \quad \text{THEN } x_{\mathrm{SUP}} \geq x_i / x_{\mathrm{INF}} \leq x_i$$

1.5 Specifications

The classification of specifications in an optimization problem is done by approaching the specifications as logical propositions. Thus, we can define a classification into two levels. The upper level is determined by the way in which the specification is stated, whereby we would have two types of specification.

- Specification stated as a simple proposition: No logical operators. It is an imposition expressed as a linear equation or inequality (we work in a linear environment).
- Specification stated as a compound proposition: Composed of simple propositions and logical operators.

1.5.1 Specifications Stated as Simple Propositions

The specifications that are formulated as simple propositions give rise to most of the constraints of optimization problems. Simple propositions can also be classified here with greater difficulty, and not exclusively, depending on the meaning or objective of the constraints or the intervening variables. In this methodology, we will define the following types of specifications stated as simple propositions:

1.5.1.1 Quantitative Selection Specification

A specification based on the variables involved. Whenever logical activities are presented on sets of elements or binary logical calculations, it is necessary to analyze the existence of this type of specification, since in many cases they are *not explicitly described in the description of the system because they are understood.*

They can be of three types:

- Upper bound
- Lower bound
- Equality

1.5.1.2 Capacity Specifications

Based on the meaning of the element data, it is a specification catalogued by some authors. It is based on availability data for elements that have a resource character in the system. The general format of the specification is as follows:

Capacity consumption $\leq$ *Capacity contribution*

Both consumption and contribution can be fixed (an attribute) or variable. The two most common formats in which it is presented in a system are:

- The specification focused on the consumption of capacity: The capacity of the resource can be partially consumed.
- The specification focused on the contribution of capacity: The capacity of the resource is used to satisfy a fixed demand for capacity.

1.5.1.3 Demand Specification

Based also on data meaning, it occurs when an element has an attribute that represents a quantity demanded of an element (individual or collective) that is necessary to cover or supply.

1.5.1.4 Bound Imposition Specifications

These are the most easily recognizable specifications in a system. It would be a typology in terms of the way of stating the specification. There are two types:

- Imposition of maximums: These impose upper bounds on variables or functions of variables.
- Imposition of minimums: These impose lower bounds on variables or functions of variables.

Both the upper or lower bound quantitative selection specifications as the specifications of capacity or demand could be within this typology. However, it is convenient to consider all typologies for their particular aspects. Here, we consider direct and explicit element boundary impositions, not corresponding to a sum of logical choices and not necessarily based on demand or availability data, but on technical criteria of the system.

1.5.1.5 Allocation or Balance Specifications

These impose values on variables or functions of variables. They also impose a balance between the inputs and outputs of a measurable element, collective or individual. Included in this typology are constraints on the distribution of stocks or those that impose equality between variable functions. As in the previous case, the sum of logical choices is excluded.

It is inevitable that certain specifications will sometimes be considered to belong to more than one typology. This is not at all inconvenient.

1.5.2 Specifications Stated as Compound Propositions

Within this group of specifications is the definition of logical calculations.

Next, we are going to present all the references that we use to model this type of proposition, which are widely explained in the book on our methodology. Table 1.4 shows the list of logical operators.

Given two propositions ϕ *and* ψ, equivalences between operators are collected in Table 1.5 representing the first three references. Table 1.6 shows references to the reduction of signs in simple propositions.

ξ is a small enough value to avoid in continuous propositions that the value less than V that the linear function could take is greater than $(V - \xi)$, in the case of $X < V$. The same applies for $X > V$.

Table 1.4 List of logical operators

Operator	Symbol	Expression
Negation	$\neg$	NOT (ϕ)
Disjunction	$\vee$	ϕ OR ψ
Conjunction	$\wedge$; &	ϕ AND ψ
Conditional	$\rightarrow$	IF ϕ THEN ψ
Biconditional	$\leftrightarrow$	ϕ IF AND ONLY IF ψ
Exclusive disjunction	$\oplus$	EITHER ϕ OR ψ

Table 1.5 Equivalences between operators

Proposition	Equivalent proposition	References
$\phi \leftrightarrow \psi$	$\phi \rightarrow \psi$ $\psi \rightarrow \phi$	f_1
$\phi \oplus \psi$	$(\phi \wedge \neg(\psi)) \vee (\neg(\phi) \wedge \psi)$	f_2
$\phi \rightarrow \psi$	$\neg\psi \rightarrow \neg\phi$	f_3

Table 1.6 Reduction of signs in simple propositions

Simple proposition	Reduction		
Sign	$X \in Z$	$X \in \mathfrak{R}$	References
$X < V$	$X \leq \lceil V \rceil - 1$	$X \leq V - \xi$	f_4
$X > V$	$X \geq \lfloor V \rfloor + 1$	$X \geq V + \xi$	f_5
$X \neq V$	$X < V$ OR $x > V$ $\Rightarrow f_4$ and $f_5 \Rightarrow$ $(X \leq \lceil V \rceil - 1)$ OR $(X \geq \lfloor V \rfloor + 1)$	$X < V$ OR $X > V$ $\Rightarrow f_4$ and $f_5 \Rightarrow$ $(X \leq V - \xi)$ OR $(X \geq V + \xi)$	f_6

Table 1.7 Increment and decrement parameters

	$V \in Z$		$V \notin Z$	
	$X \in Z$	$X \in \mathfrak{R}$	$X \in Z$	$X \in \mathfrak{R}$
R^I	1	ξ	$\lceil V \rceil - V$	ξ
R^D	1	ξ	$V - \lfloor V \rfloor$	ξ

ξ will be a small enough value.

Table 1.8 Negation operator modeling

Sign	Model	References
NOT $(X < V)$	$X \geq V$	f_8
NOT $(X \leq V)$	$X > V \Rightarrow f_5 \Rightarrow X \geq V + R^I$	f_9
NOT $(X > V)$	$X \leq V$	f_{10}
NOT $(X \geq V)$	$X < V \Rightarrow f_4 \Rightarrow X \leq V - R^D$	f_{11}
NOT $(X = V)$	$(X \leq V - R^D)$ OR $(X \geq V + R^I)$	f_{12}

Table 1.9 Conditional operator modeling with binary input proposition

IF ϕ THEN ψ					
ϕ		ψ			
Type	Sign	Type	Sign	Model	Refs.
Binary	$\alpha = 1$	Binary	$\beta = 1$	$\alpha \leq \beta$	f_{13}
Binary	$\alpha = 1$	Integer/Continuous	$X \leq V$	$X \leq V + (UB_X - V)(1 - \alpha)$	f_{14}
Binary	$\alpha = 1$	Integer/Continuous	$X \geq V$	$X \geq V\alpha + LB_X (1 - \alpha)$	f_{15}
Binary	$\alpha = 1$	Integer/Continuous	$X = V$	$X \leq V + (UB_X - V)(1 - \alpha)$ $X \geq V\alpha + LB_X (1 - \alpha)$	f_{16}

Binary expressions will always be defined with value 1. Therefore:

$$\text{For } \alpha \text{ a binary expression such that} : \alpha = 0 \Rightarrow 1 - \alpha = 1; (1 - \alpha) \text{ is still a binary expression} \quad (f7)$$

Next, in the modeling of operators, we use increment and decrement parameters with respect to the independent term V *and the type of function* X, as in the case of the reduction of signs (Table 1.7):

1.5.2.1 Modeling Operators Individually

From Tables 1.8, 1.9, 1.10, 1.11 and 1.12 the formulation of the logical operators Negation, Conditional, and Biconditional is presented when only one operator appears in the compound proposition.

Disjunction Operator ($\vee$; **OR**)

Given simple propositions ϕ_i, i=1, 2,. . ., n: ϕ_1 OR ϕ_2 OR . . . OR ϕ_n

Each proposition $\phi_i / \phi_i \in \{0, 1\}$ is defined as the generic expression $\alpha_i = 1$

Define a binary logical calculation $\omega_i \forall_i / \phi_i \in Z \vee \phi_i \in \mathfrak{R}(\forall i / \phi_i \notin \{0, 1\})$:

Table 1.10 Conditional operator modeling with integer/continuous input proposition

	IF ϕ THEN ψ						
	ϕ			ψ			
Refs.	Type	Sign	Model	Type	Sign	Model	Refs.
f_{17}	Int/Cont	$X \leq V_1$	$X \geq (V_1 + R^I)(1 - \omega) + LB_X\, \omega$	Int/Cont	$Y \leq V_2$	$Y \leq V_2 + (UB_Y - V_2)(1 - \omega)$	f_{20}
f_{18}	Int/Cont	$X \geq V_1$	$X \leq (V_1 - R^D)(1 - \omega) + UB_X\, \omega$	Int/Cont	$Y \geq V_2$	$Y \geq V_2\omega + LB_Y(1 - \omega)$	f_{21}
f_{19}	Int/Cont	$X = V_1$	$X \leq V_1 + (UB_X - V_1)(1 - \omega)$ $X \geq V_1\omega + LB_X(1 - \omega)$ $X \leq (V_1 - R^D)\delta_1 + UB_X\omega + UB_X\delta_2$ $X \geq (V_1 + R^I)\delta_2 + LB_X\omega + LB_X\delta_1$ $\delta_1 + \delta_2 = 1 - \omega$	Int/Cont	$Y = V_2$	$Y \leq V_2 + (UB_Y - V_2)(1 - \omega)$ $Y \geq V_2\omega + LB_Y(1 - \omega)$	f_{22}

Here, we make a reduction with respect to the table presented in García (2021), which is also correct, but the number of constraints in references f_{17}, f_{18} can be reduced

Table 1.11 Biconditional connective modeling with binary propositions

ϕ IF AND ONLY IF ψ					
ϕ		ψ			
Type	Sign	Type	Sign	Model	Refs.
Binary	$\alpha = 1$	Binary	$\beta = 1$	$\alpha = \beta$	f_{23}
Binary	$\alpha = 1$	Integer/Continuous	$X \leq V$	$X \leq V + (UB_X - V)(1 - \alpha)$ $X \geq (V + R^I)(1 - \alpha) + LB_X \alpha$	f_{24}
Binary	$\alpha = 1$	Integer/Continuous	$X \geq V$	$X \geq V\alpha + LB_X (1 - \alpha)$ $X \leq (V - R^D)(1 - \alpha) + UB_X \alpha$	f_{25}
Binary	$\alpha = 1$	Integer/Continuous	$X = V$	$X \leq V + (UB_X - V)(1 - \alpha)$ $X \geq V\alpha + LB_X (1 - \alpha)$ $X \leq (V - R^D)\delta_1 + UB_X \alpha + UB_X\delta_2$ $X \geq (V + R^I)\delta_2 + LB_X \alpha + LB_X\delta_1$ $\delta_1 + \delta_2 = 1 - \alpha$	f_{26}

$$\Rightarrow f_{SV} \Rightarrow \forall i/\phi_i \in Z \vee \phi_i \in \mathfrak{R} : \text{IF } \omega_i = 1 \text{ THEN} \phi_i \tag{f33}$$

Quantitative selection specification to be incorporated to the model:

$$\sum_{i/\phi_i \in \mathfrak{R} \vee \phi_i \in Z} \omega_i + \sum_{i/\phi_i \in \{0,1\}} \alpha_i \geq 1 \tag{f34}$$

Exclusive Disjunction operator (⊕; EITHER ... OR...)

Given simple propositions ϕ_i, $i = 1, 2, \ldots, n : \phi_1 \oplus \phi_2 \oplus \ldots \oplus \phi_n$

Each proposition $\phi_i/\phi_i \in \{0, 1\}$ is defined as the generic expression $\alpha_i = 1$

Define a binary logical calculation ω_i $\forall$ $i/\phi_i \in Z \vee \phi_i \in \mathfrak{R}(\forall i/\phi_i \notin \{0, 1\})$.

$$\forall i/\phi_i \in Z \vee \phi_i \in \mathfrak{R} : \omega_i = 1 \leftrightarrow \phi_i \tag{f35}$$

$$\sum_{i/\phi_i \in \mathfrak{R} \vee \phi_i \in Z} \omega_i + \sum_{i/\phi_i \in \{0,1\}} \alpha_i = 1 \tag{f36}$$

The only possibility to simplify the modeling of this operator is that we can convert this operator to the OR operator. This happens when all simple propositions involve a cost of the objective function. In that case, we can model it as an OR operator.

1.5.2.2 Modeling Compound Propositions with Various Operators

The modeling process of a compound proposition with several operators is done from the lowest level in the structure of the proposition to the highest level. The level is determined by the priority of the operators according to the structure of parentheses. The lower the level, the higher the execution priority of the operator.

Table 1.12 Biconditional connective modeling with integer/continuous propositions

	ϕ IF AND ONLY IF ψ						
	ϕ			ψ			
Refs.	Type	Sign	Model	Type	Sign	Model	Refs.
f_{27}	Int/Cont	$X \leq V_1$	$X \leq V_1 + (UB_X - V_1)(1 - \omega)$ $X \geq (V_1 + R^I)(1 - \omega) + LB_X\,\omega$	Int/Cont	$Y \leq V_2$	$Y \leq V_2 + (UB_Y - V_2)(1 - \omega)$ $Y \geq (V_2 + R^I)(1 - \omega) + LB_Y\,\omega$	f_{30}
f_{28}	Int/Cont	$X \geq V_1$	$X \geq V_1\omega + LB_X(1 - \omega)$ $X \leq (V_1 - R^D)(1 - \omega) + UB_X\,\omega$	Int/Cont	$Y \geq V_2$	$Y \geq V_2\omega + LB_Y(1 - \omega)$ $Y \leq (V_2 - R^D)(1 - \omega) + UB_Y\omega$	f_{31}
f_{29}	Int/Cont	$X{=}V_1$	$X \leq V_1 + (UB_X - V_1)(1 - \omega)$ $X \geq V_1\omega + LB_X(1 - \omega)$ $X \leq (V_1 - R^D)\delta_1 + UB_X\omega + UB_X\delta_2$ $X \geq (V_1 + R^I)\delta_2 + LB_X\omega + LB_X\delta_1$ $\delta_1 + \delta_2 = 1 - \omega$	Int/Cont	$Y{=}V_2$	$Y \leq V_2 + (UB_Y - V_2)(1 - \omega)$ $Y \geq V_2\omega + LB_Y(1 - \omega)$ $Y \leq (V_2 - R^D)\delta_3 + UB_Y\omega + UB_Y\delta_4$ $Y \geq (V_2 + R^I)\delta_4 + LB_Y\omega + LB_Y\delta_3$ $\delta_3 + \delta_4 = 1 - \omega$	f_{32}

The process will always end with a proposition that has only one type of operator and that will be modeled as defined in the previous section.

We call ψ the proposition that is part of the original compound proposition ϕ, and in which only one operator type appears. The modeling process of ψ depending on the operator is as follows:

Negation Operator (¬;NOT):
The result of the negation operator modeling replaces ψ with $\neg\psi$ in ϕ, but does not incorporate additional constraints into the model.

Disjunction Operator (∨;OR):

$$\psi = (\psi_1 \vee \psi_2 \vee \ldots \vee \psi_i \vee \ldots)$$

Each proposition $\psi_i/\psi_i \in \{0, 1\}$ is defined as the generic expression $\alpha_i = 1$.

$\forall i/\psi_i \in Z \vee \psi_i \in \mathfrak{R}(\forall i/\psi_i \notin \{0, 1\})$ we define a binary logical calculation $\omega_i : \quad \omega_I = 1 \leftrightarrow \psi_i$ (f_{35}).

The constraint that results in f_{34}:

$$\sum_{i/\psi_i \in \mathfrak{R} \vee \psi_i \in Z} \omega_i + \sum_{i/\psi_i \in \{0,1\}} \alpha_i \geq 1$$

is not incorporated as a constraint to the model, but instead replaces ψ in ϕ. With this we reduce the operators from the original proposition ϕ.

Simplifications
This was not clear from the methodology due to a misprint in the editing process. Now it is explained better:

The modeling presented guarantees a correct formulation. However, it is sometimes possible to simplify the definition of binary logic calculations ω_i. It is convenient to analyze whether it may be correct to replace that expression with one of these two expressions:

$$\forall i/\psi_i \in Z \vee \psi_i \in \mathfrak{R} : \psi_i \rightarrow \omega_i = 1 \text{ (We do not define the value 0)}$$

$$\forall i/\psi_i \in Z \vee \psi_i \in \mathfrak{R} : \psi_i \rightarrow \omega_i = 1 \text{ (We do not define the value 1)}$$

That would simplify the number of constraints in the model, but sometimes we can make mistakes by using a wrong simplification.

Let us look at a couple of examples of this.

Example 1.2

$$\text{EITHER } (x_1 \geq 10 \text{ OR } x_2 \geq 10) \text{ OR } x_3 \geq 10$$
$$x_1 \geq 0, x_2 \geq 0, x_3 \geq 0 \text{ integers}$$

In this example, we must define three binary logical calculations:

$$\omega_1 = 1 \text{ IF AND ONLY IF } x_1 \geq 10$$
$$\omega_2 = 1 \text{ IF AND ONLY IF } x_1 \geq 10$$
$$\omega_3 = 1 \text{ IF AND ONLY IF } x_1 \geq 10$$
$$\Rightarrow \text{EITHER } (\omega_1 = 1 \text{ OR } \omega_2 = 1) \text{ OR } \omega_3 = 1$$
$$\Rightarrow f_{34} \Rightarrow \text{EITHER } (\omega_1 + \omega_2 \geq 1) \text{ OR } \omega_3 = 1$$
$$\Rightarrow f_{35} \Rightarrow \omega_4 = 1 \text{ IF AND ONLY IF } \omega_1 + \omega_2 \geq 1$$
$$\Rightarrow f_{36} \Rightarrow \quad \omega_4 + \omega_3 = 1$$

Regarding the first two defined logical calculations, we see what happens with each simplification.

We remove the definition of the value 0:

Definition:

$$\text{IF } x_1 \geq 10 \text{ THEN } \omega_1 = 1$$
$$\text{IF } x_2 \geq 10 \text{ THEN } \omega_2 = 1$$

In a conditional, when the input is not satisfied, the output is free; therefore, in the expression,

$$\text{EITHER } (\omega_1 + \omega_2 \geq 1) \text{ OR } \omega_3 = 1$$

$\omega_1 + \omega_2 \geq 1$ could be satisfied; however, ($x_1 \geq 10$ OR $x_2 \geq 10$) is not satisfied. Therefore, the value 0 cannot be simplified.

We remove the definition of value 1:

Definition:

$$\text{IF } \omega_1 = 1 \text{ THEN } x_1 \geq 10$$
$$\text{IF } \omega_2 = 1 \text{ THEN } x_2 \geq 10$$

In that case, as long as ($\omega_1 + \omega_2 \geq 1$), ($x_1 \geq 10$ OR $x_2 \geq 10$) would be satisfied. But in the case that $\omega_1 = 0$ and $\omega_2 = 0$, ($\omega_1 + \omega_2 \geq 1$) would not be satisfied, but instead ($x_1 \geq 10$ OR $x_2 \geq 10$) could be satisfied since if the input is not satisfied, the output is free to be satisfied or not. Therefore, it could happen that all the expressions of the EITHER OR operator are satisfied, which is not allowed.

Conclusion: We cannot remove any value from the definition.

In this case, only the objective criterion could allow the simplification of values, and for that it is necessary to be sure that the propositions always improve or worsen the objective function and analyze its viability.

Example 1.3

$$\text{IF } (x_1 \geq 10 \text{ OR } x_2 \geq 10) \text{ THEN } x_3 \geq 10$$
$$x_1 \geq 0, x_2 \geq 0, x_3 \geq 0 \text{ integers}$$

We have the definition of the same binary logic calculations as in the previous example.

The modeling of the expression becomes:

$$\Rightarrow \text{IF } (\omega_1 + \omega_1 \geq 1) \text{ THEN } x_3 \geq 10$$

If in the logical calculations ω_1 and ω_2 we eliminate the value 1:

Definition of the value 0:

$$\text{IF } \omega_1 = 1 \text{ THEN } x_1 \geq 10$$
$$\text{IF } \omega_2 = 1 \text{ THEN } x_2 \geq 10$$

It would not be correct because we can have $\omega_1 + \omega_1 = 0$, and therefore, $x_3 \geq 10$ does not have to be satisfied; however, $(x_1 \geq 10 \text{ OR } x_2 \geq 10)$ could be satisfied.

If in the logical calculations ω_1 and ω_2, we eliminate the value 0:

Definition of value 1:

$$\text{IF } x_1 \geq 10 \text{ THEN } \omega_1 = 1$$
$$\text{IF } x_2 \geq 10 \text{ THEN } \omega_2 = 1$$

It would be correct, because whenever $(x_1 \geq 10 \text{ OR } x_2 \geq 10)$, $\omega_1 + \omega_1 \geq 1$ will be satisfied and then $x_3 \geq 10$ will be satisfied.

Important In general, if the OR operator is inside an EITHER...OR operator, we need to define the two values (only observing the objective criterion can lead to the simplification of some value). If it is in the input of an IF...THEN operator, we need to define the value 1. If it is in the output, we need to define the value 0.

Exclusive Disjunction ($\bigoplus$; EITHER... OR...)

$$\psi = \left(\psi_1 \bigoplus \psi_2 \bigoplus \ldots \bigoplus \psi_i \bigoplus \ldots\right)$$

Each proposition $\psi_i / \psi_i \in \{0, 1\}$ is defined as the generic expression $\alpha_i = 1$.

For $\forall i / \psi_i \in Z \vee \psi_i \in \mathfrak{R}(\forall i / \psi_i \notin \{0, 1\})$: we define a binary logical calculation $\omega_i : \omega_i = 1 \leftrightarrow \psi_i$

The constraint that results in f_{36}:

$$\sum_{i/\psi_i \in \Re \vee \psi_i \in Z} \omega_i + \sum_{i/\psi_i \in \{0,1\}} \alpha_i = 1$$

is not incorporated as a constraint to the model, but instead replaces ψ in ϕ. With this we reduce operators from the original proposition ϕ.

Conjunction Operator (∧;AND)

This operator had not been used individually for the obvious reasons that there was no need for any modeling exercise. However, within a proposal with more operators in an upper level, it operates in a similar way to the OR and EITHER…OR operators:

$$\psi = (\psi_1 \wedge \psi_2 \wedge \ldots \wedge \psi_i \wedge \ldots \wedge \psi_n)$$

Each proposition $\psi_i/\psi_i \in \{0,1\}$ is defined as the generic expression $\alpha_i = 1$.

$$\forall i/\psi_i \in Z \vee \psi_i \in \Re(\forall i/\psi_i \notin \{0,1\}) : \omega_i = 1 \leftrightarrow \psi_i$$

An expression ϕ_0 is created with the following format:

$$\phi_{0:} \sum_{i/\psi_i \in \Re \vee \psi_i \in Z} \omega_i + \sum_{i/\psi_i \in \{0,1\}} \alpha_i \geq n \qquad \text{(f37)}$$

Replacing ψ with ϕ_0 in ϕ.

The analysis of simplifications in the defined binary logical calculations ω_i is equivalent to the analysis of simplifications in the OR operator.

Conditional AND Biconditional Operators

The constraints resulting from operator modeling replace ψ.

Let $\pi_{1\ldots}\ \pi_r$ be simple propositions as a result of modeling the operator when there are no other operators.

Proposition (π_1 AND π_2 AND … AND π_r) replaces ψ in ϕ.

1.5.2.3 Avoid the Equal Sign in Simple Propositions

In general, in compound propositions, whenever possible, we should avoid the signs of equality in simple propositions ($X = V$). If the independent term corresponds to a lower bound of X, we can replace it with the sign $\leq$ (reference f_{LB}). Similarly, if it corresponds to an upper bound, we can work with the sign $\geq$ (reference f_{UB}).

On the other hand, we will also use the same simplification when the objective criterion does not improve, in any case, by decreasing (f_{LB}) or increasing (f_{UB}) the value of equality V, even if V does not correspond to a lower or upper bound, respectively.

1.5.2.4 Distributive Law Between Propositions and Other Equivalences

Given propositions ϕ, ψ, and σ:

$$\phi \vee (\psi \wedge \sigma) \equiv (\phi \vee \psi) \wedge (\phi \vee \sigma) \tag{f38}$$

$$\phi \wedge (\psi \vee \sigma) \equiv (\phi \wedge \psi) \vee (\phi \wedge \sigma) \tag{f39}$$

It is also possible to divide conditional propositions when they are at the highest level of the compound proposition.

Given propositions $\phi_1, \phi_2, \ldots, \phi_n, \phi_m$:

$$\begin{aligned} &\text{IF } \phi_1 \vee \phi_2 \vee \ldots \vee \phi_n \text{ THEN } \phi_m \Rightarrow \\ &\Rightarrow \text{IF } \phi_1 \text{ THEN} \phi_m \\ &\Rightarrow \text{IF } \phi_2 \text{ THEN} \phi_m \\ &\Rightarrow \ldots \\ &\text{IF } \phi_n \text{ THEN} \phi_m \end{aligned} \tag{f40}$$

On the other hand:

$$\begin{aligned} &\text{IF}\phi_m \text{ THEN } \phi_2 \wedge \phi_2 \wedge \ldots \wedge \phi_n \Rightarrow \\ &\text{IF}\phi_m \text{ THEN } \phi_1 \\ &\text{IF}\phi_m \text{ THEN } \phi_2 \\ &\Rightarrow \ldots \\ &\text{IF}\phi_m \text{ THEN } \phi_n \end{aligned} \tag{f41}$$

1.5.3 *Data in Propositions*

Let us call the data propositions with the term P_{At} whether they are one or several attributes joined by operators.

First, we distinguish in Table 1.13 the case of data propositions at the upper level of the proposition:

In the case of data propositions at a lower level of the compound proposition, we use f_1 y f_2 to substitute the biconditional operator and the operator EITHER OR and operate as follows:

If the conditional operator does not exist in the upper level:

If necessary, the distributive law (f_{38}) is applied until a union of propositions of the following form is obtained:

Table 1.13 Data proposition modeling at the upper level of the proposition

Proposition	Modelling	Reference
$\phi \vee P_{At}$	∀element/NOT(P_{At}): ϕ	f_{42}
$\phi \wedge P_{At}$	∀element/P_{At}: ϕ	f_{43}
$\phi \oplus P_{At}$	∀element/NOT(P_{At}): ϕ ∀element/P_{At}: NOT (ϕ)	f_{44}
IF P_{At} THEN ϕ	∀element/P_{At}: ϕ	f_{45}
P_{At} IF AND ONLY IF ϕ	∀element/P_{At}: NOT (ϕ) ∀element/NOT (P_{At}): ϕ	f_{46}[a]

Let ϕ be a proposition of variables
[a]There was a typo in García (2021)

$$(\phi_1 \text{ OR } P_{\text{At1}}) \text{ AND } (\phi_2 \text{OR } P_{\text{At2}}) \text{ AND} \ldots$$

For each compound proposition united with the operator AND, the following specification is created:

$$f_{42} \Rightarrow \forall \text{element/NOT}(P_{\text{At1}}) : \phi_1$$

If the conditional operator exists in the upper level:
If necessary, apply f_{39} until you obtain a proposal of the following form:

$$\text{IF}\phi_1 \text{ OR } P_{\text{At1}}\text{OR } (\phi_2 \text{ AND } P_{\text{At2}}) \text{ OR } (\phi_3 \text{ OR } P_{\text{At3}}) \text{ OR} \ldots \text{THEN}\psi$$

$$f_{40} \Rightarrow \text{IF}\phi_1 \text{ THEN } \psi \Rightarrow \ldots$$

$$\text{IF } P_{\text{At1}} \text{ THEN } \psi \quad \Rightarrow f_{45} \Rightarrow \forall \text{element}/P_{\text{At1}} : \psi$$

$$\begin{aligned} \text{IF } (\phi_2 \text{ AND } P_{\text{At2}}) \text{ THEN}\psi \Rightarrow & \forall \text{ element}/P_{\text{At2}} \\ & : \text{IF}\phi_2 \text{THEN } \psi \text{IF } (\phi_3 \text{ OR } P_{At3}) \text{ THEN } \psi \\ & \Rightarrow \forall \text{ element/NOT}(P_{At3}) : \text{IF } \phi \text{ THEN}\psi \end{aligned}$$

$$\Rightarrow \forall \text{element}/P_{At3} : \psi$$

1.6 Objective Criterion

For the objective function, we present three cases in which the cost of a variable is not a fixed value, but a variable value depending on the value that the variable takes. These three cases can also be applied in situations where the cost of the variable is not part of the objective, but is part of a specification. The modeling process is the same. This is also reflected in the characteristics of the different problems presented in the book.

The cases that may arise are the following.

1. The cost of variable x depends on the range of values in which the variable falls.
2. The cost of variable x depends on the value that another variable takes.
3. The cost depends on the deviation of the variable with respect to a reference value.

On the other hand, we are going to point out at the end of the section the balance functions, which, although they do not imply a variable cost of variables, are common in modeling problems and are based on the use of bounds calculations.

1.6.1 Cost According to the Interval of Values

With a variable x, we define a set of n intervals $(U_{i-1}, U_i]$ $i = 1\ldots n$, $U_i \geq 0$ $i = 0\ldots n$, and a cost c_i associated with each interval).

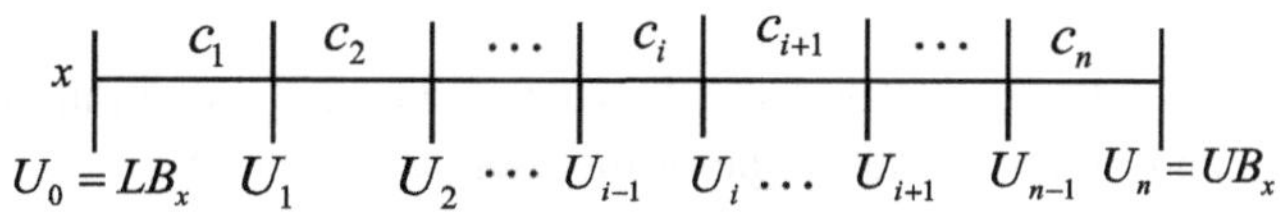

After a modeling process, the result is the following:

Defined variables:

$$\alpha_i = \begin{cases} 1 & \text{if } x \in [U_{i-1} + B, U_i] \\ 0 & \text{otherwise} \end{cases}$$

x_i: the same type as x.

i=1. . .n;

Constraints:

$$x = \sum_{i=1}^{n} x_i$$

$$\forall i : x_i \leq U_i \alpha_i$$

$$\sum_{i=1}^{n} \alpha_i = 1 \text{ We could also use } \sum_{i=1}^{n} \alpha_i \leq 1$$

$$\forall i : x \geq (U_{i-1} + B)\ \alpha_i$$

$$\forall i : x \leq U_i \alpha_i + \text{UB}_x (1 - \alpha_i)$$

These last two can also be replaced by:

$$\forall i : x_i \geq (U_{i-1} + B)\ \alpha_i$$

$$B = \begin{cases} 0 & i = 1 \\ 1 & i > 1; x \text{ integer} \\ \xi(\text{small enough value}) & i > 1; x \text{ continuous} \end{cases}$$

Cost term for the objective function or for a constraint:

$$\sum_{i=1}^{n} c_i x_i$$

1.6.2 Cost According to the Value of Another Variable

Although the variable that determines the cost is integer or continuous, it will be reduced to depend on the value of one or more binary variables.

$$\beta = 1 \Rightarrow \text{Cost of } x : c_1$$

$$\beta = 0 \Rightarrow 1 - \beta = 1 \Rightarrow \text{Cost of } x : c_2$$

$$\text{Cost of variable } \beta : C$$

Variables:

$$x_i \geq 0 \text{ of the same type as } x.$$

Constraints:

$$x_1 \leq \mathrm{UB}_x \beta$$

$$x_2 \leq \mathrm{UB}_x (1 - \beta)$$

$$x = x_1 + x_2$$

Cost term for the objective function o for a constraint:

$$C\beta + c_1 x_1 + c_2 x_2$$

1.6.3 Costs Depending on the Deviation of the Variable

In some situations, a reference value can be imposed on the values of a variable so that we can be interested in the approach of the variable to that reference value or we are interested in distancing from it. The distance from this reference is not imposed as a specification in the problem, but the variable has freedom, penalizing or rewarding the deviation from the reference data in the objective function.

The bonuses or penalties imposed affect the units that deviated from the reference data.

The following table summarizes all the possibilities that may arise (Table 1.14):

1.6.3.1 Penalty for Excess

Given

U: Reference value (data or variable)
p: unit penalty (data)

Affected variable: x
Auxiliary variables:

x_d: Deviated units for the default of x over U
x_e: Deviated units for excess of x over U

Constraints:

$$x + x_d - x_e = U$$

Cost term for the objective function o for a constraint:

$$px_e$$

(x would enter in the function with its base cost)

1.6.3.2 Bonus for Excess

Given

Table 1.14 Cases of deviation

Deviation	Repercussion	Section
Excess	Penalty	3.1
	Bonus	3.2
Default	Penalty	3.3
	Bonus	3.4

U: Reference value (data o variable)
b: Unit bonus (data)

Affected variable: x
Auxiliary variables:

x_d: Deviated units for default of x over U
x_e: Deviated units for excess of x over U

Binary logical calculations:

$$\alpha_d : \alpha_d = 1 \text{ IF AND ONLY IF } x_d > 0$$
$$\alpha_e : \alpha_e = 1 \text{ IF AND ONLY IF } x_e > 0$$

Constraints:

$$x + x_d - x_e = U$$
$$x_d \leq U^* \alpha_d$$
$$x_e \leq (\text{UB}_x - U^*) \alpha_e$$
$$\alpha_d + \alpha_e \leq 1$$

* If the reference value U is a variable, we have to use another upper bound for x_d and x_e in the modeling process, UB_x for example:

$$x + x_d - x_e = U$$
$$x_d \leq UB_x \alpha_d$$
$$x_e \leq UB_x \alpha_e$$
$$\alpha_d + \alpha_e \leq 1$$

Cost term for the objective function o for a constraint:

$bx_e$is incorporated

x would enter in the function with its base cost.

1.6.3.3 Penalty for Default

Given

U: Reference value (data or variable)
p: Unit penalty (data)

Affected variable: x
Auxiliary variables:

x_d: Deviated units for default of x over U
x_e: Deviated units for excess of x over U

Constraint:

$$x + x_d - x_e = U$$

Cost term de for the objective function o for a constraint:

$$px_d$$

x would enter in the function with its base cost.

1.6.3.4 Bonus for Default

Given

U: Reference value (data or variable)
p: unit penalty (data)

Affected variable: x
Auxiliary variables:

x_d: Deviated units for default of x over U
x_e: Deviated units for excess of x over U

Binary logical calculations:

$$\alpha_d : \alpha_d = 1 \text{ IF AND ONLY IF } x_d > 0$$
$$\alpha_e : \alpha_e = 1 \text{ IF AND ONLY IF } x_e > 0$$

Constraints:

$$x + x_d - x_e = U$$
$$x_d \leq U^* \alpha_d$$
$$x_e \leq (\text{UB}_x - U^*)\alpha_e$$
$$\alpha_d + \alpha_e \leq 1$$

* If the reference value U is a variable, we have to use another upper bound for x_d and x_e in the modeling process, for example, UBx.

Cost term for the objective function o for a constraint:

$$bx_d$$

x would enter in the function with its base cost.

1.6.4 Balance Functions

In certain cases, we are interested in balancing the values that a series of variables take, both as a constraint and as an objective function.

Let $x_1, x_2, \ldots, x_i, \ldots, x_n$ be the set of variables to balance.

Balance Strategies:

Maximize the minimum value.

Lower bound calculation:

$$x_{\mathrm{MIN}} : \text{the minimum value of variables } x_i.$$

$$\forall i : x_i \geq x_{\mathrm{MIN}}$$

Objective function:

$$\text{Max } x_{\mathrm{MIN}}$$

Minimize the Maximum Value.

Upper bound calculation:

$$x_{\mathrm{MAX}} : \text{the maximum value of the variables } x_i.$$

$$\forall i : x_{\mathrm{MAX}} \geq x_i$$

Objective function:

$$\text{Min } x_{\mathrm{MAX}}$$

Minimize the Difference Between the Minimum and Maximum Values.

Bounds calculations:

$$x_{\mathrm{MIN}} : \text{the minimum value of the variables } x_i.$$

$$x_{\mathrm{MAX}} : \text{the maximum value of the variables } x_i.$$

$$\forall i : x_i \geq x_{\mathrm{MIN}}$$

$$\forall i : x_{\mathrm{MAX}} \geq x_i$$

Objective function:

$$\text{Min } x_{\mathrm{MAX}} - x_{\mathrm{MIN}}$$

(As a constraint, it would be to impose an upper bound on $x_{\mathrm{MAX}} - x_{\mathrm{MIN}}$)

Reference

García JM (2021) Modelling in mathematical programming: methodology and technics. Springer. https://doi.org/10.1007/978-3-030-57250-1

Chapter 2
Building and Solving Models with LINGO

2.1 Introduction

"LINGO is a comprehensive tool designed to make solving mathematical optimization models easy and efficient. LINGO provides a completely integrated package that includes a powerful language for expressing optimization models, a full-featured environment for building and editing problems, and a set of fast built-in solvers capable of efficiently solving most classes of optimization models" (LINDO Systems Inc. 2021). The primary features of LINGO to be presented in this handbook include the description of its algebraic modeling language and the options for input and output data.

LINGO supports a powerful set-based modeling language that allows users to express mathematical programming models efficiently and compactly.

LINGO allows you to build models that pull information directly from text files, databases, and spreadsheets (see Fig. 2.1). Similarly, LINGO can output solution information directly into a database or spreadsheet, making it easier for you to generate reports in the application of your choice. Complete separation of the model and the data improves the maintenance and scalability of the model.

All information about LINGO tools, versions, and downloadable applications is available at lindo.com.

The objective of this chapter is to learn the algebraic modeling language of the LINGO libraries in order to solve our models or any other, using all valid data inputs and outputs, always trying to present the chapter in an easy and practical way. In the LINGO manual (LINDO Systems Inc. 2021), you can find more extensive information on the functions used, with other functionalities that have not been necessary for our models.

LINGO allows two ways to implement the model:

– Extended form: The model is written in an extended format, without the use of sets and, therefore, without the use of subscripts. It is a valid format for small

J. M. García Sánchez, *Building and Solving Mathematical Programming Models*,
International Series in Operations Research & Management Science 329,
https://doi.org/10.1007/978-3-030-97626-2_2

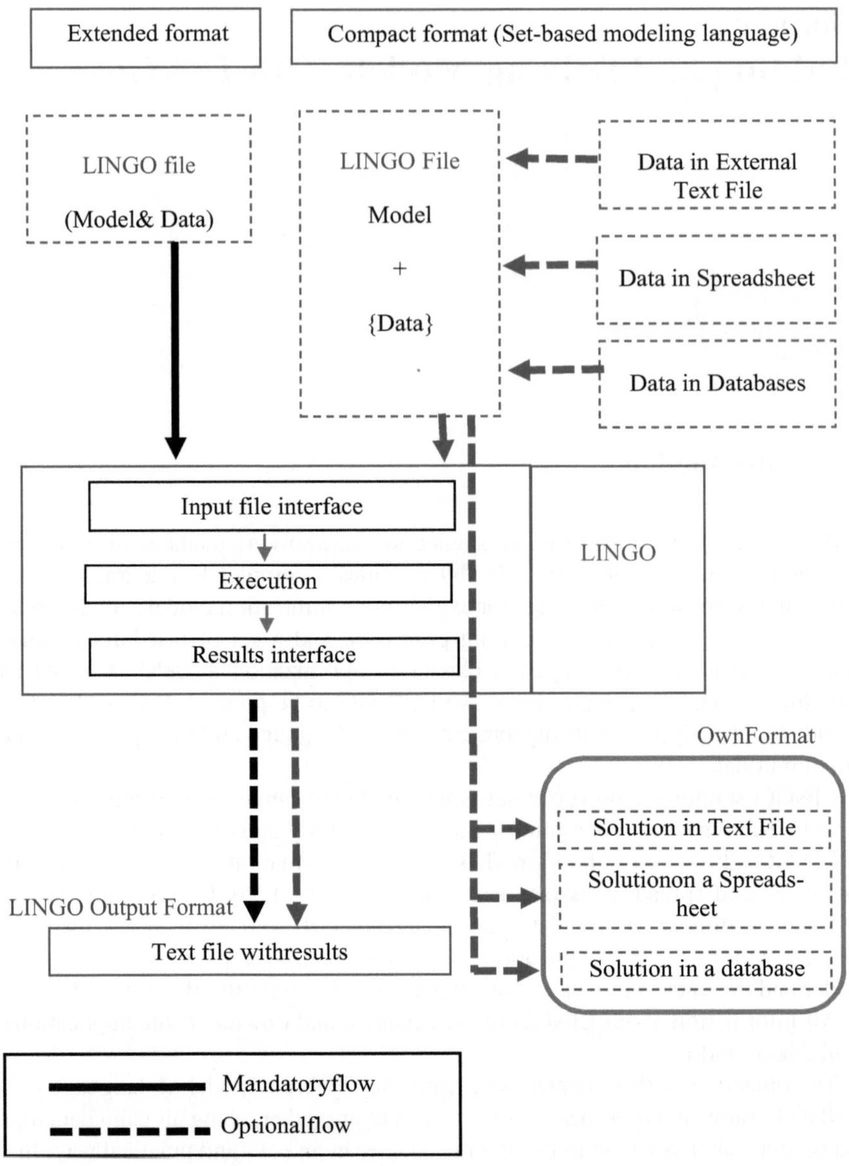

Fig. 2.1 Working framework in LINGO

models written manually or for the previous use of a computer program that automatically generates the extended model.

– Compact form: We make use of the concept of set for the introduction of the model and the data.

In extended form, the data of the model are necessarily integrated with the model. The model and the data are indivisible. An extended form illustration is shown in Example 2.1.

Example 2.1 Extended form

$$
\begin{aligned}
&\text{Max } x_1 + x_2 + y \\
&\text{s.t.} \\
&x_1 \leq 2 \\
&x_1 + y \leq 3 \\
&y + 2x_2 \leq 4 \\
&x_1 \geq 0 \\
&x_2 \geq 0 \text{ Integer} \\
&y \text{ Binary}
\end{aligned}
$$

In compact form, we divide the model and the data, in the case that the data are associated with sets of elements. For the data associated with elements that do not form a set, we can integrate them with the model as in the extended version or also define them independently of the sets.

For non-integrated data, in the compact version there are four possibilities:

- Enter data in its own section (DATA Section) within the file where the model is defined (LINGO File).
- Enter the model data in an external text file. The text file is referenced from the model file, in the DATA section.
- Enter the data of the model into a spreadsheet. The spreadsheet is referenced from the model file in the DATA section.
- Enter the data of the model into a database. The database is referenced from the model file, in the DATA section.

Similarly, in the compact version, we can export the results to an external file. When we talk about our own format in data export, we mean that we can send the solution with the coding that we consider most appropriate. A very simple example to reference this is presented in Example 2.2.

Example 2.2 Coding Illustration to Show the Solution

Given a system for assigning operators to machines.: 6 operators and 6 machines. Ignoring attributes and specifications, we propose the following table of elements and the following decision variables (Table 2.1).

Decision Activities

Assign Operators (*i*) to Machines (*j*)

$\forall i, j : \alpha_{ij} = 1$ if operator i is assigned to machine j; 0 otherwise.

After solving the problem, the optimal solution defines the following assignment.

Table 2.1 Table of elements for the illustration

			Data				
Elements	Set	QN	Name	Parameter	Type	Belonging	Value
Operators	$i = 1\ldots6$	I					
Machines	$j = 1\ldots6$	I					

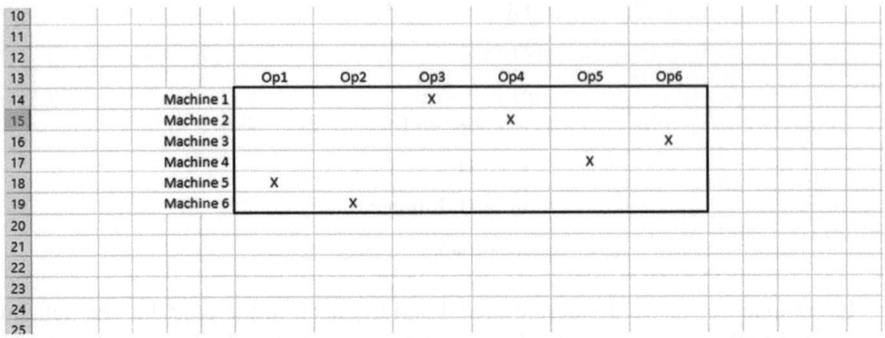

Fig. 2.2 Solution representation in a spreadsheet

$$
\begin{array}{llllll}
\alpha_{11}=0 & \alpha_{12}=0 & \alpha_{13}=0 & \alpha_{14}=0 & \alpha_{15}=1 & \alpha_{16}=0 \\
\alpha_{21}=0 & \alpha_{22}=0 & \alpha_{23}=0 & \alpha_{24}=0 & \alpha_{25}=0 & \alpha_{26}=1 \\
\alpha_{31}=1 & \alpha_{32}=0 & \alpha_{33}=0 & \alpha_{34}=0 & \alpha_{35}=0 & \alpha_{36}=0 \\
\alpha_{41}=0 & \alpha_{42}=1 & \alpha_{43}=0 & \alpha_{44}=0 & \alpha_{45}=0 & \alpha_{46}=0 \\
\alpha_{51}=0 & \alpha_{52}=0 & \alpha_{53}=0 & \alpha_{54}=1 & \alpha_{55}=0 & \alpha_{56}=0 \\
\alpha_{61}=0 & \alpha_{62}=0 & \alpha_{63}=1 & \alpha_{64}=0 & \alpha_{65}=0 & \alpha_{66}=0
\end{array}
$$

But in my output text file, I want a solution format expressed as:

Operator 1 assigned to machine 5
Operator 2 assigned to machine 6
Operator 3 assigned to machine 1
Operator 4 assigned to machine 2
Operator 5 assigned to machine 4
Operator 6 assigned to machine 3

Or if in a spreadsheet I want to represent the solution in a table with the format shown in Fig. 2.2.

2.2 Basic Format: The Extended Format

The basic structure for an extended-format model is presented in Fig. 2.3.

1. {Header}

---

2. {Title}
3. {Objective function}
4. Problem constraints
5. {Constraints associated with the type of variables}

---

6. {End}

{}: Optional content.

↑↓: Any order

1. Header:
FORMAT: MODEL:

2. Title:
FORMAT: TITLE Model title;

3. Objective Function:

FORMAT: {[Name]} (MAX/MIN) = Cost_1*VariableName_1 + Cost_2*Variable-Name_2 + … + Cost_n*VariableName_n ;

LINGO allows the option to name the objective function. If it does, it must be enclosed in square brackets []. For variable names, we can use up to 64 characters, starting with a letter (A-Z). We can use letters, numbers, or the underscore. It is not case sensitive.

Example: [O.F.] MIN = x1 + 2*x2 + 3*x3 – 5*x4 ;

4. Constraints:
FORMAT: {[Constraint Name]} TechicalData_1*VariableName_1 + Techical-Data_2*VariableName_2 + … + TechicalData_n*VariableName_n SIGN Independent_Term ;
SIGNs: (<; <= ; >; >=; =)
Example: x1 + x2<=1;
LINGO internally performs the conversion from < to <=, or from > to >=, if you have chosen those signs.

5. Constraints associated with the type of variables:
By default, all variables are Continuous >=0.
To define a variable as an integer:
FORMAT: @GIN(*VariableName*);
To define a variable as binary:
FORMAT: @BIN(*VariableName*);
To define a variable as free (without sign restrictions). It can also be defined as an integer.
FORMAT: @FREE(*VariableName*);

6. END:
FORMAT: END

Fig. 2.3 Basic structure for an extended format

Other Considerations
Blank lines are ignored.

Comments are preceded by the symbol "!" and at the end of the comment ";"

Example 2.3 Extended form in LINGO
Model:

$$\begin{aligned}&\text{Max } x_1 + x_2 + y\\&\text{s.t.}\\&x_1 \leq 2\\&x_1 + y \leq 3\\&y + 2x_2 \leq 4\\&x_1 \geq 0\\&x_2 \geq 0 \text{ Integer}\\&y \text{ Binary}\end{aligned}$$

In LINGO:

```
MODEL:
MAX = x1 + x2 + y ;
 x1 <=2 ;  !x1 cannot be greater than 2;
 x1 + y <=3;
 y + 2*x2 <=4;
@GIN(X2);
@BIN(y);
END
```

2.3 Compact Set-Based Format

When we express a model in a compact form, it is necessary to use sets. The sets are directly equivalent to the sets of elements in the problem, with one difference. LINGO distinguishes two types of sets:

- Primitive sets: They collect the sets of elements of the problem and their variables and attributes, own or shared with elements that are not in a set.
- Derived sets: They are formed by combining primitive sets. They collect the data and/or variables shared between elements of primitive sets.

Elements that do not form sets do not need to be defined, although they can be defined as a set of one element. The attributes of these elements or shared attributes with elements that do not form a set (i.e., attributes without subscript) can be integrated with the model or defined directly in the DATA section. Variables where only elements without set participate, i.e., variables without subscripts, are used directly in the expressions of the constraints and the objective function.

LINGO has a section for the definition of the sets and two sections for the data:

- **SETS Section**: The elements contained in the sets are defined, as well as the names of attributes and variables associated with them.
- **DATA Section**: The attributes values associated with sets are defined, as well as any data without subscripts (not associated with sets).
- **CALC Section**: Section offered by LINGO for the creation of calculated data if it is not already calculated externally. In the same way, values for attributes that already had information can be recalculated in this section.

To illustrate the compact set-based language, we are going to work with a version of the transportation problem that is ideal to illustrate the three sections and the implementation of constraints and objective.

Example 2.4 A Transportation Problem (Table 2.2)

A company has m warehouses where the units of its product are located. Each warehouse A_i ($i = 1...10$) has K_i units.

There is a set of customers ($j = 1...12$) with a demand D_j for units of the product. The company has to supply the demand of the 12 customers from the warehouses. The cost of shipping a product from each Ai warehouse ($i = 1...10$) to each customer C_j ($j = 1...12$) is estimated in c_{ij}. If the sum of customer's costs exceeds 20, high-cost customer must receive units from Warehouse 1. We use this last specification to illustrate performing a data calculation.

Specifications

- Capacity of each warehouse
- Supply of the demand of each customer
- Imposition to receive units from Warehouse 1 to high-cost customers.

Objective

Minimize shipping cost

To illustrate the CALC Section, we are going to calculate the data of the HCC (high-cost customer) binary attribute in that section.

Table 2.2 Table of elements of the transportation problem

			Data				
Elements	Set	QN	Name	Parameter	Type	Belonging	Value
Warehouses	$i = 1...10$	I_U	Stock	K_i	I	S	...
			Cost	c_{ij}	C ($)	S	...
Customers	$j = 1...12$	I_U	Demand	D_j	I	S	...
			HCC	CCj	B	W	...
				c_{ij}			
Product	–	C_D		K_i; D_j; c_{ij}			

Decision Activities

Ship products from warehouses (*i*) to customers (*j*)

$\forall i, j : x_{ij} =$ Number of product units shipped from Warehouse *i* to Customer *j*.

Specifications

Capacity of each warehouse (Capacity specification)

$$\forall i : \sum_{j=1}^{12} x_{ij} \leq K_i$$

Demand of each client (Demand specification)

$$\forall j : \sum_{i=1}^{10} x_{ij} \geq D_j$$

Imposition of shipping from Warehouse 1 if you are a high-cost customer (Imposition of lower bound).

$$\forall j / CC_j = 1 : x_{1j} \geq 1$$

Complete Model

$$\text{Min} \sum_{i=1}^{10} \sum_{j=1}^{12} c_{ij} x_{ij}$$

$$\text{s.t.}$$

$$\forall i : \sum_{j=1}^{12} x_{ij} \leq K_i$$

$$\forall j : \sum_{i=1}^{10} x_{ij} \geq D_j$$

$$\forall j / CC_j = 1 : x_{1j} \geq 1$$

$$\forall i, \forall j : x_{ij} \geq 0$$

2.3.1 SETS Section

Format

SETS:

Definition of Sets

ENDSETS

Table 2.3 Element declaration

Format	Example	Elements of the set
1...n	1...10	1, 2, 3, 4, 5, 6, 7, 8, 9 ,10
letterM...letterN	a...g	a, b, c, d, e, f, g
stringM...stringN	Pen1...Pen4	Pen1, Pen2, Pen3, Pen4
dayM...dayN	MON...FRI	MON, TUE, WED, THU, FRI
monthM...monthN	OCT...JAN	OCT, NOV, DEC, JAN
monthYearM...monthYearN	OCT2001...JAN2002	OCT2001, NOV2001, DEC2001, JAN2002

Primitive Set Definition

SetName {/ Element_List /} {: {Attribute_List,} {Variable_List}};

The list of elements (*Element_List*), the list of attributes(*Attribute_List*) and variables (*Variables_List*) are separated by commas. When the list of elements is not defined, the set name refers to a single element (it can collect the case of elements that are not part of any set). As we have commented, it is not mandatory to do so. Its data can be defined directly in the DATA section and its variables can be used directly in the model.

LINGO offers various possibilities to simplify the list of elements in a set. They are collected in Table 2.3:

For our illustration of the transportation problem, the set of warehouses corresponds to a primitive set, and the only argument of its own or shared with elements outside of sets is its capacity (K_i). The definition of the set could be in various ways:

```
Warehouses / Warehouse1, Warehouse2, Warehouse3, Warehouse4,
Warehouse5, Warehouse6, Warehouse7, Warehouse8, Warehouse9,
Warehouse10 /: K;
```

More simplified:

```
Warehouses /1..10/: K;
```

In Sect. 2.3.2 (DATA section), we will see the format to incorporate data from external files. Among the data that can be incorporated are also those related to the elements of a set.

Derived Set Definition

Derived sets are necessary whenever we have data shared between sets or variables with more than one subscript. The format is as follows:

SetName (Parent_set_list) {/ Element_List /} {:{Attribute_List,} {Variable_List}};

The *parent_set_list* is a list of previously defined sets separated by commas. For simplicity, we can assume that these sets are primitive sets, although derived sets are also allowed.

The derived sets are formed by the combinations of the elements of all the primitive sets that define them. The association is always done in the following order:

Let the primitive set S_1 be formed by n_1 elements; let the set S_2 be formed by n_2 elements. The derived set formed by (S_1, S_2) orders the combinations as follows: (1, 1) (1, 2)... (1, n2) (2, 1) (2, 2)... (2, n2)... (n1, 1) (n1, 2)... (n1, n2)

If more primitive sets were combined, the order of the combinations would follow the same logic. For three primitive sets S_1, S_2, S_3:

(1, 1, 1) (1, 1, 2) ... (1, 1, n_3)... (1, 2, 1) (1, 2, 2)... (1, 2, n_3)...

For our illustration, we need to combine the primitive sets of warehouses and customers, previously declared, to be able to declare the variables and cost data.

We have previously declared the primitive sets with their data not shared with other sets:

```
Warehouses / 1..10/: K;
Customers / 1..12/: D, CC;
```

For the declaration of the derived set, we can use any name:

```
WC (Warehouses, Customers): c, x;
```

If certain combinations cannot be considered in the system, LINGO allows only the valid combinations to be declared. For our illustration, if only the following combinations were allowed:

Warehouse 1 ⇒ *Customer* 1, *Customer* 2, *Customer* 3;
Warehouse 2 ⇒ *Customer* 4, *Customer* 5, *Customer* 6, *Customer* 7, *Customer* 8;
Warehouse 3 ⇒ *Customer* 1, *Customer* 9, *Customer* 10, *Customer* 11, *Customer* 12;
Warehouse 4 ⇒ *Customer* 1, *Customer* 4, *Customer* 6, *Customer* 9, *Customer* 11;

The declaration would be made as follows:

```
WC (Warehouses, Customers) /1 1, 1 2, 1 3, 2 4, 2 5, 2 6, 2 7, 2 8, 3 1, 3 9, 3
10, 3 11, 3 12, 4 1, 4 4, 4 6, 4 9 4 11/: c, x;
```

I discourage this type of implementation. It is more convenient to allow all the combinations and, later in the constraints, to cancel the variables of the not allowed combinations. This makes it easier to write the constraints in LINGO than if we have only declared the subset of valid combinations. Section 2.3.4 presents this case.

2.3.2 *DATA Section*

Format
DATA:
Attribute values
ENDDATA

LINGO allows for as many DATA sections as we want in the LINGO file.

As we discussed in the LINGO framework, we have four possibilities to define attribute values:

- Integrated data in the model file.
- Defining the link to a text file where the values are found.
- Defining the link to a spreadsheet file.
- Defining the link to a database file.

In the cases of external files, we will also illustrate how to incorporate, from that file, data referring to the elements of a set, since we can read from an external file the number of elements of a primitive set or the allowed combinations of a derived set.

2.3.2.1 Integrated Data in the Model File

In the DATA section, the values of the attributes are defined sequentially.

DATA:
AttributeName = Value_1 {,} Value_2 {,} …{,} Value_n;
ENDDATA

We can separate the values by commas or spaces.

Combined data entry is also allowed. We can see it directly in Example 2.5.

Example 2.5 Introduction of Attribute Values
Let a set, defined as SET1, consist of three elements. Two attributes, A and B, are associated with the set.

```
SETS:
  SET1 /1..3/: A,B;
ENDSETS
```

Values can be entered attribute by attribute or in combination:

Attribute by attribute:

```
DATA:
 A = 3, 2, 3;
  B = 5 5 8;
ENDDATA
```

In combination:

```
DATA:
A, B = 3, 5, 2, 5, 3, 8;
ENDDATA
```

For our transportation problem

```
SETS:
Warehouses /1..10/: K;
Customers /1..12/: D, CC;
WC (Warehouses, Customers): c, x;
ENDSETS
DATA:
  K = 40, 50, 30, 45, 100, 34, 15, 12, 18, 50;
  D = 20, 25, 25, 30, 10, 10 ,16, 23, 25, 35, 30, 13;
 c = 1, 4, 3, 2, 5, 5, 7, 9, 1, 2 , 3, 3, 1, 4, 3, 2, 5, 5, 5, 5, 1, 2 , 3, 3,4,
4, 3, 2, 5, 5, 7, 9, 1, 2 , 3, 2, 3, 1, 4, 3, 2, 5, 4, 7, 9, 1, 2 , 3, 2,5, 4, 3,
5, 5, 5, 7, 7, 1, 2 , 3, 3,  1, 4, 3, 2, 3, 5, 7, 9, 1, 2 , 1, 1,7, 4, 3, 6, 5,
5, 7, 2, 1, 2 , 7, 1, 1, 4, 3, 2, 5, 2, 7, 6, 1, 2 , 2, 3,3, 4, 5, 7, 5, 5, 7, 3, 1,
2 , 5, 3, 1, 4, 3, 2, 5, 2, 7, 6, 1, 2 , 2, 4, 9, 4, 3, 2, 5, 5, 7, 1, 1, 2 , 3, 3,
1, 4, 3, 2, 1, 5, 7, 7, 1, 2 , 3, 3;
ENDDATA
```

2.3.2.2 Data in External Text File

The link is established in the DATA section in the following format:

DATA:
AttributeName = @FILE(*'File_Path & File_Name'*);
ENDDATA

The data within the file is loaded sequentially, ending the loading of the attribute data with the symbol ~ (Alt + 126). In this way, we can insert as many attributes into the file as we want. LINGO reads each attribute sequentially until it finds a ~ symbol.

Within the text file we can include comments with the same format as in the LINGO file (! Comment;). These comments can serve to head the values of each attribute and thus have the information in the text file better referenced. It is totally optional.

Let us see how our example (transportation problem) would look like with the data in a text file.

```
DATA:
K = @FILE('c:\Files\TPData.txt');
D= @FILE('c:\Files\TPData.txt');
c= @FILE('c:\Files\TPData.txt');
ENDDATA
```

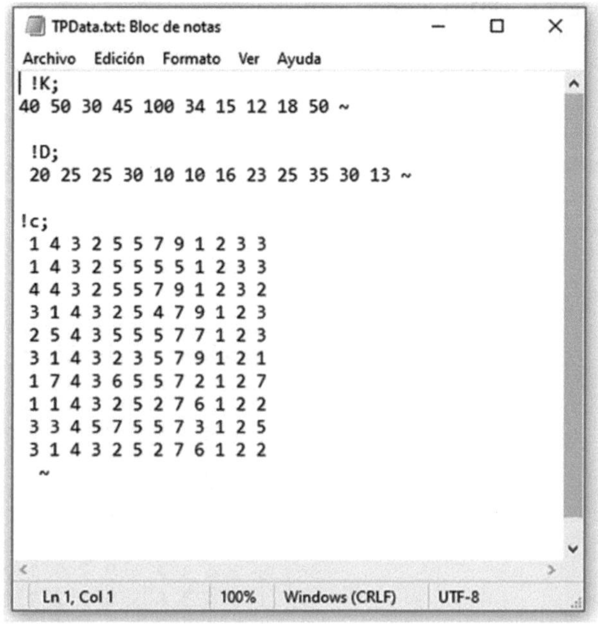

```
!K;
40 50 30 45 100 34 15 12 18 50 ~

 !D;
 20 25 25 30 10 10 16 23 25 35 30 13 ~

!c;
 1 4 3 2 5 5 7 9 1 2 3 3
 1 4 3 2 5 5 5 5 1 2 3 3
 4 4 3 2 5 5 7 9 1 2 3 2
 3 1 4 3 2 5 4 7 9 1 2 3
 2 5 4 3 5 5 5 7 7 1 2 3
 3 1 4 3 2 3 5 7 9 1 2 1
 1 7 4 3 6 5 5 7 2 1 2 7
 1 1 4 3 2 5 2 7 6 1 2 2
 3 3 4 5 7 5 5 7 3 1 2 5
 3 1 4 3 2 5 2 7 6 1 2 2
  ~
```


Fig. 2.4 TPData.txt

And the file "TPData.txt" presents the format collected in Fig. 2.4:

We can store data related to the elements of the set in the external files, so that they can be read from the LINGO file, just as we have read the attribute data. The format for linking the data is the same. For text files, we could modify the TPData.txt file to include, as an illustration, the number of elements in each primitive set (Fig. 2.5).

For the data in Fig. 2.5, we can read the number of warehouses and the number of customers in a DATA section before the SETS section:

```
MODEL:
TITLE Example;
DATA:
 n=@FILE('c:\Files\TPData.txt');
 m=@FILE('c:\Files\TPData.txt');
ENDDATA
SETS:
 Warehouses/1..n/: K;
 Customers/1..m/:D,CC;
 WC(Warehouses, Customers):x,c;
ENDSETS
DATA:
 K=@FILE('c:\Files\TPData.txt');
 D=@FILE('c:\Files\TPData.txt');
```

Fig. 2.5 TPData.txt with elements information

```
TPData.txt - Notepad
File Edit Format View Help
!n;
10~

!m;
12~

 !K;
40 50 30 45 100 34 15 12 18 50 ~

 !D;
 20 25 25 30 10 10 16 23 25 35 30 13 ~

!c;
 1 4 3 2 5 5 7 9 1 2 3 3
 1 4 3 2 5 5 5 5 1 2 3 3
 4 4 3 2 5 5 7 9 1 2 3 2
 3 1 4 3 2 5 4 7 9 1 2 3
 2 5 4 3 5 5 5 7 7 1 2 3
 3 1 4 3 2 3 5 7 9 1 2 1
 1 7 4 3 6 5 5 7 2 1 2 7
 1 1 4 3 2 5 2 7 6 1 2 2
 3 3 4 5 7 5 5 7 3 1 2 5
 3 1 4 3 2 5 2 7 6 1 2 2
  ~
```

```
c=@FILE('c:\Files\TPData.txt');
ENDDATA
```

2.3.2.3 Data in Spreadsheet File

We based on the Microsoft Excel spreadsheet from Microsoft Corporation.

Format
DATA:
AttributeName = @OLE('*File_Path & File_Name*', *RangeName*);
ENDDATA

RangeName defines the name given in the spreadsheet to the range of cells where the values are found. This task is done in the text box at the top left of the spreadsheet. Range names are not case sensitive. If the Excel file is open, it is not necessary to indicate the path of the file.

For our example:

```
DATA:
K = @OLE('c:\Files\TPSheet.xlsx',Capacity);
```

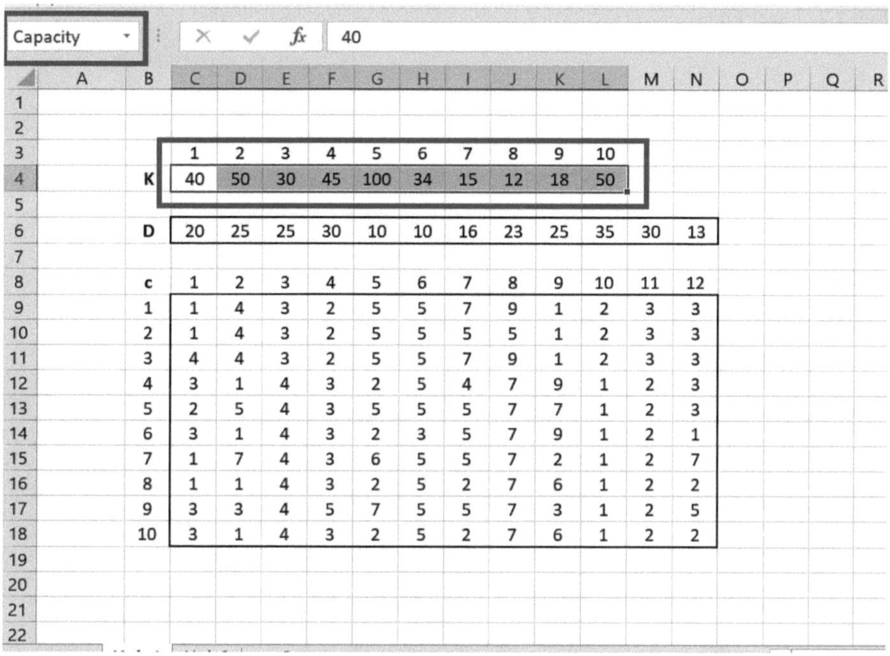

B	C	D	E	F	G	H	I	J	K	L	M	N
	1	2	3	4	5	6	7	8	9	10		
K	40	50	30	45	100	34	15	12	18	50		
D	20	25	25	30	10	10	16	23	25	35	30	13
c	1	2	3	4	5	6	7	8	9	10	11	12
1	1	4	3	2	5	5	7	9	1	2	3	3
2	1	4	3	2	5	5	5	5	1	2	3	3
3	4	4	3	2	5	5	7	9	1	2	3	3
4	3	1	4	3	2	5	4	7	9	1	2	3
5	2	5	4	3	5	5	5	7	7	1	2	3
6	3	1	4	3	2	3	5	7	9	1	2	1
7	1	7	4	3	6	5	5	7	2	1	2	7
8	1	1	4	3	2	5	2	7	6	1	2	2
9	3	3	4	5	7	5	5	7	3	1	2	5
10	3	1	4	3	2	5	2	7	6	1	2	2

Fig. 2.6 TPSheet.xlsx

```
D = @OLE('c:\Files\TPSheet.xlsx',Demand);
c = @OLE('c:\Files\TPSheet.xlsx',Costs);
ENDDATA
```

And the spreadsheet file "TPSheet.xlsx" is shown in Fig. 2.6:

In the image, you can see how the range C4: L4 has been labeled with the name "Capacity." Similarly, the cell range C6: N6 was labeled with the name "Demand" and the range C9: N18 was labeled with "Costs."

LINGO allows you to declare the data on the same command line. Simply enter the attribute names and ranges in the same order.

```
K,D,c =@OLE('c:\Files\TPSheet.xlsx', Capacity, Demand, costs);
```

For the information of elements, it operates the same as with text files. Below, we illustrate the case of the transportation problem in which only a subset of combinations between warehouses and customers are admissible. We are going to read from the spreadsheet not only the number of warehouses and customers, but also the allowed combinations and their costs.

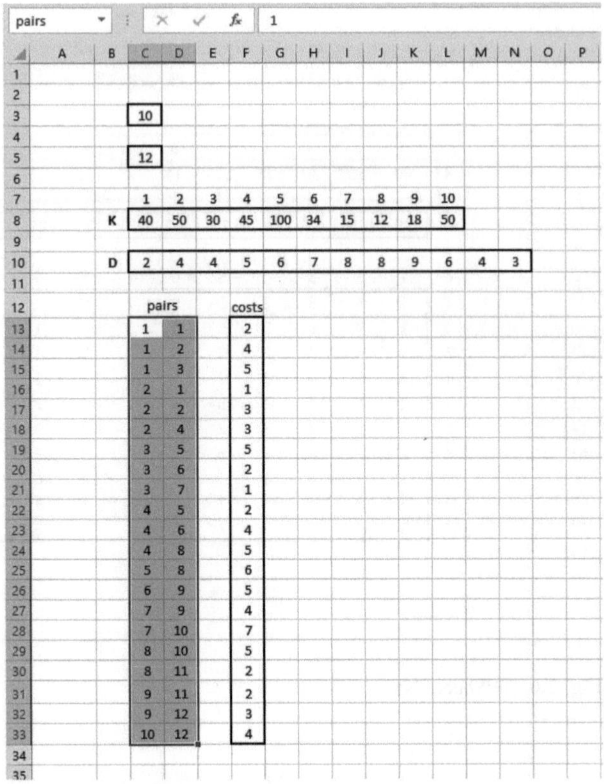

Fig. 2.7 Excel File: TPSheet1.xlsx

Example 2.6 Transportation Problem with Admissible Combinations Subset (Fig. 2.7)

In the LINGO File:

```
DATA:
 n=@OLE('c:\Files\TPSheet1.xlsx', n);
 m=@OLE('c:\Files\TPSheet1.xlsx', m);
ENDDATA
SETS:
 Warehouses/1..n/: K,A,B;
 Customers/1..m/:D,CC;
 WC( Warehouses, Customers)/ @OLE('c:\Files\TPSheet1.xlsx', pairs)/:
x,c;
ENDSETS
DATA:
 K =@OLE('c:\Files\TPSheet1.xlsx', capacity);
 D =@OLE('c:\Files\TPSheet1.xlsx', demand);
 c =@OLE('c:\Files\TPSheet1.xlsx',costs);
ENDDATA
```

As we pointed out in Sect. 2.3.1, it is easier to declare all combinations and, later, in the constraints, to prohibit the illegal combinations by nulling the variables.

2.3.2.4 Data in the Database File

To link the data in a database with the attribute values, we need to create an ODBC data source with the database. In Windows 10, this is done following the next steps. For other systems, the procedure is similar.

1. Open the ODBC data source manager:
   Run Start ⇒ Windows Administrative Tools ⇒ ODBC Data Sources

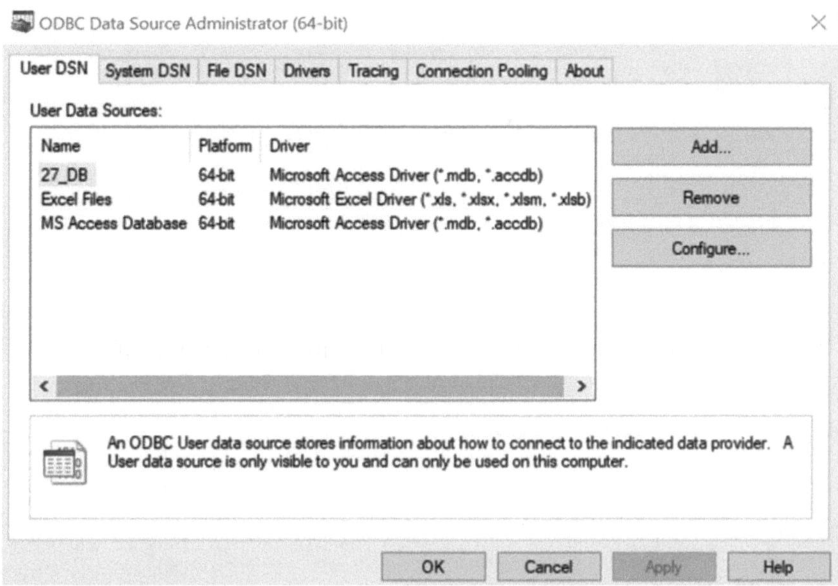

2. Press Add Button: Select Driver (Microsoft Access Driver for us) and pulse Finish

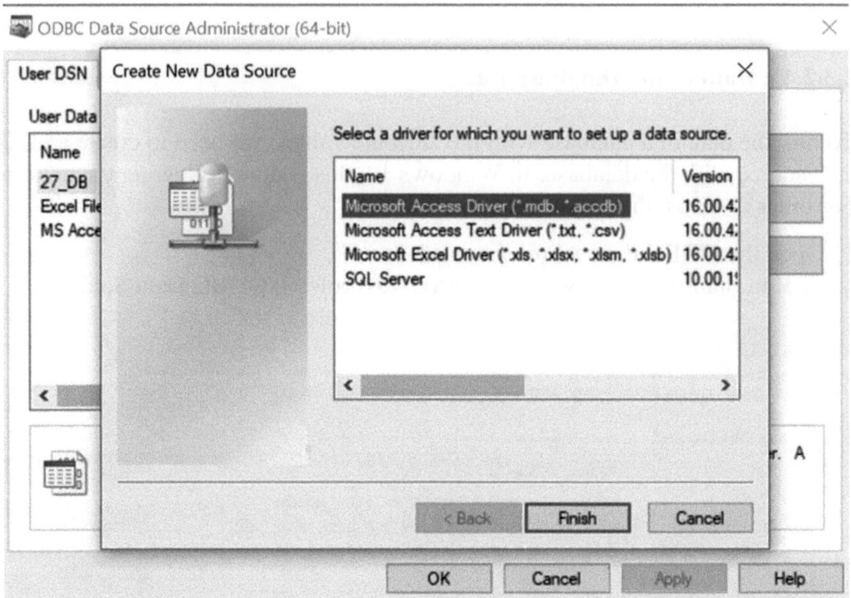

3. Name the data source and select the database that contains the information.

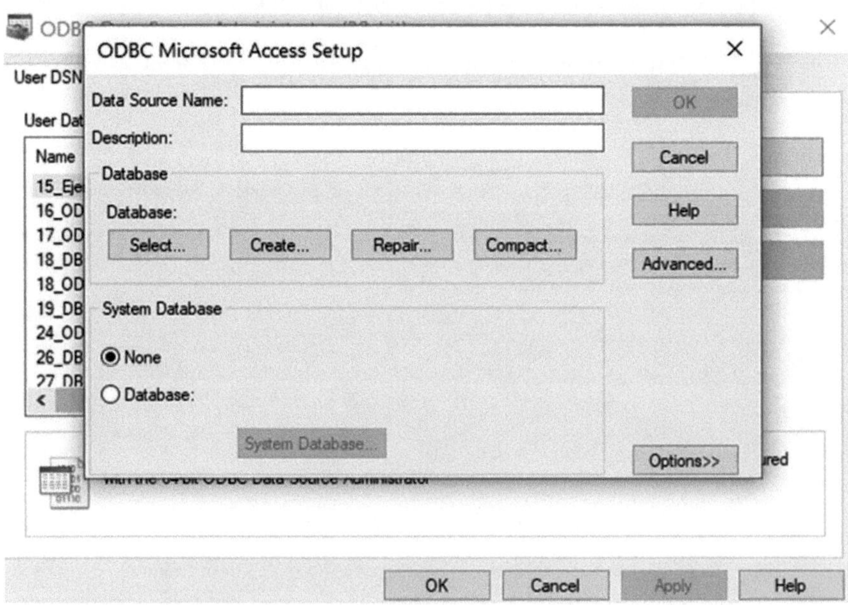

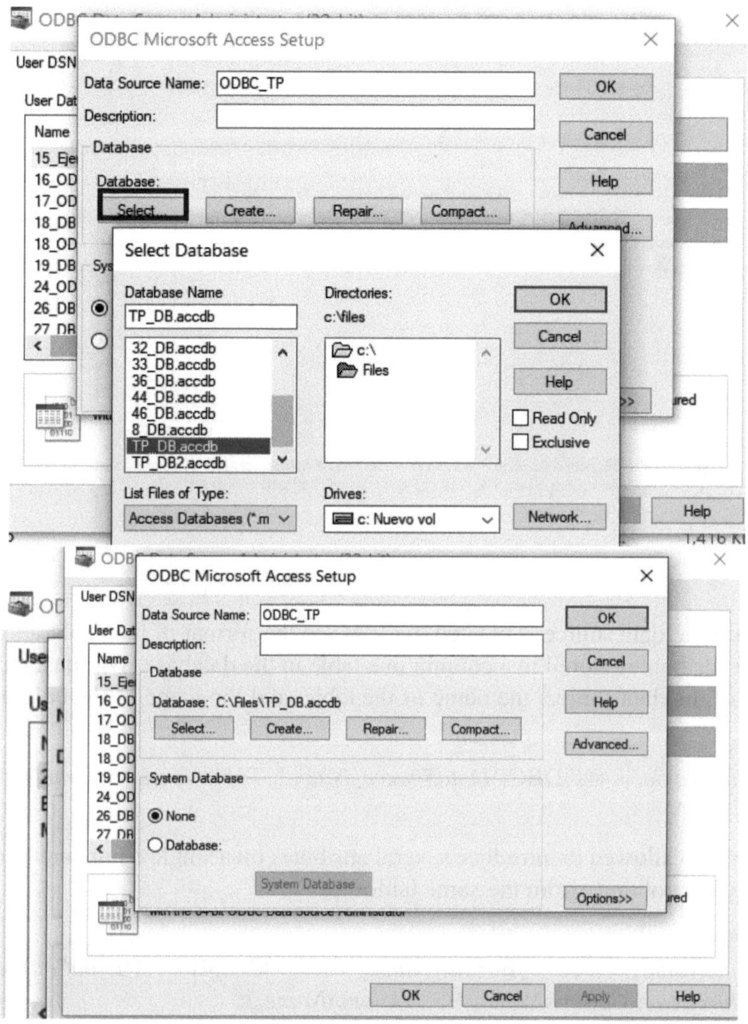
ODBC Microsoft Access Setup
Data Source Name: ODBC_TP
Description:
Database
Database:
Select...
Create...
Repair...
Compact...
OK
Cancel
Help
Select Database
Database Name
TP_DB.accdb
32_DB.accdb
33_DB.accdb
36_DB.accdb
44_DB.accdb
46_DB.accdb
8_DB.accdb
TP_DB.accdb
TP_DB2.accdb
Directories:
c:\files
c:\
Files
OK
Cancel
Help
Read Only
Exclusive
List Files of Type:
Access Databases (*.m
Drives:
c: Nuevo vol
Network...
ODBC Microsoft Access Setup
Data Source Name: ODBC_TP
Description:
Database
Database: C:\Files\TP_DB.accdb
Select...
Create...
Repair...
Compact...
OK
Cancel
Help
Advanced...
System Database
None
Database:
System Database
Options>>
OK
Cancel
Apply
Help

4. The new data source has been added to the list of existing data sources.

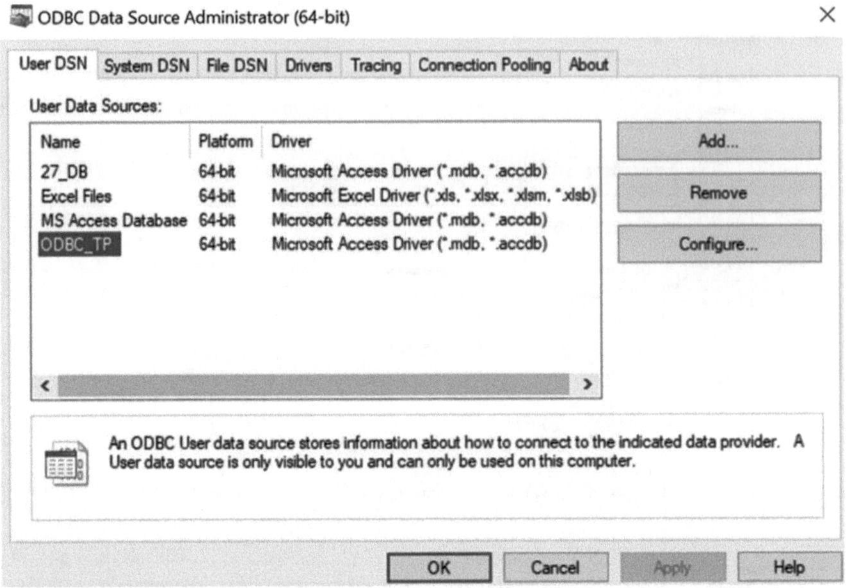

Once the data source is created, we can see the format of the declaration:

An attribute is stored in a column in a table in the database. We must indicate the name of the data source, the name of the table, and the name of the column:

DATA:
AttributeName = @ODBC('*DataSource_Name*', *TableName*, *ColumnName*);
ENDDATA

It is also allowed to introduce several attributes on a single command line, as long as they are columns from the same table:

DATA:
$AttributeName_1$, . . ., $AttributeName_n$ = @ODBC ('*DataSource_Name*', *TableName*, $ColumnName_1$, . . ., $ColumnName_n$);
ENDDATA

For our example:

Before creating the data source, we have created the database. The structure of the database is shown below. It has three tables (warehouses, customers, costs). The three tables are shown in Fig. 2.8.

The ID columns that index the tables are not required. They have been introduced to have a relational database. In either case, LINGO reads the information in order of how it was entered.

The format for reading would be the following.

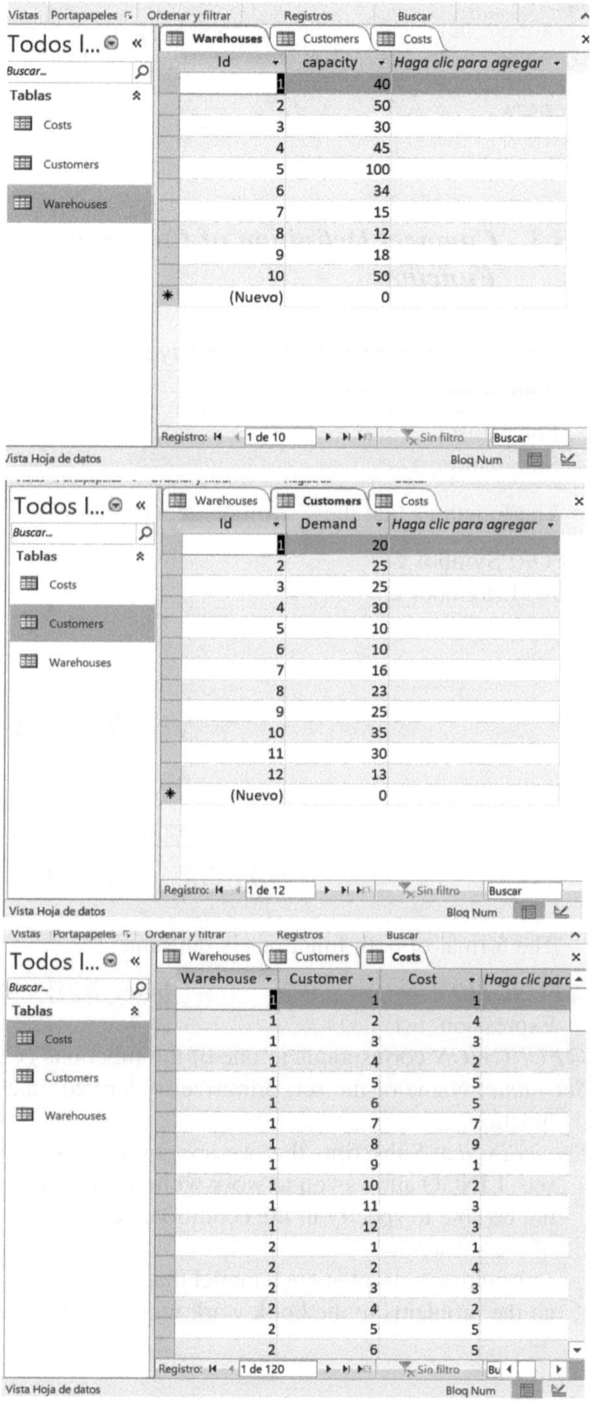

Id	capacity
1	40
2	50
3	30
4	45
5	100
6	34
7	15
8	12
9	18
10	50
(Nuevo)	0

Id	Demand
1	20
2	25
3	25
4	30
5	10
6	10
7	16
8	23
9	25
10	35
11	30
12	13
(Nuevo)	0

Warehouse	Customer	Cost
1	1	1
1	2	4
1	3	3
1	4	2
1	5	5
1	6	5
1	7	7
1	8	9
1	9	1
1	10	2
1	11	3
1	12	3
2	1	1
2	2	4
2	3	3
2	4	2
2	5	5
2	6	5

Fig. 2.8 Warehouses, customers, and costs

```
DATA:
 K =@ODBC(ODBC_TP,'Warehouses','capacity');
 D =@ODBC(ODBC_TP,'Customers','demand');
 c =@ODBC(ODBC_TP,'Costs','cost');
ENDDATA
```

2.3.3 *Compact Definition of Constraints and Objective Function*

For the compact definition, we must pay attention to the two mathematical symbols of compaction in a model:

- For All: ∀
- Summation: Σ

The functions that these symbols represent are:

@FOR: Symbol ∀
@SUM: Symbol Σ

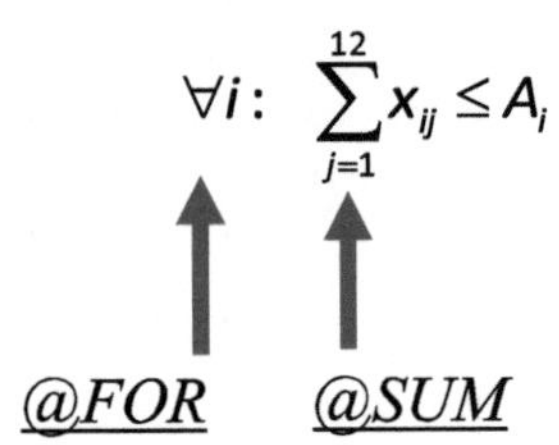

The format of both functions is the same:

@FUNCTION (*Set_name* {(*Set_index_list*)} {| |Conditional_qualifier}: Expression_list) ;

@FUNCTION corresponds to one of the functions (@FOR, @SUM).

Set_name: Name of the set, primitive or derived, that we cover in the For or Sum function.

Set_index_list: Subscripts that we are going to use to reference the elements of the sets. LINGO allows you to work without subscripts, although in that case we will not be able to specify in the conditional qualifier a condition regarding the value of the subscripts.

Looking to implement a model that is as close to mathematical representation, all the problems in the book work with subscripts.

Conditional_qualifier: It constitutes the "such that" (/) that can be applied in either of the two functions. It is used to limit the elements of the set to which the function is applied. The signs used in the conditional expression are defined as follows:

Sign	Representation in LINGO
$=$	#EQ#
$\neq$	#NE#
$\geq$	#GE#
$>$	#GT#
$<$	#LT#
$\leq$	#LE#
Y	#And#
O	#Or#

Expression_list: They are the mathematical constraints applied in the case of FOR, or the function to be added in the case of SUM. It has the same format as in the extended version. For @FOR, expression_list can contain multiple expressions, separated by;. The @SUM function supports only one expression.

Every restriction, as in the extended version, ends with;.
As an illustration, we develop the model of our transportation problem.

$$\forall i : \sum_{j=1}^{12} x_{ij} \leq K_i$$

```
@FOR(Warehouses(i):@SUM(Customers(j): x(i,j))<= K(i)) ;
```

FUNCTION	(	Set_name	(set_index_list)	:	Expression								)
@FOR	(	Warehouses	(i)	:	@SUM	(	Customers	(j)	:	x(i,j)	)	<= K(i)	)
					FUNCTION	(	Set_name	(set_index_list)	:	expression	)		

Without the use of subscripts:

```
@FOR(Warehouses:@SUM(Customers: x)<=K) ;
```

$$\forall j : \sum_{i=1}^{10} x_{ij} \geq D_j$$

```
@FOR(Customers(j):@SUM(Warehouses(i):x(i,j))>=D(j));
```

$$\forall j/\mathrm{CC}_j = 1 : x_{1j} \geq 1$$

```
@FOR(Customers(j)|CC(j)#EQ#1:x(1,j)>=1);
```

$$\forall i, \forall j : x_{ij} \geq 0$$

It is not necessary to declare the variables as integers since the unimodularity of the coefficient matrix ensures an integer solution in the optimum of the problem. Without declaring the type of variables, LINGO interprets them as continuous variables. In case their declaration as integer variables had been necessary, the format would have been:

```
@FOR(Warehouses(i):@FOR(Customers(j):@GIN(x(i,j))));
```

Objective function:

$$\text{Min} \sum_{i=1}^{10} \sum_{j=1}^{12} c_{ij} x_{ij}$$

```
MIN = @SUM(Warehouses(i):@SUM(Customers(j):c(i,j)*x(i,j)));
```

The function that adds the elements of the derived set would have been equivalent, since it collects all combinations of customers with warehouses.

```
MIN = @SUM(WC(i,j):c(i,j)*x(i,j));
```

The next section discusses some drawbacks in this type of equivalency.

2.3.4 *Incompatibilities in the Use of Primitive Sets Instead of Derived Sets*

As we have just shown, the primitive sets that constitute a derived set produce an alternative in the definition of expressions, being equivalent to the same function to loop through each primitive set in a chained way or to loop through the derived set. As an example, we present the objective function of our illustration that we just presented in the previous section.

However, if a subset of valid combinations is specified in the definition of the derived set, then the option of looping through the primitive sets is not a valid option.

Suppose the model in Example 2.6 where a subset of valid combinations for warehouses and customers was defined:

```
MODEL:
DATA:
 n=@OLE('c:\Files\TPSheet1.xlsx', n);
 m=@OLE('c:\Files\TPSheet1.xlsx', m);
ENDDATA
SETS:
 Warehouses/1..n/: K,A,B;
 Customers/1..m/:D,CC;
 WC( Warehouses, Customers)/@OLE('c:\Files\TPSheet1.xlsx', pairs)/:
x,c;
ENDSETS
DATA:
 K =@OLE('c:\Files\TPSheet1.xlsx', capacity);
 D =@OLE('c:\Files\TPSheet1.xlsx', demand);
 c =@OLE('c:\Files\TPSheet1.xlsx',costs);
ENDDATA
```

On this model, posing the constraints and objective function in the format of looping through the primitive sets, as we did in the previous illustrations, would not be correct:

```
@FOR(Warehouses(i): @SUM(Customers(j): x(i,j))<=K(i));
@FOR(Customers(j): @SUM(Warehouses(i): x(i,j))>=D(j));
MIN = @SUM(Warehouses(i):@SUM(Customers(j):c(i,j)*x(i,j)));
```

It would not be valid because the sums would loop through undeclared combinations. An error message is displayed: Subscript out of range (Fig. 2.9).

Therefore, the way to define the above constraints and the objective function should be as follows:

```
@FOR(Warehouses(i): @SUM(WC(i,j): x(i,j)) <= K(i));
@FOR(Customers(j): @SUM(WC(i,j): x(i,j)) >= D(j));
MIN = @SUM(WC(i,j):c(i,j)*x(i,j));
```

In this way, it only considers the variables $x(i, j)$ and the costs $c(i, j)$ defined in the derived set WC.

2.3.5 CALC Section

Section dedicated to the attributes that are defined as calculated data and, therefore, their values are not obtained from any external file nor are they integrated into the LINGO file. Their values are calculated from the values of other attributes.

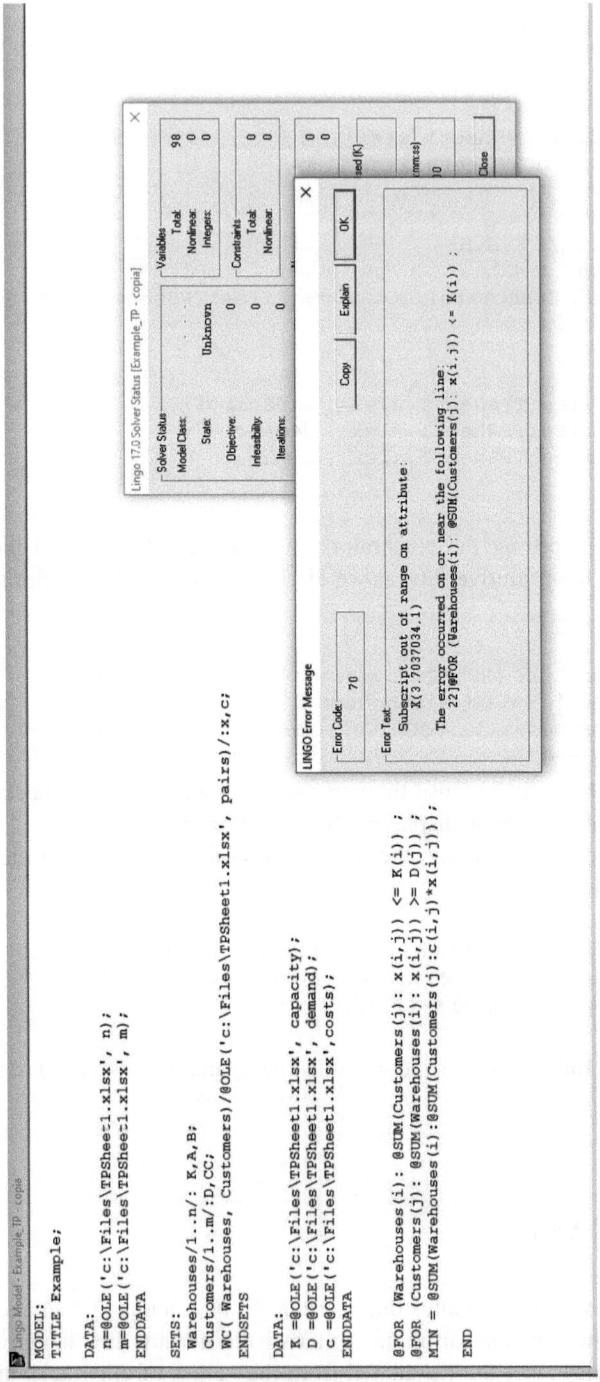

Fig. 2.9 Error message

Format
CALC:
Calculated data
ENDCALC

This section may require the use of the @FOR and @SUM functions. The @FOR function is used if we need to calculate an attribute for the elements of a set. The @SUM function will be necessary if it is necessary to perform the sum of any attribute of the elements of a set. It may also be necessary to use other predefined LINGO functions. A comprehensive set of functions can be found in the LINGO manual. Here, we present the most important ones:

@MAX
This returns the maximum value of the *expression* taken over the set *setname*:
@MAX(*setname* [(*set_index_list*) [| *cond_qualifier*]]: *expression*)

@MIN
This returns the minimum value of *expression* taken over the set *setname*:
@MIN(*setname* [(*set_index_list*) [| *cond_qualifier*]]: *expression*)

Example 2.7 @MAX and @MIN Illustration
If we want to collect the highest and lowest value of the demand:

```
MaxDemand = @MAX(Customers(j):D(j));
MinDemand = @MIN(Customers(j):D(j));
```

If we want to collect the highest shipping cost from each customer:

We need to include the attribute in the SETS section, since it is associated with the elements of a set.

```
SETS:
 Warehouses/1..10/: K,A,B;
 Customers/1..12/:D, CC, MaxCost;
 WC( Warehouses, Customers):x,c;
ENDSETS
DATA:
 K =@OLE('c:\Files\TPSheet.xlsx', capacity);
 D =@OLE('c:\Files\TPSheet.xlsx', demand);
 c =@OLE('c:\Files\TPSheet.xlsx', costs);
ENDDATA
CALC:
@FOR(Customers(j):MaxCost(j)=
@MAX(Warehouses(i):c(i,j)));
ENDCALC
```

LINGO allows us to use the @MAX and @MIN functions with variables, but in that case it does not internally linearize the function and makes the model non-linear. For this reason, I do not recommend its use to calculate maximums and minimums on a set of variables. For that, it is better to make a logical proposition to obtain the value. In the book, only these functions will be used on the data to calculate maximum or minimum values.

@IF

@IF(*logical_condition*, *true_result*, *false_result*)

The @IF function evaluates the *logical_condition* and, if true, returns *true_result*. Otherwise, it returns *false_result*.

Example 2.8 Illustration of the Function @IF

In our example, we have a calculated attribute, the CC_j data of binary type:

$CC_j = 1$ if the sum of the costs of customer j is greater than 20; 0 otherwise.

Mathematically:

$$\forall j: \text{ IF } \sum_{i=1}^{10} c_{ij} > 20 \text{ THEN } CC_j = 1; \text{OTHERWISE } CC_j = 0$$

In LINGO:

```
CALC:
@FOR(Customers (j): CC(j)= @IF(@SUM(Warehouses(i):c(i,j))
#GT#20,1,0));
ENDCALC
```

2.4 Execution of the Solving Process

The model solving process is executed from the LINGO-> Solve Tools menu.

There is also a shortcut icon for the same function on the toolbar (Fig. 2.10).

During model resolution, LINGO presents a resolution status window (Fig. 2.11) that presents information on whether admissible solutions have been found, the best solution found, the optimum bound, model type, model size, execution time, and some other secondary data.

Prior to execution, in the LINGO menu itself, in the Options section, the resolution parameters can be set, referring to the type of resolution algorithm, the branching strategies, tolerances, etc. The default options are usually well determined; no changes are necessary.

Only when the model is very large, it may be convenient to expand the memory used in the *Model Generator* tab (Fig. 2.12). Sometimes the amount of memory available for the internal generation of the resolution process is a reason for an error in the execution.

Fig. 2.10 Toolbar

Fig. 2.11 Resolution status window

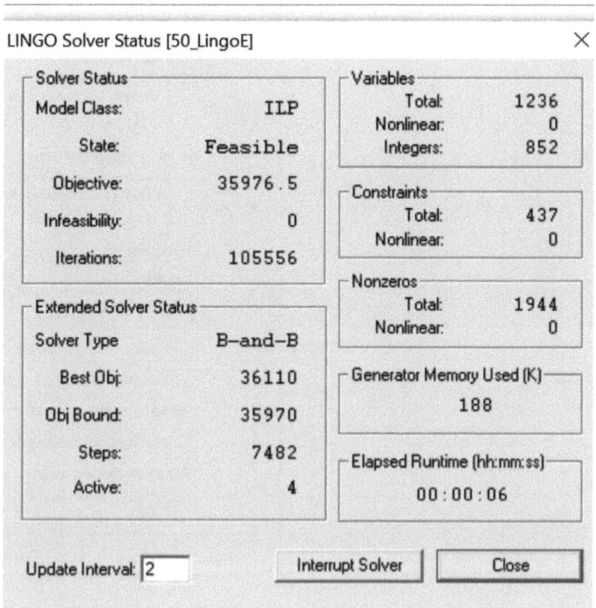

2.5 LINGO Solution Report

LINGO always reports a text file with the resolution results. However, it may be more interesting for us to create the format for the results.

The report file that LINGO creates has the structure shown in Fig. 2.13.

Most relevant information:

- Value of the objective function
- Execution time in seconds.
- Model class (more relevant):
  - LP: Linear programming
  - ILP: Integer linear programming
  - PILP: Pure integer linear programming
  - MILP: Mixed integer linear programming
  - QP: Quadratic programming
  - NLP: Non-linear programming

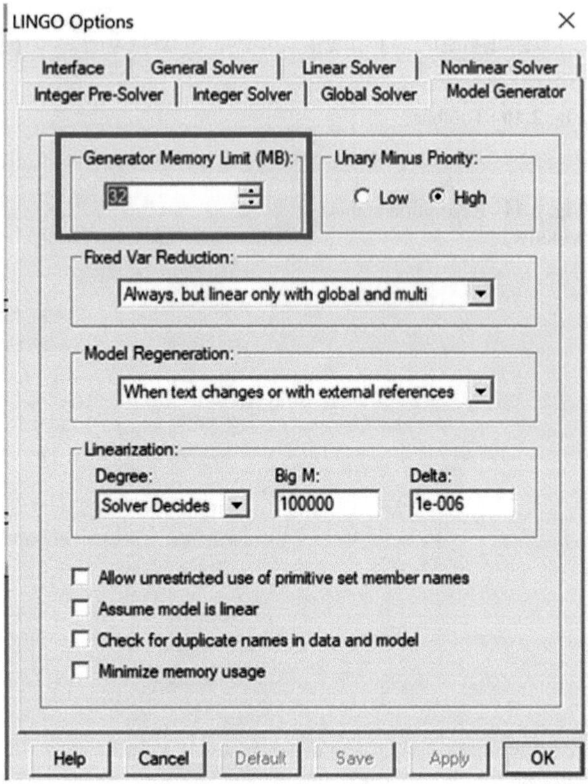

Fig. 2.12 Generator memory limit

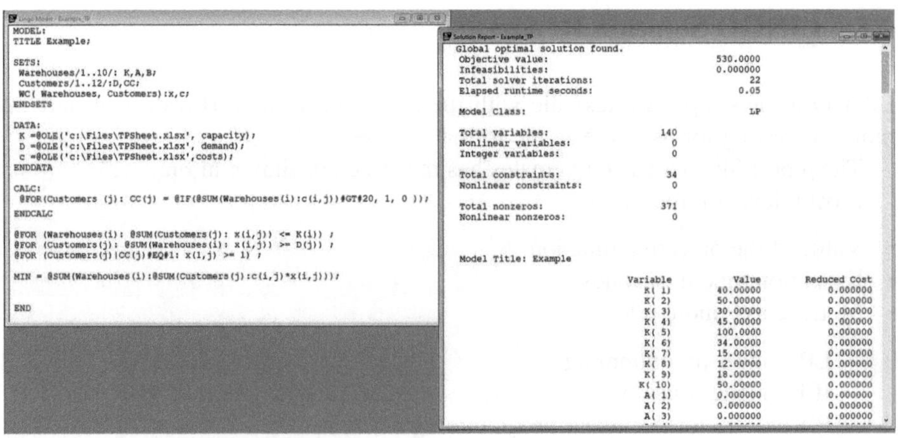

Fig. 2.13 LINGO report file

- Total variables
- Total constraints

In the same file, LINGO reports the value of all data and variables of the problem, including their cost in the objective function. Also, in the last lines it shows the slack of the constraints and value of the dual price:

2.6 Output Data Format

As for input data, LINGO offers three possibilities:

– Write the results in an external text file.
– Write the results in a spreadsheet file.
– Write the results into a database.

The format is similar to the use of these resources for the input data. It should always be done in a DATA section.

2.6.1 Output Data in External Text Files

Formats
DATA:
@TEXT('*Filepath & Filename*') = @WRITE(obj_1{, . . ., obj_n}) ;
@TEXT('*Filepath & Filename*') = @WRITEFOR (*setname*{ (*set_index_list*) {| *cond_qualifier*}}: obj_1{, . . ., obj_n});
ENDDATA

By object obji, we can understand any text string, data, variable, or function.
A widely used function in printing data to a text file is:

@NEWLINE (*n*)

Insert *n* blank lines into the output file.
Let us consider several examples of printing results to a file.

Example 2.9 Writing Output Data in a Text File

```
DATA:
@TEXT("c:\Files\Out_TP.txt")=@WRITE("Results ", ":",@NEWLINE(1));
@TEXT("c:\Files\Out_TP.txt")=@WRITE("Customer 1's data is: "
,"Demand=", D(1), ", Cost with Warehouse 1= ", C(1,1), @NEWLINE(1));
@TEXT("c:\Files\Out_TP.txt")=@WRITEFOR(Warehouses(i)|i#GT#1: "Cost
with Warehouse ", i, "= ", C(i,1),@NEWLINE(1));
ENDDATA
```

Output
Results:

Customer 1's data is: Demand=20, Cost with Warehouse 1 = 1
Cost with Warehouse 2= 1
Cost with Warehouse 3= 4
Cost with Warehouse 4= 3
Cost with Warehouse 5 = 2
Cost with Warehouse 6= 3
Cost with Warehouse 7= 1
Cost with Warehouse 8= 1
Cost with Warehouse 9 = 3
Cost with Warehouse 10= 3

```
DATA:
@TEXT("c:\Files\Out_TP.txt")=@WRITE("Shipments to each Customer:",
@NEWLINE(1));
@TEXT("c:\Files\Out_TP.txt")=@WRITEFOR(Customers(j):"Customer ", j,
": ", @WRITEFOR(Warehouses(i): "Warehouse " ,i,"= ", X(i,j),"; "),
@NEWLINE(1));
ENDDATA
```

Result
Shipments to each Customer:
Customer 1: A1= 1; A2= 10; A3= 0; A4= 0; A5= 0; A6= 0; A7= 9; A8= 0; A9= 0; A10= 0;
Customer 2: A1= 1; A2= 0; A3= 0; A4= 7; A5= 0; A6= 13; A7= 0; A8= 0; A9= 0; A10= 4;
Customer 3: A1= 1; A2= 18; A3= 6; A4= 0; A5= 0; A6= 0; A7= 0; A8= 0; A9= 0; A10= 0;
Customer 4: A1= 30; A2= 0; A3= 0; A4= 0; A5= 0; A6= 0; A7= 0; A8= 0; A9= 0; A10= 0;
Customer 5: A1= 1; A2= 0; A3= 0; A4= 9; A5= 0; A6= 0; A7= 0; A8= 0; A9= 0; A10= 0;
Customer 6: A1= 1; A2= 0; A3= 0; A4= 0; A5= 0; A6= 9; A7= 0; A8= 0; A9= 0; A10= 0;
Customer 7: A1= 1; A2= 0; A3= 0; A4= 0; A5= 0; A6= 0; A7= 0; A8= 12; A9= 0; A10= 3;
Customer 8: A1= 1; A2= 22; A3= 0; A4= 0; A5= 0; A6= 0; A7= 0; A8= 0; A9= 0; A10= 0;
Customer 9: A1= 1; A2= 0; A3= 24; A4= 0; A5= 0; A6= 0; A7= 0; A8= 0; A9= 0; A10= 0;
Customer 10: A1= 0; A2= 0; A3= 0; A4= 0; A5= 35; A6= 0; A7= 0; A8= 0; A9= 0; A10= 0;
Customer 11: A1= 1; A2= 0; A3= 0; A4= 29; A5= 0; A6= 0; A7= 0; A8= 0; A9= 0; A10= 0;

Customer 12: A1= 1; A2= 0; A3= 0; A4= 0; A5= 0; A6= 12; A7= 0; A8= 0; A9= 0; A10= 0;

```
DATA:
@TEXT("c:\Files\Out_TP.txt")=@WRITE("We now write the solution in
another format:",@NEWLINE(1));
@TEXT("c:\Files\Out_TP.txt")=@WRITEFOR(Customers(j):"Customer ", j,
": ",@NEWLINE(1),
@WRITEFOR(Warehouses(i):@IF(x(i,j)#GT#0,"Warehouse ",''),@IF(x(i,
j)#GT#0,i,''),
@IF(x(i,j)#GT#0,': ',''), @IF(x(i,j)#GT#0, x(i,j),''),@IF(x(i,j)
#GT#0," uds. ",'')),@NEWLINE(1));
ENDDATA
```

Result

We now write the solution in another format:

Customer 1:

Warehouse 1: 1 uds. Warehouse 2: 10 uds. Warehouse 7: 9 uds.

Customer 2:

Warehouse 1: 1 uds. Warehouse 4: 7 uds. Warehouse 6: 13 uds. Warehouse 10: 4 uds.

Customer 3:

Warehouse 1: 1 uds. Warehouse 2: 18 uds. Warehouse 3: 6 uds.

Customer 4:

Warehouse 1: 30 uds.

Customer 5:

Warehouse 1: 1 uds. Warehouse 4: 9 uds.

Customer 6:

Warehouse 1: 1 uds. Warehouse 6: 9 uds.

Customer 7:

Warehouse 1: 1 uds. Warehouse 8: 12 uds. Warehouse 10: 3 uds.

Customer 8:

Warehouse 1: 1 uds. Warehouse 2: 22 uds.

Customer 9:

Warehouse 1: 1 uds. Warehouse 3: 24 uds.

Customer 10:

Warehouse 5: 35 uds.

Customer 11:

Warehouse 1: 1 uds. Warehouse 4: 29 uds.

Customer 12:

Warehouse 1: 1 uds. Warehouse 6: 12 uds.

2.6.2 Output Data in Spreadsheet Files

Formats

DATA:

@OLE(‘*PathName & FileName*’, *RangeName*$_1$,. . ., *RangeName*$_n$)= @WRITE (*obj*$_1${, . . ., *obj*$_n$}) ;

@OLE(‘*PathName & FileName*’, *RangeName*$_1$,. . ., *RangeName*$_n$)= @WRITEFOR (*setname* {(*set_index_list*)} {| *cond_qualifier*}: *obj*$_1${, . . ., *obj*$_n$});

ENDDATA

Previously, we had to create the range in the spreadsheet. Information is transferred with the @WRITE and @WRITEFOR functions, just like text files.

Example 2.10 Writing Output Data in a Spreadsheet for the Transportation Problem

We prepare in the spreadsheet a range of 120 cells to store the solution. We call the range solution (Fig. 2.14).

In the LINGO file, we create a DATA section to print the values shipped from each warehouse to each customer.

```
DATA:
@OLE("c:\Files\TPSheet.xlsx",'Solution') = @WRITEFOR(WC(i,j):x(i,
j));
ENDDATA
```

The result is showed in Fig. 2.15.

2.6.3 Output Data in Database Files

Formats

DATA

@ODBC(*data_source_name*, ‘*table_name*’, ‘*column_name*$_1$’,.., ‘*column_name*$_n$’,) = @WRITE(*obj*$_1${, . . ., *obj*$_n$}) ;

@ODBC(*data_source_name*, ‘*table_name*’, ‘*column_name*$_1$’,. . ., ‘*column_name*$_n$’)= @WRITEFOR (*setname* {(*set_index_list*)} {| *cond_qualifier*}: *obj*$_1${, . . ., *obj*$_n$});

ENDDATA

To print information in a database from LINGO, the only requirement is that the records on which the information is to be printed are created. LINGO does not create the record. This is shown in the following illustration.

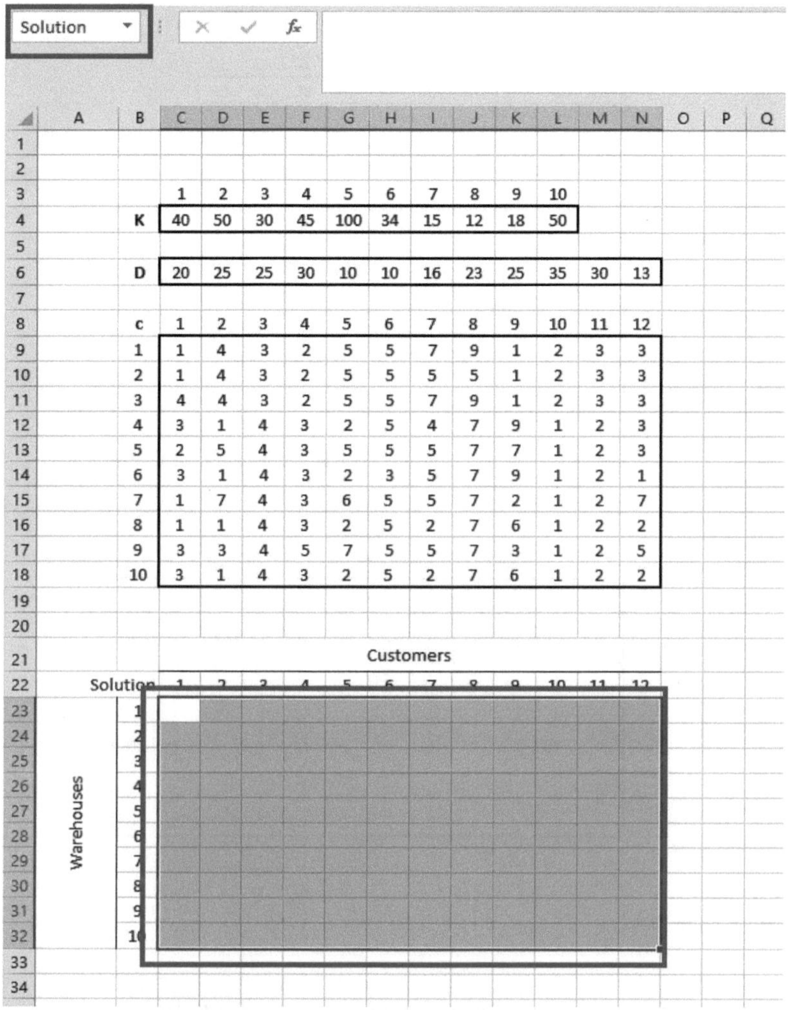

Fig. 2.14 Range solution in the spreadsheet

Example 2.11 Writing Output Data in a Database for the Transportation Problem

To illustrate, we introduce a table in the database "*TP_DB.accdb*" to save the solution (Fig. 2.16).

We have created the 120 records where the information will be registered (there are 120 variables x_{ij}). The columns to record are empty. We will print the warehouse, the customer, and the quantity shipped x_{ij}.

The DATA section for this would be the following:

	1	2	3	4	5	6	7	8	9	10		
K	40	50	30	45	100	34	15	12	18	50		
D	20	25	25	30	10	10	16	23	25	35	30	13

c	1	2	3	4	5	6	7	8	9	10	11	12
1	1	4	3	2	5	5	7	9	1	2	3	3
2	1	4	3	2	5	5	5	5	1	2	3	3
3	4	4	3	2	5	5	7	9	1	2	3	3
4	3	1	4	3	2	5	4	7	9	1	2	3
5	2	5	4	3	5	5	5	7	7	1	2	3
6	3	1	4	3	2	3	5	7	9	1	2	1
7	1	7	4	3	6	5	5	7	2	1	2	7
8	1	1	4	3	2	5	2	7	6	1	2	2
9	3	3	4	5	7	5	5	7	3	1	2	5
10	3	1	4	3	2	5	2	7	6	1	2	2

		Customers											
	Solution	1	2	3	4	5	6	7	8	9	10	11	12
Warehouses	1	1	1	1	30	1	1	1	1	1	0	1	1
	2	10	0	18	0	0	0	0	22	0	0	0	0
	3	0	0	6	0	0	0	0	0	24	0	0	0
	4	0	7	0	0	9	0	0	0	0	0	29	0
	5	0	0	0	0	0	0	0	0	0	35	0	0
	6	0	13	0	0	0	9	0	0	0	0	0	12
	7	9	0	0	0	0	0	0	0	0	0	0	0
	8	0	0	0	0	0	0	12	0	0	0	0	0
	9	0	0	0	0	0	0	0	0	0	0	0	0
	10	0	4	0	0	0	0	3	0	0	0	0	0

Fig. 2.15 Output data

```
DATA:
@ODBC(ODBC_TP,'Solution','warehouse','customer','x')=
@WRITEFOR(WC(i,j):i,j,x(i,j));
ENDDATA
```

The code fills the solution in the *solution* table (Fig. 2.17):

Fig. 2.16 *Solution* table

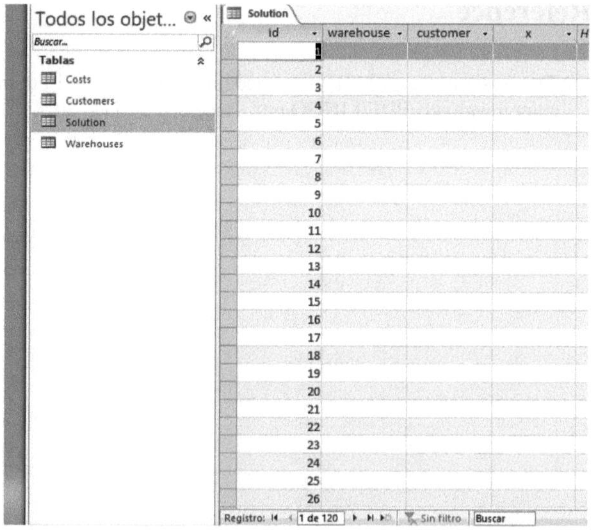

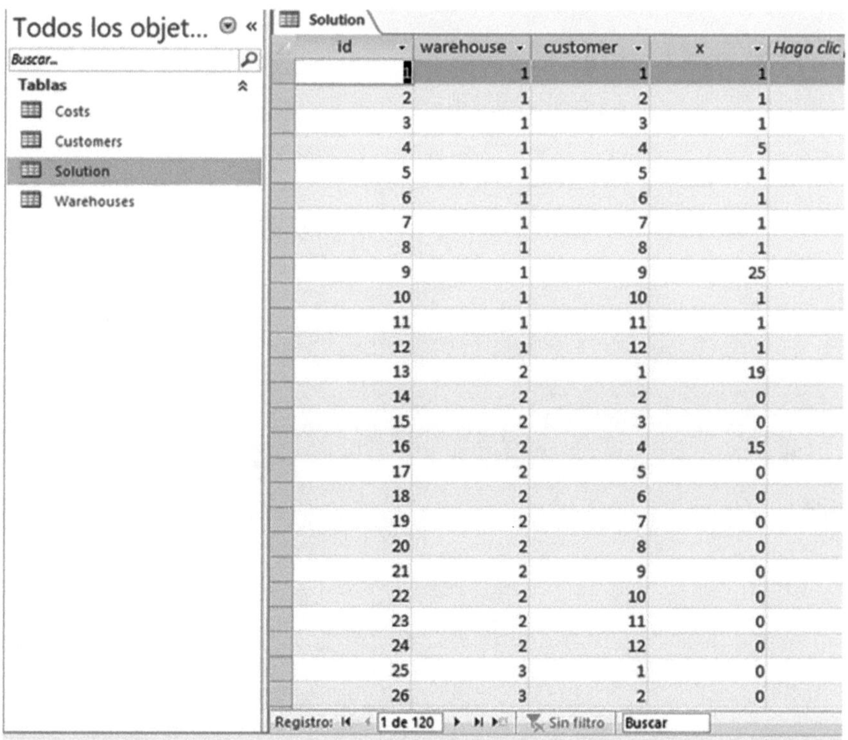

id	warehouse	customer	x
1	1	1	1
2	1	2	1
3	1	3	1
4	1	4	5
5	1	5	1
6	1	6	1
7	1	7	1
8	1	8	1
9	1	9	25
10	1	10	1
11	1	11	1
12	1	12	1
13	2	1	19
14	2	2	0
15	2	3	0
16	2	4	15
17	2	5	0
18	2	6	0
19	2	7	0
20	2	8	0
21	2	9	0
22	2	10	0
23	2	11	0
24	2	12	0
25	3	1	0
26	3	2	0

Fig. 2.17 Printed solution

Reference

LINDO Systems Inc. (2021) The modeling language and optimizer. LINGO. https://www.lindo.com/downloads/PDF/LINGO.pdf. Accessed 31 Dec 2021

Chapter 3
Low-Level Problems

3.1 Introduction

Firstly, we present the specific characteristics that each problem covers with respect to the methodology, in addition to the following information:

- Number of versions
- Difficulty in data handling
- Location of input and output data

We will also show the input data and results of the Lingo execution.
Problems:

Number	Problem
1	Product manufacturing
2	Fruit sale
3	Securities portfolio
4	Capital investments
5	Number partitioning
6	Installation of parking spaces for electric Cars
7	Independent edge set

Versions and Elements:

		Elements				
Problem	Versions	Collective elements	Measurable individual elements	Grouping of items	Upper bound in the number of elements	Data handling
1	1		✓			Low
2	1	✓				Low
3	1	✓				Low

(continued)

J. M. García Sánchez, *Building and Solving Mathematical Programming Models*, International Series in Operations Research & Management Science 329,
https://doi.org/10.1007/978-3-030-97626-2_3

Problem	Versions	Elements				
		Collective elements	Measurable individual elements	Grouping of items	Upper bound in the number of elements	Data handling
4	1		✓			Low
5	2					Low
6	1	✓				Low
7	2					Medium

Calculations:

Problem	Binary logical calculations	Not-binary logical calculations	Bounds calculations	Conditioned bounds calculations
1				
2				
3				
4				
5			✓	
6			✓	
7				

Specifications:

Problem	QSS[a]	Capacity	Demand	Bound imposition	Allocation/assignment
1		✓		✓	
2			✓		
3		✓		✓	✓
4					✓
5	✓			✓	✓
6		✓		✓	✓
7	✓				

[a]*QSS* quantitative selection specifications

Problem	Flow balance	PLS[a]—One operator	PLS—Several operators	PLS—Attributes	Variable costs cases
1					
2					
3					
4	✓				
5					
6					
7					

[a]*PLS* propositional logic specifications

Objective criterion, input data and output data:

	Objective criterion		Input data				
Problem	Calculations	Particular cases	Integrated data	Text file	Spreadsheet	Database	Output data
1			✓				✓
2			✓				✓
3			✓				✓
4			✓				
5	✓				✓		✓
6	✓		✓				
7						✓	✓

3.2 Problem 1: Product Manufacturing

A company has 1000 tons of mineral B1, 2000 tons of mineral B2, and 500 tons of B3. From these materials, products A1, A2, and A3 can be obtained. The company wants to determine the quantity of each product that must be manufactured in order to obtain the maximum economic benefit from the operation.

Next, the necessary quantity of each mineral is detailed to obtain 1 ton of each product and the benefit obtained with each one (Table 3.1).

The company must also take into account:

- that it should not manufacture more than 10 tons of A2, since there is not as much demand in the market.
- that there is a company that buys mineral B2 at a price of $20/ton.

Maximize the company's profit.

Table 3.1 Products information

Product	B1	B2	B3	Benefits ($/T.)
A1	5	10	10	300
A2	5	8	5	200
A3	10	5	0	50

Modelling

List of elements in Table 3.2.

We include the sale price of mineral B2 in the table. To do this, we assign a null value to the price of the other minerals. In a similar way, we proceed with the production cap of the products.

Anyway, we could also have omitted those data in the table and used the values directly in the problem specifications.

Table 3.2 Problem 1: table of elements

Elements	Set	QN	Data				
			Name	Parameter	Type	Belonging	Value
Minerals	$i = 1 \ldots 3$ {B1, B2, B3}	I_U	Necessary quantity	Q_{ij}	C (Tons)	S	. . .
			Capacity	K_i	C (Tons)	W	{1000, 2000, 500}
			Price	P_i			{–, 20, –}
Products	$j = 1 \ldots 3$ {A1, A2, A3}	I_M	Profit	Pr_i	C ($)	W	{300, 200, 50}
			Production Limit	T_j	C (Tons)	W	{–, 10, –}
				Q_{ij}			

Decision Activities
Manufacture products (*j*)

$$\forall j : x_j = \text{Amount of product } j \text{ manufactured}$$

Sell minerals (i): Event: *i* = 2

$$y_2 = \text{Amount of mineral 2 sold}$$

Specifications

- Capacity of each mineral (capacity specification):

$$\forall i/i \neq 2 : \sum_{j=1}^{3} q_{ij} x_j \leq K_i \tag{1.1}$$

$$i = 2 : \sum_{j=1}^{3} q_{2j} x_j + y_2 \leq K_2 \tag{1.2}$$

- Do not manufacture more than 10T of product A2 (upper bound imposition)

$$j = 2 : x_2 \leq T_2 \tag{1.3}$$

Objective Function
Maximize profits:

$$\text{Max} \sum_{j=1}^{3} Pr_j x_j + p_2 y_2 \tag{1.4}$$

Complete Model

$$\text{Max} \quad \sum_{j=1}^{3} Pr_j x_j + p_2 y_2 \tag{1.4}$$

s.t.

$$\forall i / i \neq 2 : \sum_{j=1}^{3} q_{ij} x_j \leq K_i \tag{1.1}$$

$$i = 2 : \sum_{j=1}^{3} q_{2j} x_j + y_2 \leq K_2 \tag{1.2}$$

$$x_2 \leq T_2 \tag{1.3}$$

$$x_j \geq 0; y_2 \geq 0$$

Implementation in Lingo

```
MODEL:
SETS:
Minerals/1..3/: K,y;
Products/1..3/: Pr,T,x;
M_P(Minerals,Products): q;
ENDSETS
DATA:
k=1000 2000 500;
Pr= 300 200 50;
T= 0 10 0;
q= 5 5 10 10 8 5 10 5 0;
p2= 20;
ENDDATA

!1.1;
MAX = @SUM(Products(j): Pr(j)*x(j)) + p2*y(2);

!1.2;
@FOR(Minerals(i)|i#NE#2: @SUM(Products(j): q(i,j)*x(j))<=K(i));

!1.3;
@FOR(Minerals(i)|i#EQ#2: @SUM(Products(j):q(i,j)*x(j))+y(2)<=K
(i));

!1.4;
x(2)<=T(2);

DATA:
 @TEXT("c:\Files_\Out1.txt") = @WRITEFOR(Products(j):'Product ', j, "
= " , x(j),@NEWLINE(1));
 @TEXT("c:\Files_\Out1.txt") = @WRITE(@NEWLINE(1));
 @TEXT("c:\Files_\Out1.txt") = @WRITE('OBJECTIVE FUNCTION VALUE:',
@SUM(Products(j):Pr(j)*x(j))+ p2*y(2));
ENDDATA

END
```

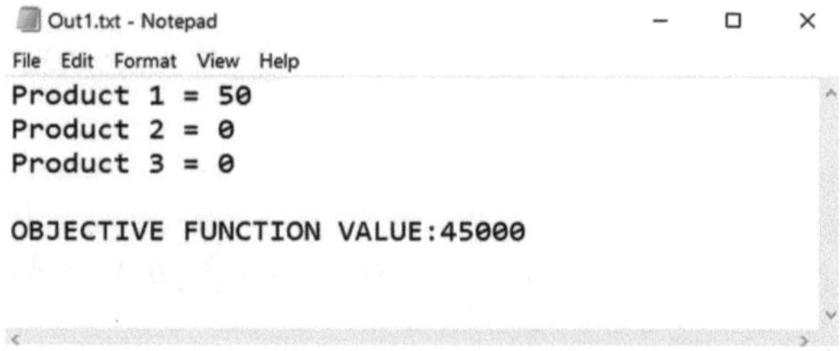

Fig. 3.1 Output file [C:\Files_\Out1.txt]

Solution
The solution is shown in Fig. 3.1.

3.3 Problem 2: Fruit Sale (Salazar 2021)

A fruit merchant needs 16 boxes of oranges, 5 boxes of bananas, and 20 boxes of apples. He uses wholesalers who are in a position to meet his needs, but only sell the fruit in full containers. Wholesaler A sends in each container 8 boxes of oranges, 1 of bananas, and 2 of apples. Wholesaler B sends 2 boxes of oranges, one of bananas, and 7 of apples in each container. Knowing that wholesaler A is 150 km away and wholesaler B is 300 km away, calculate how many containers we will have to buy from each wholesaler, in order to save time and minimizing the distance (each container means a trip).

Modeling
List of elements in Table 3.3.

Table 3.3 Problem 2–Table of elements

			Data				
Elements	Sct	QN	Name	Parameter	Type	Belonging	Value
Fruit boxes	$i = 1...3$ {Oranges, Bananas, Apples}	C_D	Demand	A_i	I	W	{16, 5, 20}
			Boxes in wholesaler container	CCM_{ij}	I	S	–
Wholesalers	$j = 1, 2$ {A, B}	I_U	Distance	Dj	C (Km)	W	{150, 300}
				CCM_{ij}			
Containers	–	C_I		CCM_{ij}			

The number of boxes carried by a container from each wholesaler is shared by these three elements, regardless of whether the container's participation is auxiliary. The methodology prepares the problem to provide flexibility in the case of increased features. That would be the case where, for example, there are two types of containers. In that case, we would have no doubt about the need to share the attribute with the set of container types.

Decision Activities

Buy containers from wholesalers (*j*)

$$\forall j : x_j = \text{Number of containers purchased from the wholesaler } j. \quad (2.1)$$

Specifications

– Demand for fruit boxes (demand specification):

Contribution ≥ Demand: As the unit contributions are different from 1, the specification is with a sign of ≥.

$$\forall i : \sum_{j=1}^{2} \text{CM}_{ij} x_j \geq A_i \quad (2.2)$$

Objective Function

Minimize distances:

$$\text{Min} \sum_{j=1}^{2} D_j x_j \quad (2.3)$$

Complete Model

$$\text{Min} \sum_{j=1}^{2} D_j x_j \quad (2.3)$$

s.t.

$$\forall i : \sum_{j=1}^{2} \text{CM}_{ij} x_j \geq A_i \quad (2.2)$$

$$\forall j : x_j \geq 0 \text{ Integers} \quad (2.1)$$

Implementation in Lingo

```
MODEL:
SETS:
Boxes/1..3/: A;
Wholesalers/1..2/:D, x;
```

```
C_M(Boxes,Wholesalers): CM;
ENDSETS

DATA:
A= 16 5 20;
D = 150 300;
CM= 8 2 1 1 2 7 ;
ENDDATA
!2.3; Min = @SUM(Wholesalers(j): D(j)*x(j));
!2.2; @FOR(Boxes(i): @SUM(Wholesalers(j): CM(i,j)*x(j))>= A(i));
!2.1; @FOR(Wholesalers(j):@GIN(x(j)));

DATA:
 @TEXT("c:\Files_\Out2.txt")= @WRITEFOR(Wholesalers(j):'Purchased
containers from ', j, " = " , x(j),@NEWLINE(1));
 @TEXT("c:\Files_\Out2.txt")= @WRITE(@NEWLINE(1));
 @TEXT("c:\Files_\Out2.txt")= @WRITE('OBJECTIVE FUNCTION VALUE:',
@SUM(Wholesalers(j): D(j)*x(j)));
ENDDATA
END
```

Solution

The solution is shown in Fig. 3.2.

3.4 Problem 3: Securities Portfolio (Castillo et al. 2002)

An investor owns shares in various securities. More specifically, he owns b shares of the Ai securities, $i = 1, 2, \ldots, m$. The current prices of these securities are v_i. Consider that it is possible to predict the dividends to be paid at the end of the year that begins and the final prices of the different securities; that is, A_i will pay d_i and will have a new price w_i.

Our goal is to adjust our portfolio, that is, the number of shares in each security that will be bought and sold at the current price, so that dividends are maximized.

Certain conditions must be ensured that a well-balanced portfolio must satisfy, such as:

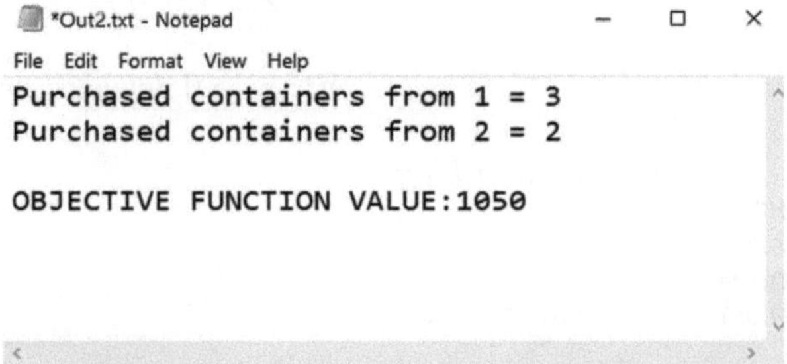

Fig. 3.2 Output file [C:\Files_\Out2.txt]

The total capital of the portfolio should not change with the adjustment; that is, the total capital that is bought is the same as that is sold.

The portfolio should avoid being overly dependent on any one value. This condition can be established by requiring that the capital associated with any specific security, after the purchase and sale adjustment, represents at least a certain percentage *r*% of the current total capital of the portfolio.

The total capital in the future must be at least the current capital plus a certain percentage *s*% of the capital currently invested.

Modeling

List of elements in Table 3.4.

Table 3.4 Problem 3: table of elements

			Data				
Elements	Set	QN	Name	Parameter	Type	Belonging	Value
Shares of securities	$i = 1 \ldots m$	C_I	Available shares	b_i	I	W	...
			Current value	v_i	C ($)	W	...
			Dividends	d_i	C ($)	W	...
			Future value	w_i	C ($)	W	...
System	–	I_U	Minimum percentage	r	C (%)	W	...
			Total capital percentage	s	C (%)	W	...

Decision Activities

Purchase shares of securities (*i*)

$$x_i : \text{Number of shares of the security } i \text{ purchased.} \tag{3.1}$$

Sell shares of securities (*i*)

$$y_i : \text{Number of shares of the security } i \text{ sold} \tag{3.2}$$

Auxiliary Calculation

Shares of a security at the end of the year:

$$\forall i : z_i = \text{Number of shares of the security } i \text{ after the buy-sell process}$$

$$\forall i : z_i = b_i + x_i - y_i \tag{3.3}$$

Specifications

– The available shares act as a cap on the amount that can be sold (capacity specification):

$$\forall i : y_i \leq b_i \tag{3.4}$$

- Total capital purchased = Total capital sold (balance specification):

$$\sum_{i=1}^{m} v_i x_i = \sum_{i=1}^{m} v_i y_i \tag{3.5}$$

- The capital of each security is at least a percentage r of the total capital (lower bound imposition):

$$\forall i : v_i z_i \geq \frac{r}{100} \sum_{j=1}^{m} v_j z_j \tag{3.6}$$

- The total future capital must be at least the current capital plus a percentage s of it (lower bound imposition):

$$\sum_{i=1}^{m} w_i z_i \geq \sum_{i=1}^{m} v_i z_i + \frac{s}{100} \sum_{i=1}^{m} v_i z_i \tag{3.7}$$

Objective Function
Maximize dividends:

$$\text{Max} \sum_{i=1}^{m} d_i z_i \tag{3.8}$$

Complete Model

$$\text{Max} \sum_{i=1}^{m} d_i z_i \tag{3.8}$$

s.t.

$$\forall i : z_i = b_i + x_i - y_i \tag{3.3}$$

$$\forall i : y_i \leq b_i \tag{3.4}$$

$$\sum_{i=1}^{m} v_i x_i = \sum_{i=1}^{m} v_i y_i \tag{3.5}$$

$$\forall i : v_i z_i \geq \frac{r}{100} \sum_{j=1}^{m} v_j z_j \tag{3.6}$$

$$\sum_{i=1}^{m} w_i z_i \geq \sum_{i=1}^{m} v_i z_i + \frac{s}{100} \sum_{i=1}^{m} v_i z_i \tag{3.7}$$

$$x_i, y_i \geq 0 \text{ integers} \qquad (3.1,\ 3.2)$$

$$z_i \geq 0$$

Implementation in Lingo

We assume a model with 20 values and data collected in Table 3.5 (data integrated in the lingo file):

$r = 3\%$; $s = 5\%$;

Table 3.5 Input data

i	b_i	v_i	d_i	w_i
1	12	7	0.8	9
2	120	7	0.5	9
3	50	7	0.5	4
4	40	5	0.8	10
5	150	7	0.2	6
6	100	7	2.3	8
7	25	7	1.3	5
8	30	5	1.9	6
9	200	8	1.1	11
10	120	9	0.8	5
11	70	4	0.3	6
12	70	3	0.5	4
13	50	2	0.2	1
14	100	2	0.3	1
15	15	1	0.1	1.5
16	25	3	1.1	7.7
17	50	6	0.9	10
18	50	5	0.7	5.5
19	60	6	2.3	5.5
20	40	9	1.3	10

```
MODEL:
SETS:
Securities/1..20/: b,v,d,w,x,y,z;
ENDSETS
DATA:
r=3;
s=5;
b=12 120 50 40 150 100 25 30 200 120 70 70 50 100 15 25 50 50 60 40;
v= 7 7 7 5 7 7 7 5 8 9 4 3 2 2 1 3 6 5 6 9;
d= 0.8 0.5 0.5 0.8 0.2 2.3 1.3 1.9 1.1 0.8 0.3 0.5 0.2 0.3 0.1 1.1 0.9 0.7 2.3
1.3;
w= 9 9 4 10 6 8 5 6 11 5 6 4 1 1 1.5 7.7 10 5.5 5.5 10;
ENDDATA

!3.8; Max = @SUM(Securities(i):d(i)*x(i));

!3.3; @FOR(Securities(i): z(i)=b(i)+x(i)-y(i));
!3.4; @FOR(Securities(i): y(i)<=b(i));
!3.5; @SUM(Securities(i):v(i)*x(i))=@SUM(Securities(i):v(i)*y(i));
!3.6; @FOR(Securities(i):v(i)*z(i)>=(r/100)* @SUM(Securities(j):v
(j)*z(j)));
!3.7; @SUM(Securities(i):w(i)*z(i))>=@SUM(Securities(i):v(i)*z(i))
+(s/100)* @SUM(Securities(i):v(i)*z(i));
!3.1; @FOR(Securities(i):@GIN(x(i)));
!3.2;@FOR(Securities(i):@GIN(y(i)));

DATA:
@TEXT("c:\Files_\Out3.txt")= @WRITEFOR(Securities(i): 'Security ',
i, ': purchased shares ', " = " , x(i), '; sold shares ', ' = ' ,y(i),
@NEWLINE(1));
@TEXT("c:\Files_\Out3.txt")= @WRITE(@NEWLINE(1));
@TEXT("c:\Files_\Out3.txt")= @WRITE('OBJETIVE FUNCTION VALUE:',@SUM
(Securities(i):d(i)*x(i)));
ENDDATA
END
```

Solution

It is shown in Fig. 3.3.

3.5 Problem 4: Capital Investments (D'Auria 2008)

We have \$2200 available to invest over the next 5 years. At the beginning of each year, we can invest part of the money in 1-year or 2-year deposits. One-year deposits pay 5% interest, while 2-year deposits pay 11% at the end of the 2 years. In addition, at the beginning of the second year, it is possible to invest money in 3-year bonds of company A, which have a total return of 17%. We raise the model to achieve that after 5 years the capital is as large as possible.

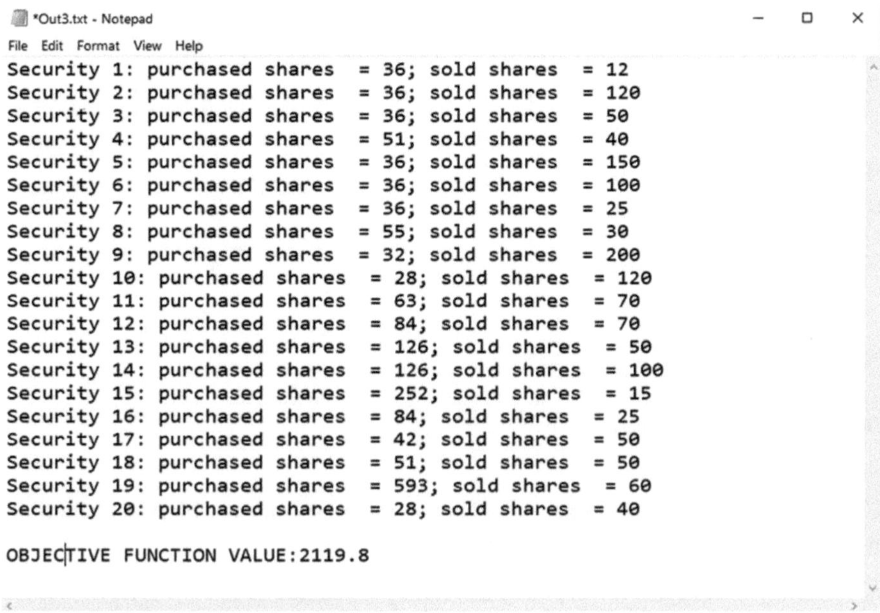

```
Security 1: purchased shares  = 36; sold shares  = 12
Security 2: purchased shares  = 36; sold shares  = 120
Security 3: purchased shares  = 36; sold shares  = 50
Security 4: purchased shares  = 51; sold shares  = 40
Security 5: purchased shares  = 36; sold shares  = 150
Security 6: purchased shares  = 36; sold shares  = 100
Security 7: purchased shares  = 36; sold shares  = 25
Security 8: purchased shares  = 55; sold shares  = 30
Security 9: purchased shares  = 32; sold shares  = 200
Security 10: purchased shares  = 28; sold shares  = 120
Security 11: purchased shares  = 63; sold shares  = 70
Security 12: purchased shares  = 84; sold shares  = 70
Security 13: purchased shares  = 126; sold shares  = 50
Security 14: purchased shares  = 126; sold shares  = 100
Security 15: purchased shares  = 252; sold shares  = 15
Security 16: purchased shares  = 84; sold shares  = 25
Security 17: purchased shares  = 42; sold shares  = 50
Security 18: purchased shares  = 51; sold shares  = 50
Security 19: purchased shares  = 593; sold shares  = 60
Security 20: purchased shares  = 28; sold shares  = 40

OBJECTIVE FUNCTION VALUE:2119.8
```

Fig. 3.3 Output file [C:\Files_\Out3.txt]

Modeling

List of elements in Table 3.6.

Table 3.6 Problem 4: table of elements

Elements	Set	QN	Data				
			Name	Parameter	Type	Belonging	Value
System	-	I_U	Available money	D	C ($)	S	$2200
Years (periods)	$t = 1 \ldots 5$	I_U					...
Deposits	$j = 1, 2$	I_U	Interest	Ij	C (%)	W	{5%, 11%}
Three-year bonds	–	I_U	Yield	R	C (%)	W	{17%}
Money	–	I_M		D			

Decision Activities + Auxiliary Variables

The bond element appears only at $t = 2$ (Fig. 3.4), since it is the only period where there is a possibility of flow input, that is, of investing in bonds. In the same way, the 2-year deposit ($j = 2$) is no longer used at $t = 4$ as it has no application in 5-year planning.

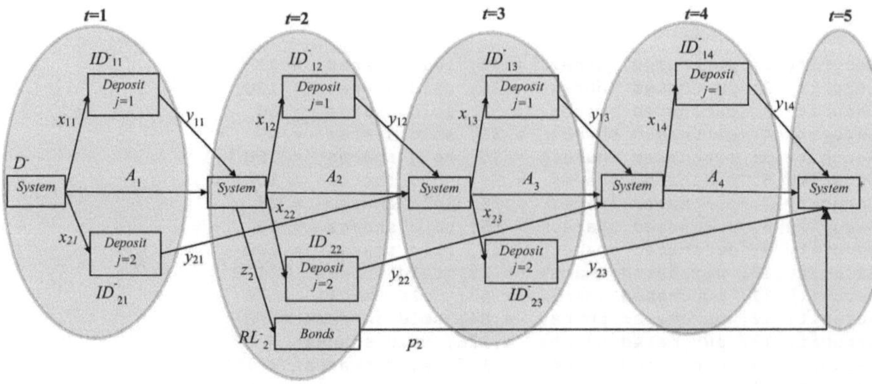

Fig. 3.4 Graph representing the activity

The definition of variables (decision variables and auxiliary calculations) extracted from the graph of Fig. 3.4 would be:

x_{jt} = Money invested in the deposit *j* in year *t*. *j* = 1.2; (*t* = 1 . . . 4/*j* = 1); (*t* = 1. . .3/ *j* = 2)
z_2 = Money invested in obligations in year 2
A_t = Money available at the end of year *t*. *t* = 1 . . . 5
ID^-_{jt} = Interest on deposit *j* in year *t*. *j* = 1.2; *t* = 1 . . . 4/*j* = 1; *t* = 1 . . . 3/*j* = 2
RL^-_2 = Yield on obligations in the second year
y_{jt} = Money deposited on deposit *j* in year *t*. *j* = 1.2; *t* = 1. . .4/*j* = 1; *t* = 1. . .3/*j* = 2
p_2 = Money received from investment in bonds

From all these variables, we could consider x_{jt} and z_2 as decision activities. The rest of the variables (ID^-_{jt}, RL^-_2, A_t, y_{jt}, p_2) would be defined as auxiliary calculations, obtained from two sources:

– As system specifications referring to the calculation of interest and returns on deposits and obligations (assignment specifications):

$$\forall j,t : \mathrm{ID}_{jt} = \frac{I_j}{100} x_{jt} \tag{4.1}$$

$$\mathrm{RL}_2 = \frac{R}{100} z_2 \tag{4.2}$$

– From the flow balance equations of the graph (balance specifications):

System Node:

$$t = 1 : D = \sum_{j=1}^{2} x_{j1} + A_1 \quad \Rightarrow A_1 = D - \sum_{j=1}^{2} x_{j1} \tag{4.3}$$

$$t = 2 : A_1 + y_{11} = \sum_{j=1}^{2} x_{j2} + z_2 + A_2 \quad \Rightarrow A_2 = A_1 + y_{11} - \sum_{j=1}^{2} x_{j2} - z_2 \tag{4.4}$$

$$t = 3 : A_2 + y_{12} + y_{21} = \sum_{j=1}^{2} x_{j3} + A_3 \quad \Rightarrow A_3 = A_2 + y_{12} + y_{21} - \sum_{j=1}^{2} x_{j3} \tag{4.5}$$

$$t = 4 : A_3 + y_{13} + y_{22} = x_{14} + A_4 \quad \Rightarrow A_4 = A_3 + y_{13} + y_{22} - x_{14} \tag{4.6}$$

$$t = 5 : A_4 + y_{14} + y_{23} + p_2 = A_5 \tag{4.7}$$

Deposits:

$$\forall j, t : x_{jt} + \mathrm{ID}_{jt} = y_{jt} \tag{4.8}$$

Bonds:

$$z_2 + \mathrm{RL}_2 = p_2 \tag{4.9}$$

There is no other specification in the system.
Objective Function
Maximize capital at the end of planning:

$$\text{Max } A_5 \tag{4.10}$$

Complete Model

$$\text{Max } A_5 \tag{4.10}$$

$$\text{s.t.}$$

$$\forall j, t : \mathrm{ID}_{jt} = \frac{I_j}{100} x_{jt} \tag{4.1}$$

$$\mathrm{RL}_2 = \frac{R}{100} z_2 \tag{4.2}$$

$$D = \sum_{j=1}^{2} x_{j1} + A_1 \tag{4.3}$$

$$A_1 + y_{11} = \sum_{j=1}^{2} x_{j2} + z_2 + A_2 \tag{4.4}$$

$$A_2 + y_{12} + y_{21} = \sum_{j=1}^{2} x_{j3} + A_3 \tag{4.5}$$

$$A_3 + y_{13} + y_{22} = x_{14} + A_4 \tag{4.6}$$

$$A_4 + y_{14} + y_{23} + p_2 = A_5 \tag{4.7}$$

$$\forall j, t : x_{jt} + \mathrm{ID}_{jt} = y_{jt} \tag{4.8}$$

$$z_2 + \mathrm{RL}_2 = p_2 \tag{4.9}$$

$$\forall j, t : x_{jt} \geq 0; y_{jt} \geq 0$$

$$\forall j, t : \mathrm{IL}_{jt} \geq 0$$

$$\forall t : A_t \geq 0$$

$$z_2 \geq 0; \mathrm{RL}_2 \geq 0; p_2 \geq 0$$

Implementation in Lingo

```
MODEL:
SETS:
Deposits/1..2/: I;
Years/1..5/:A;
D_Y(Deposits,Years)/1 1, 1 2,1 3, 1 4,2 1, 2 2, 2 3/: x,ID,y;
ENDSETS
DATA:
D=2200;
I=5,11;
R=17;
ENDDATA
!4.10; Max = A(5);
!4.1; @FOR(D_Y(j,t): ID(j,t)=(I(j)/100)*x(j,t));
!4.2; RL2=(R/100)*z2;
!4.3; D = @SUM(Deposits(j):x(j,1))+ A(1);
!4.4; A(1)+ y(1,1) =@SUM(Deposits(j):x(j,2)) + z2 + A(2);
!4.5; A(2)+ y(1,2)+ y(2,1) =@SUM(Deposits(j):x(j,3)) + A(3);
!4.6; A(3)+ y(1,3)+ y(2,2) =x(1,4) + A(4);
!4.7; A(4)+ y(1,4)+ y(2,3)+p2 = A(5);
!4.8; @FOR(D_Y(j,t):x(j,t)+ID(j,t)=y(j,t));
!4.9; z2+ RL2= p2;
END
```

Solution

The solution is shown in Fig. 3.5.

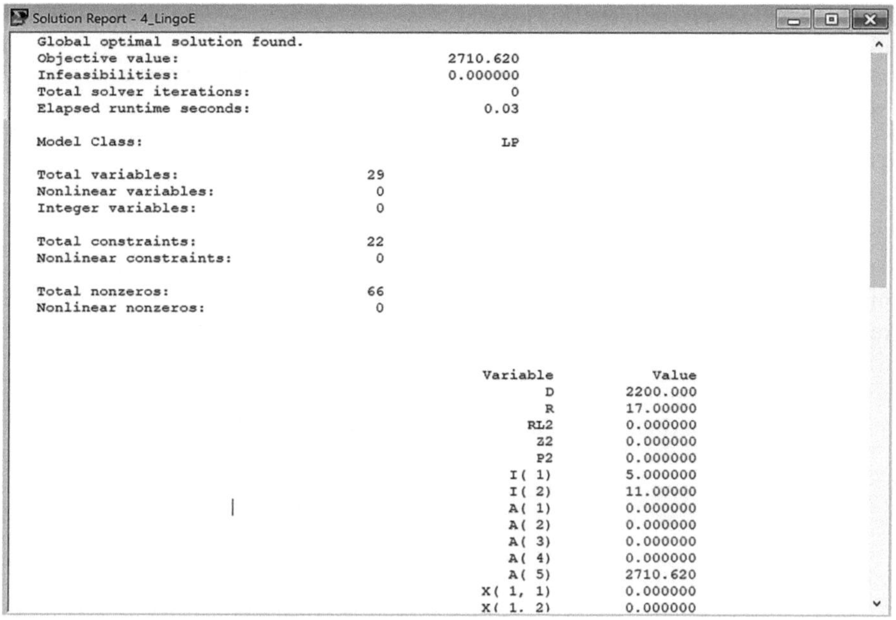

```
Solution Report - 4_LingoE
Global optimal solution found.
Objective value:                              2710.620
Infeasibilities:                              0.000000
Total solver iterations:                             0
Elapsed runtime seconds:                          0.03

Model Class:                                        LP

Total variables:                      29
Nonlinear variables:                   0
Integer variables:                     0

Total constraints:                    22
Nonlinear constraints:                 0

Total nonzeros:                       66
Nonlinear nonzeros:                    0

                                Variable           Value
                                       D        2200.000
                                       R        17.00000
                                     RL2        0.000000
                                      Z2        0.000000
                                      P2        0.000000
                                  I( 1)         5.000000
                                  I( 2)         11.00000
                                  A( 1)         0.000000
                                  A( 2)         0.000000
                                  A( 3)         0.000000
                                  A( 4)         0.000000
                                  A( 5)         2710.620
                               X( 1, 1)         0.000000
                               X( 1, 2)         0.000000
```

Fig. 3.5 Problem 4: LINGO solution report

3.6 Problem 5: Number Partitioning (Korf 1998)

Given a set of n elements, each represents (owns) a known integer. It is about dividing the elements into two sets (assigning elements to sets) in such a way as to balance the sum of numbers in each set.

Modeling

List of elements in Table 3.7.

Table 3.7 Problem 5: table of elements

			Data				
Elements	Set	QN	Name	Parameter	Type	Belonging	Value
Elements	$i = 1 \ldots n$	I_U	Number	n_i	I	W	...
Sets	$j = 1, 2$	I_U					

Decision Activities

Assign elements (*i*) to sets (*j*)

$$\forall i,j : \alpha_{ij} = 1 \quad \text{if the element } i \text{ is assigned to the set } j; 0 \text{ otherwise} \qquad (5.1)$$

Specifications

– Quantitative selection specification:

Activity	Elements selecting	Selectable elements	Selection	Constraints
α_{ij}	Elements $i = 1 \ldots n$	Sets $j = 1, 2$	=1	$\forall i : \sum_{j=1}^{2} \alpha_{ij} = 1$ (5.2)
	Sets $j = 1, 2$	Elements $i = 1 \ldots n$	–	–

Objective Criterion

Balance the total sum of numbers in the sets. We will see two versions. The first, more systematic, uses dimension calculations to balance. The second version, more reduced, applies the idea of presetting the order in the value of the sum of each set.

Version 1:

Auxiliary Calculation:

We define an auxiliary calculation to collect the sum of elements of each set.

$$\forall j : \quad x_j = \text{sum of the numbers of the elements in the set } j$$

$$\forall j : \quad x_j = \sum_{i=1}^{n} n_i \alpha_{ij} \tag{5.3}$$

An equilibrium of the variables x_j is requested; we apply the use of upper/lower bound variables to carry out the equilibrium:

Upper/Lower Bound Calculation:

$$x_{\mathrm{MIN}} = \text{Lower bound of } x_j \text{ values}$$

$$\forall j : x_{\mathrm{MIN}} \leq x_j \tag{5.4}$$

$$x_{\mathrm{MAX}} = \text{Upper bound of } x_j \text{ values}$$

$$\forall j : x_{\mathrm{MAX}} \geq x_j \tag{5.5}$$

Objective Function:

$$\text{Min } x_{\mathrm{MAX}} - x_{\mathrm{MIN}} \tag{5.6}$$

Version 2:

Since we can always determine that one of the sets is greater than or equal to the other, we impose that the sum of the first set is $\geq$ that of the second set:

Bounds imposition:

$$x_1 \geq x_2 \tag{5.7}$$

Objective Function:

Now we simply minimize the difference of the two variables:

$$\text{Min } x_1 - x_2 \tag{5.8}$$

Complete Model (Version 2)

$$\text{Min } x_1 - x_2 \tag{5.8}$$

$$\text{s.t.}$$

$$\forall i : \sum_{j=1}^{2} \alpha_{ij} = 1 \tag{5.2}$$

$$\forall j : \quad x_j = \sum_{i=1}^{n} n_i \alpha_{ij} \tag{5.3}$$

$$x_1 \geq x_2 \tag{5.7}$$

$$\forall i, j : \alpha_{ij} \text{ binaries} \tag{5.1}$$

$$\forall j : x_j \geq 0$$

The two versions are extendable to any number of sets. In the case of version 2, the bounds imposition proceeds as follows when we consider m sets:

$$x_1 \geq x_2 \geq x_3 \ldots \geq x_m$$

In the objective function:

$$\text{Min } x_1 - x_m$$

Implementation in Lingo

We assume a model with 20 elements and the following data:

$n = \{7\ 8\ 2\ 5\ 7\ 1\ 5\ 5\ 9\ 9\ 4\ 3\ 2\ 2\ 1\ 3\ 6\ 3\ 11\ 12\}$

These data are stored in an Excel file. In the results section, we can view the file (Fig. 3.6).

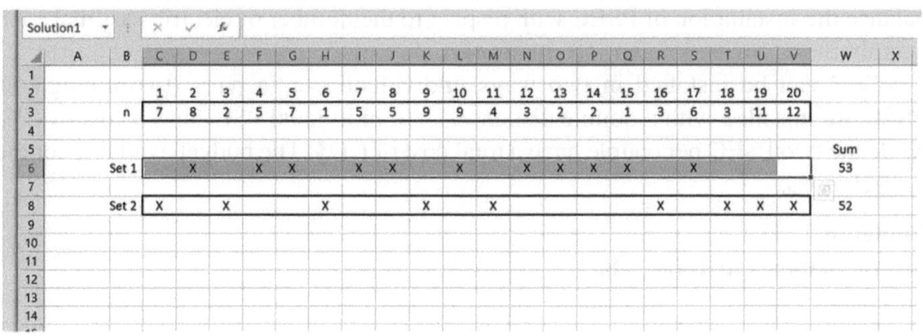

	C	D	E	F	G	H	I	J	K	L	M	N	O	P	Q	R	S	T	U	V	W
	1	2	3	4	5	6	7	8	9	10	11	12	13	14	15	16	17	18	19	20	
n	7	8	2	5	7	1	5	5	9	9	4	3	2	2	1	3	6	3	11	12	
																					Sum
Set 1		X		X	X		X	X		X		X	X	X	X		X				53
Set 2	X		X			X			X		X					X		X	X	X	52

Fig. 3.6 Input and output file [C:\Files_\5_Excel.xlsx]

```
MODEL:
SETS:
Elements/1..20/: n;
PSet/1..2/:x;
E_S(Elements,PSet):Alpha;
ENDSETS
DATA:
n=@OLE("c:\Files_\5_Excel.xlsx",'n');
ENDDATA
!5.8; Min = x(1)-x(2);
!5.2; @FOR(Elements(i): @Sum(Pset(j):Alpha(i,j))=1);
!5.3;
@FOR(Pset(j): x(j)=@Sum(Elements(i): n(i)*Alpha(i,j)));
!5.7; x(1)>=x(2);
!5.1; @FOR(E_S(i,j):@BIN(Alpha(i,j)));

DATA:
@OLE("c:\Files_\5_Excel.xlsx","Solution1")
=@WRITEFOR(E_S(i,j)|j#EQ#1:@IF(Alpha(i,1)#EQ#1,'X',''));
@OLE("c:\Files_\5_Excel.xlsx","Solution2")
=@WRITEFOR(E_S(i,j)|j#EQ#2:
@IF(Alpha(i,2)#EQ#1,'X',' '));
 @OLE("c:\Files_\5_Excel.xlsx","VSolution1") = @WRITE(x(1));
 @OLE("c:\Files_\5_Excel.xlsx","VSolution2") = @WRITE(x(2));
ENDDATA
END
```

Solution

The solution is shown in Fig. 3.6.

3.7 Problem 6: Installation of Parking Spaces for Electric Cars

Let a city be divided into 18 neighborhoods. The city council wants to install parking spaces for electric cars (PSEC) in the city and the approach it wants to make is to balance the installation of PSEC with respect to the number of electric cars that each neighborhood owns (known information).

PSEC can be installed in two ways: individually or in pairs (two contiguous PSEC are installed with a double dispenser). An individual PSEC has a cost of *CI*\$, while the two PSEC per couple have a total cost of *CD*\$. The budget available for the project is *P*\$.

Modeling

List of elements in Table 3.8.

Table 3.8 Problem 6: table of elements

			Data				
Elements	Set	QN	Name	Parameter	Type	Belonging	Value
Neighborhoods	$i = 1 \ldots 18$	I_U	Electric cars	V_i	I	W	...
Individual PSEC	–	C_I	Cost	CI	C($)	W	–
PSEC couples	–	C_I	Cost	CD	C($)	W	–
System	–	I_U	Budget	P	C($)	W	–

Decision Activities

Install individual PSEC in neighborhoods (*i*)

$$x_i = \text{Number of individual PSEC installed in neighborhood } i. \tag{6.1}$$

Install PSEC couples in Neighborhoods (*i*)

$$y_i = \text{Number of PSEC couples installed in neighborhood } i. \tag{6.2}$$

Specifications

– Budget available (capacity specification):

$$\sum_{i=1}^{18} \mathrm{CI}x_i + \mathrm{CD}y_i \leq P \tag{6.3}$$

Objective Criterion

Balance installed PSEC with respect to the number of electric cars. For this case, we are simply going to use as balance the difference between electric cars and installed PSEC. In other environments, it may be more convenient to balance ratios.

Auxiliary Calculation

A simple auxiliary calculation to collect the number of PSEC installed in each neighborhood:

$$\forall i: \ z_i = \text{Number of PSEC installed in neighborhood } i.$$

$$z_i = x_i + 2y_i \tag{6.4}$$

Lower and Upper Bound Calculations

$$x_{\text{MIN}} = \text{Lower bound of the differences between electric cars and installed PSEC}(V_i - z_i)$$

$$\forall i: x_{\text{MIN}} \leq V_i - z_i \tag{6.5}$$

x_{MAX} = Upper bound of the differences between electric cars and installed PSEC

$$\forall i : V_i - z_i \leq x_{MAX} \quad (6.6)$$

Objective Function
Balance differences

$$\text{Min } x_{MAX} - x_{MIN} \quad (6.7)$$

Complete Model

$$\text{Min } x_{MAX} - x_{MIN} \quad (6.7)$$

s.t.

$$\sum_{i=1}^{18} \text{CI}x_i + \text{CD}y_i \leq P \quad (6.3)$$

$$z_i = x_i + 2y_i \quad (6.4)$$

$$\forall i : x_{MIN} \leq V_i - z_i \quad (6.5)$$

$$\forall i : V_i - z_i \leq x_{MAX} \quad (6.6)$$

$$\forall i : x_i \geq 0 \text{ integers} \quad (6.1)$$

$$\forall i : y_i \geq 0 \text{ integers} \quad (6.2)$$

$$\forall i : z_i \geq 0$$

$$x_{MIN} \geq 0$$

$$x_{MAX} \geq 0$$

Implementation in Lingo
Problem data:

Neighborhoods	1	2	3	4	5	6	7	8	9	10	11	12	13	14	15	16	17	18
Electric cars	14	24	5	8	19	3	11	17	6	7	5	9	3	6	10	12	4	8

CI = 3000
CD = 5600
P = 250,000€

```
MODEL:
SETS:
Neighborhoods/1..18/:V,x,y,z;
ENDSETS
DATA:
 V= 14 24 5 8 19 3 11 17 6 7 5 9 3 6 10 12 4 8;
 CI= 3000;
 CD= 5600;
```

```
 P=250000;
ENDDATA

!6.7; Min = xMAX - xMIN;
!6.3; @SUM(Neighborhoods(i):CI*x(i))+@SUM(Neighborhoods(i):CD*y
(i))<=P;
!6.4; @FOR(Neighborhoods(i):z(i)=x(i)+2*y(i));
!6.5; @FOR(Neighborhoods(i):xMIN<= z(i)/ V(i));
!6.5; @FOR(Neighborhoods(i):V(i)-z(i)<=xMAX);
!6.1; @FOR(Neighborhoods(i):@GIN(x(i)));
!6.2; @FOR(Neighborhoods(i):@GIN(y(i)));

END
```

Solution

The solution is shown in Fig. 3.7.

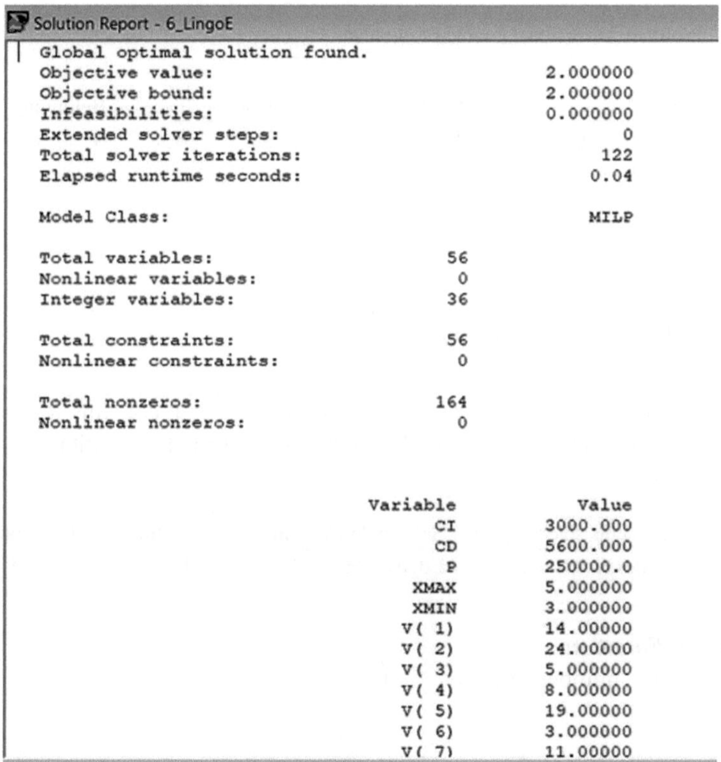

```
Solution Report - 6_LingoE
 Global optimal solution found.
 Objective value:                              2.000000
 Objective bound:                              2.000000
 Infeasibilities:                              0.000000
 Extended solver steps:                               0
 Total solver iterations:                           122
 Elapsed runtime seconds:                          0.04

 Model Class:                                      MILP

 Total variables:                     56
 Nonlinear variables:                  0
 Integer variables:                   36

 Total constraints:                   56
 Nonlinear constraints:                0

 Total nonzeros:                     164
 Nonlinear nonzeros:                   0


                                Variable           Value
                                      CI        3000.000
                                      CD        5600.000
                                       P        250000.0
                                    XMAX        5.000000
                                    XMIN        3.000000
                                   V( 1)        14.00000
                                   V( 2)        24.00000
                                   V( 3)        5.000000
                                   V( 4)        8.000000
                                   V( 5)        19.00000
                                   V( 6)        3.000000
                                   V( 7)        11.00000
```

Fig. 3.7 Problem 6—LINGO solution report

3.8 Problem 7: Independent Edge Set (Alan 1985)

Given an undirected graph G (N, E), we try to obtain a set with the greatest number of unconnected edges (two edges are connected when they share a node).

Modeling

We are going to make the same model with two different configurations of the graph information. In the first case, the relationship between nodes and edges is expressed in binary form. In the second case, the nodes of each edge are referenced by two ordinal attributes. From this latest version, a Lingo implementation will be made with a data source in an accdb database.

Version 1

Table of Elements

For this problem, we use the table of elements collected in Table 3.39 for the undirected graphs defined in Garcia (2021). It is represented in Table 3.9.

Table 3.9 Problem 7 (version 1): table of elements

			Data				
Elements	Set	QN	Name	Parameter	Type	Belonging	Value
Nodes	$i = 1 \ldots n$	I_U	Node_Edge	A_{ik}	B	S	...
Edges	$k = 1 \ldots m$	I_U		A_{ik}			

Decision Activities

Select edges (*k*)

$\forall k : \alpha_k = 1$ if the edge k is selected; 0 otherwise.

Specifications

- For each node, there cannot be more than one edge that contains it; that is, two edges that contain the same node cannot be selected (quantitative selection specification):

Activity	Elements selecting	Selectable elements	Selection	Constraints
α_k	Nodes ($i = 1 \ldots n$)	Edges ($k/A_{ik} = 1$)	≤ 1	$\forall i : \sum_{k/A_{ik}=1} \alpha_k \leq 1$

Objective Function

Maximize the number of edges:

$$\text{Max} \sum_{i=1}^{m} \alpha_k$$

Version 2

Now the nodes of an edge are referenced by an ordinal. List of elements in Table 3.10.

Table 3.10 Problem 7 (version 2): table of elements

Elements	Set	QN	Data				
			Name	Parameter	Type	Belonging	Value
Edges	$k = 1 \ldots m$	I_U	Node1	$n1_k$	O	W	...
			Node2	$n2_k$	O	W	...
Nodes	$i = 1 \ldots n$	I_U					...

Decision Activities

Select edges (*k*)

$$\forall k : \alpha_k = 1 \text{ if the edge } k \text{ is selected; 0 otherwise.} \tag{7.1}$$

Specifications

– Quantitative selection specification:

Activity	Elements selecting	Selectable elements	Selection	Constraints
α_k	Nodes ($i = 1 \ldots n$)	Edges ($k/n1_k = i$ OR $n2_k = i$)	≤ 1	$\forall i, \sum_{k/n1_k=i \ OR \ n2_k=i} \alpha_k \leq 1$ (7.2)

Objective Function

$$\text{Max} \sum_{i=1}^{m} \alpha_k \tag{7.3}$$

Complete Model

$$\text{Max} \sum_{i=1}^{m} \alpha_k \tag{7.3}$$

s.t.

$$\forall i, \sum_{k/n1_k=i \ OR \ n2_k=i} \alpha_k \leq 1 \tag{7.2}$$

$$\forall k : \alpha_k \text{ binaries} \tag{7.1}$$

Implementation in Lingo

The data and results are collected in a database, represented in Figs. 3.8 and 3.9.

Fig. 3.8 Problem 7: input data

Edges | System

Id	n	m	Hc
1	9	10	
* (Nuevo)	0	0	

Edges | System

Id	Edge	Node1	Node2	Selected
3	1	1	2	☐
4	2	1	3	☐
5	3	2	4	☑
6	4	1	5	☑
7	5	1	6	☐
8	6	3	6	☑
9	7	4	7	☐
10	8	4	8	☐
11	9	7	8	☐
12	10	8	9	☑
* (Nuevo)	0	0	0	☐

Fig. 3.9 Problem 7: input and output data

```
MODEL:
DATA:
 n= @ODBC('7_ODBC','System','n');
 m= @ODBC('7_ODBC','System','m');
ENDDATA
SETS:
Nodes/1..n/;
Edges/1..m/:Alpha,N1,N2;
ENDSETS
DATA:
 N1 =@ODBC('7_ODBC','EDGES','NODE1');
 N2 =@ODBC('7_ODBC','EDGES','NODE2');
ENDDATA

!7.3; MAX = @SUM(Edges(k):Alpha(k));
!7.2; @FOR(Nodes(i):@SUM(Edges(k)|N1(k)#EQ#i#OR#N2(k)#EQ#i: Alpha
(k))<=1);
!7.1; @FoR(Edges(k):@BIN(Alpha(k)));
DATA:
@ODBC('7_ODBC','EDGES','Selected') =
@WRITEFOR(Edges(k):Alpha(k));
ENDDATA
END
```

Solution

The solution is shown in Fig. 3.9.

References

Alan G (1985) Matchings. In: Algorithmic graph theory. Cambridge University Press, Cambridge

Castillo E, Conejo A, Pedregal P (2002) Formulación y resolución de modelos de programación matemática en ingeniería y ciencia. Universidad de Castilla-La Mancha, Ciudad Real. isbn:978-84-600-9751-8

D'Auria B (2008) Programación Lineal. http://halweb.uc3m.es/esp/Personal/personas/bdauria/0809/IO_II/2008-10-20L_T2_II_82.pdf. Accessed 31 Dec 2021

García JM (2021) Modelling in mathematical programming: methodology and technics. Springer, Cham. https://doi.org/10.1007/978-3-030-57250-1

Korf RE (1998) A complete anytime algorithm for number partitioning. Artificial Intell 106(2): 181–203

Salazar B (2021) Investigación de Operaciones. Ejercicios de Programación Lineal. https://www.ingenieriaindustrialonline.com/investigacion-de-operaciones/ejercicios-de-programacion-lineal/. Accessed 31 Dec 2021

Chapter 4
Low-Medium Level Problems

4.1 Introduction

Problems:

Number	Problem
8	Power generation planning
9	Purchase of processing machines
10	Project selection
11	Assignment of workers to centers
12	Warehouse rental
13	Article and position planning
14	Online sale
15	Plot cultivation
16	Production and distribution
17	Production of divisible tasks
18	Selection of nodes and edges

Versions and elements:

		Elements				
Problem	Versions	Collective elements	Measurable individual elements	Grouping of items	Upper bound in the number of elements	Data handling
8	2		✓			Medium
9	1				✓	Medium
10	1					Low
11	2	✓		✓		Low
12	2	✓	✓			Low

(continued)

J. M. García Sánchez, *Building and Solving Mathematical Programming Models*, International Series in Operations Research & Management Science 329,
https://doi.org/10.1007/978-3-030-97626-2_4

Problem	Versions	Elements				
		Collective elements	Measurable individual elements	Grouping of items	Upper bound in the number of elements	Data handling
13	1					Medium
14	1	✓				Medium
15	1		✓			Medium
16	1	✓				Medium
17	1		✓			Low
18	1					Medium

Calculations:

Problem	Binary logical calculations	Not-binary logical calculations	Bounds calculations	Conditioned bounds calculations
8			✓	
9	✓			
10				
11				
12	✓			
13	✓			
14	✓			
15	✓			
16	✓			
17	✓		✓	
18				

Specifications:

Problem	QSS[a]	Capacity	Demand	Bound imposition	Allocation/assignment
8		✓	✓	✓	
9	✓	✓			
10	✓	✓			
11	✓	✓		✓	
12		✓			✓
13	✓	✓			
14		✓			
15	✓	✓		✓	
16		✓	✓		
17	✓				✓
18	✓				

[a]*QSS* quantitative selection specifications

Problem	Flow balance	PLS[a]—One operator	PLS—Several operators	PLS—Attributes	Variable costs cases
8					
9					
10		✓	✓		
11					
12		✓			
13		✓		✓	
14		✓			
15		✓			
16	✓	✓			
17		✓			
18		✓			

[a]*PLS* propositional logic specifications

Objective criterion, input data and output data:

	Objective criterion		Input data				
Problem	Calculations	Particular cases	Integrated data	Text file	Spreadsheet	Database	Output data
8	✓			✓			
9					✓		
10			✓				
11					✓		✓
12	✓		✓				
13					✓		✓
14					✓		
15				✓			
16	✓					✓	
17	✓					✓	
18				✓			✓

4.2 Problem 8. Power Generation Planning (Sarker and Newton 2007)

A long-term power generation plan considers the operational schedule of existing power generating units and the installation schedule of new plants to meet electricity demand for a number of future years, at the minimum possible cost. The electricity demand is described by a load duration curve (LDC), as shown in Fig. 4.1. The area under the LDC represents the requirements of electricity production in megawatt-hours. To keep the model simple, such as linear, the LDC is approximated by a piecewise constant function shown as a histogram (or step function) in Fig. 4.1. Bar

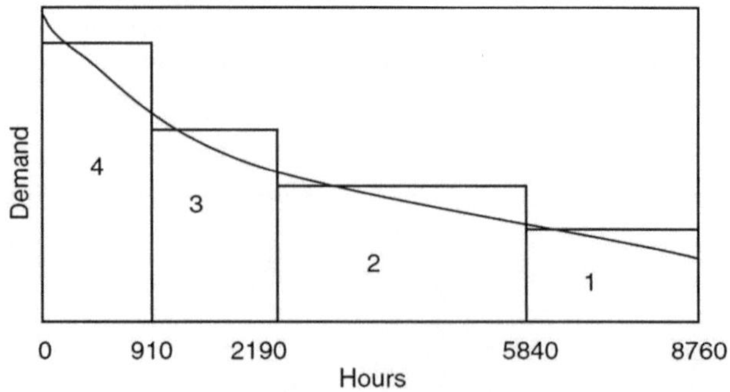

Fig. 4.1 Load duration curve (LDC)

1 refers to the minimum load, bar 4 refers to the peak load, and two intermediate bars represent the middle-load demand. In any company or country, a number of different types of technology, such as steam turbines, gas turbines, hydro, combined cycle, and diesel cycle, are used for electricity generation. These power generation units require different inputs, installation costs, and variable costs. In addition, they have different technical lives and follow different maintenance policies.

The planning problem is to find the operating schedule of existing units, the new generation units to be installed in the future, and the installation schedule of new units by minimizing the sum of installation and operating costs for a given number of years while not violating technical and financial constraints.

Once a new generating unit with a given capacity is added, that unit will be maintained throughout the planning horizon. Additionally, more capacity can be added in a future period. For example, suppose that the model has decided to install a new plant of type $i = 1$ with 50 MW capacity in period 1, it will exist in the system until the technical life of this unit expires. Note that the duration of the planning horizon, considered in this problem, is shorter than the technical life of any new power generating unit available on the market.

On the other hand, the administration has a preference between the different types of technology, that is, between the different generating units in the case of new units. This preference is described by a percentage factor of the minimum that each new unit must have over the total energy installed in the new units.

Modeling

The system is simplified and does not consider each energy production plant individually, but rather globally considers all plants of a type of energy as an energy generation unit. Since there are also no data referring to the plants as items of a collective element, but rather the data refer globally to the set of all the plants of a generation type, we will consider each generating unit as an individual element.

The new generating units will also be considered as individual elements for the same simplification of the problem. They are considered measurable because the energy capacity to be installed is measured, which in a more specific scenario would correspond to the volume of facilities, quantity of equipment and resources, etc.

Considering the existing units and the new units as differentiated elements gives greater flexibility to the problem (Version 1), but we could consider a second version in which we unify these two elements into one (Version 2).

The data with which the system works are as follows:

- Variable costs of generating units (identical for existing and new units) depending on the year ($/MWh)
- The annual availability factor of each generating unit, also identical for new and existing ones.
- The fixed costs of installing capacity in new units ($/MWh). Dependent on the year.
- The capacity already installed in the existing units.
- The maximum energy of each annual load bar and the duration of each bar.
- The percentage of preference for each new unit.

Table 4.1 Problem 8 (version 1): table of elements

			Data				
Elements	Set	QN	Name	Parameter	Type	Belonging	Value
Existing power generating units	$i = 1\ldots n$	I_U	Installed capacity	CE_{it}	C(MW)	S	–
			Annual variable costs	VC_{it}	C ($/MWh)	S	–
			Annual availability factor	AF_{it}	C	S	–
New power generating units	$j = 1\ldots n$	I_M	Annual fixed costs	FC_{jt}	C ($/MWh)	S	–
			Annual variable costs	VCN_{jt}	C ($/MWh)	S	–
			Annual availability factor	AFN_{jt}	C	S	–
			Preference percentage	W_j	C(%)	S	–
Load bar	$p = 1\ldots 4$	I_U	Maximum energy demanded in year t	PD_{pt}	C(MW)	S	–
			Duration	TD_p	C(h)	W	–
Year	$t = 1\ldots L$	I_U	CE_{it}; PD_{pt}; VC_{it}; VCN_{jt}; FC_{jt}; AF_{it}; AFN_{jt}				
Power	–	I_M	CE_{it}; PD_{pt}				

Version 1

Version presented as resolution in Sarker and Newton (2007). List of elements in Table 4.1.

Decision Activities

Install energy capacity every year (*t*) in new power generation units (*j*)

$\forall j : x_{jt}$ = Energy capacity installed in the power generating unit j

Consume energy from existing units (*i*) during each load bar (*p*) and each year (*t*)

$\forall i, p, t : y_{ipt}$ = Amount of energy used by the generating unit i during the load bar p in the year t (*in* Sarker and Newton 2007, *would be* OE_{ipt}).

Consume energy from the new units (*j*) during each load bar (*p*) and each year (*t*)

$\forall j, p, t : z_{jpt}$ = Amount of energy used from the new generating unit j during the load bar p in the year t (*in* Sarker and Newton 2007, would be OX_{ipt}).

Specifications

- Energy demand (demand specification): Based on the PD_{pt} attribute, that is, on the maximum amount of energy demanded in each load bar and each year:

$$\forall p, t : \sum_{i=1}^{n} y_{ipt} + \sum_{j=1}^{n} z_{jpt} \geq PD_{pt} \tag{8.1}$$

- Energy capacity of existing and new units (capacity specification): Based on the CE_{it} attribute for existing units and multiplying by the availability factor to obtain the net capacity to be taken into account. For new units, it is equivalent, but the contribution of energy capacity is variable, x_{jt}.

$$\forall i, p, t : y_{ipt} \leq AF_{it}CE_{it} \tag{8.2}$$

$$\forall j, p, t : z_{jpt} \leq AFN_{jt}x_{jt} \tag{8.3}$$

- Maintain capacity of new units installed (lower bound imposition): It imposes that the capacity in new units is maintained or can be increased:

$$\forall j, t < L : x_{jt} \leq x_{jt+1} \tag{8.4}$$

- Preference for installing capacity in new units (upper bound imposition):

To collect the maximum installed capacity of each new unit, Sarker and Newton (2007) define a variable from a calculation of the upper bound on the installed capacity each year as follows.

Upper Bound Calculation

$\forall j : CF_j$ = Upper bound of the installed capacity each year of each new unit j

Constraints defining the calculation : $\forall j, t : CF_j \geq x_{jt}$

With the calculation, the minimum percentage of each new unit j is imposed on the total new installed capacity:

$$\forall j : \mathrm{CF}_j \geq \frac{W_j}{100}\left(\sum_{j'=1}^{n} \mathrm{CF}_{j'}\right) \tag{8.5}$$

Using upper or lower bound calculations to collect the maximum or minimum of a set of values, it is viable as long as we are sure that the specifications of the system or the objective function push the value of the bound toward the maximum (case of using upper bounds) or the minimum (case of lower bounds). And this occurs in this case thanks to the objective function.

However, since there is a specification that requires that the installed capacity never decreases from 1 year to the next (8.4), it would be enough to simply adjust CF_j to the value of the capacity in the last period. In that way, really CF_j does not have to be an upper bound calculation, would simply be an auxiliary calculation with a value equal to the installed capacity in the last period, or directly having used x_{jL} as CF_j. That is, we remove constraint (8.5) and remove the definition of the variable CFj and simply impose the following:

$$\forall j : x_{jL} \geq \frac{W_j}{100}\left(\sum_{j'=1}^{n} x_{j'L}\right) \tag{8.6}$$

Objective Function

Minimize costs:

$$\text{Costs} = \text{Variables costs} + \text{Fixed costs;}$$

$$\text{Variable costs} : \sum_{p=1}^{P} \mathrm{TD}_p \sum_{i=1}^{n}\sum_{t=1}^{L} \mathrm{VC}_{it} y_{ipt} + \sum_{p=1}^{P} \mathrm{TD}_p \sum_{j=1}^{n}\sum_{t=1}^{L} \mathrm{VCN}_{jt} z_{jpt}$$

$$\text{Fixed costs} : \sum_{j=1}^{n}\sum_{t=1}^{L} \mathrm{FC}_{jt} x_{jt}$$

$$\text{Min} \sum_{p=1}^{P} \mathrm{TD}_p \sum_{i=1}^{n}\sum_{t=1}^{L} \mathrm{VC}_{it} y_{ipt} + \sum_{p=1}^{P} \mathrm{TD}_p \sum_{j=1}^{n}\sum_{t=1}^{L} \mathrm{VCN}_{jt} z_{jpt} + \sum_{j=1}^{n} \times \sum_{t=1}^{L} \mathrm{FC}_{jt} x_{jt} \tag{8.7}$$

Version 2

In this second version, we are going to create a smaller model, with fewer variables. We will assume as an element associated with the units only the types of power generation units, without including the aspect of whether they are existing or new (Table 4.2). We will associate with this set of elements all the data associated with the existing and new units of version 1. The rest is the same.

Table 4.2 Problem 8 (version 2): table of elements

Elements	Set	QN	Data				
			Name	Parameter	Type	Belonging	Value
Power generating units	$i = 1\ldots n$	I_M	Installed capacity	CE_{it}	C (MW)	S	–
			Annual fixed costs	FC_{it}	C(€/MW)	S	–
			Annual variable costs	VC_{it}	C(€/MW h)	S	–
			Annual availability factor	AF_{it}	C	S	–
			Preference percentage	W_i	C(%)	S	–
Load bars	$p = 1\ldots 4$	I_U	Maximum energy demanded in year t	PD_{pt}	C	S	–
			Duration	TD_p	C	W	–
Years	$t = 1\ldots L$	I_U	K_{it}; PD_{pt}; VC_{it}; FC_{it}; AF_{it}				
Energy	–	I_M	K_{it}; PD_{pt}				

Decision Activities

For decision activities, we distinguish the action of consuming energy from each unit in each bar and each year. And with respect to the installation of capacity, we apply it to each unit and every year, subsequently imposing that the installed capacity is at least that which each unit already has (CE_{it}).

Install energy capacity every year (*t*) in power generating units (*j*)

$$\forall i : x_{it} = \text{Energy capacity installed in the power generating unit } j$$

Consume energy from the generating units (*i*) during each load bar (*p*) and each year (*t*)

$$\forall i, p, t : y_{ipt} = \text{Amount of energy used by the generator unit } i \text{ during the load bar } p \text{ in year } t$$

Specifications

– Energy demand specification (demand specification):

$$\forall p, t : \sum_{i=1}^{n} y_{ipt} \geq PD_{pt} \tag{8.8}$$

– Energy capacity of units (capacity specification):

$$\forall i, p, t : y_{ipt} \leq AF_{it}(CE_{it} + x_{it}) \tag{8.9}$$

– Maintain installed capacity (upper/lower bound imposition):

$$\forall i, t < L : x_{it} \leq x_{it+1} \tag{8.10}$$

– Preference for installing capacity in new units (upper bound imposition):

$$\forall i : x_{iL} \geq \frac{W_i}{100}\left(\sum_{i'=1}^{n} x_{i'L}\right) \tag{8.11}$$

Objective Function
Minimize costs.

$$\text{Costs} = \text{Variable Costs} + \text{Fixed Costs}$$

$$\text{Variable costs} : \sum_{p=1}^{P} \mathrm{TD}_p \sum_{i=1}^{n} \sum_{t=1}^{L} \mathrm{VC}_{it} y_{ipt}$$

$$\text{Fixed costs} : \sum_{i=1}^{n} \sum_{t=1}^{L} \mathrm{FC}_{it} x_{it}$$

$$\text{Min} \sum_{p=1}^{P} \mathrm{TD}_p \sum_{i=1}^{n} \sum_{t=1}^{L} \mathrm{VC}_{it} y_{ipt} + \sum_{i=1}^{n} \sum_{t=1}^{L} \mathrm{FC}_{it} x_{it} \tag{8.12}$$

Complete Model

$$\text{Min} \sum_{p=1}^{P} \mathrm{TD}_p \sum_{i=1}^{n} \sum_{t=1}^{L} \mathrm{VC}_{it} y_{ipt} + \sum_{i=1}^{n} \sum_{t=1}^{L} \mathrm{FC}_{it} x_{it} \tag{8.12}$$

s.t.

$$\forall p, t : \sum_{i=1}^{n} y_{ipt} \geq \mathrm{PD}_{pt} \tag{8.8}$$

$$\forall i, p, t : y_{ipt} \leq \mathrm{AF}_{it}(\mathrm{CE}_{it} + x_{it}) \tag{8.9}$$

$$\forall i, t < L : x_{it} \leq x_{it+1} \tag{8.10}$$

$$\forall i : x_{iL} \geq \frac{W_i}{100}\left(\sum_{i'=1}^{n} x_{i'L}\right) \tag{8.11}$$

$$\forall i, t : x_{jt} \geq 0$$

$$\forall i, p, t : y_{ipt} \geq 0$$

$$\forall j : \mathrm{CF}_j \geq 0$$

Implementation in LINGO

We are going to use the data collected in the text file "8_File.txt" (Fig. 4.2).

```
MODEL:
SETS:
Units/1..@FILE('C:\Files_\8_File.txt')/:W,CF;
Bars/1..@FILE('C:\Files_\8_File.txt')/:TD;
Periods/1..@FILE('C:\Files_\8_File.txt')/;
UP(Units,Periods):CE,FC,VC,AF,x;
BP(Bars,Periods):PD;
UBP(Units,Bars,Periods):y;
ENDSETS
DATA:
W= @FILE('C:\Files_\8_File.txt');
TD = @FILE('C:\Files_\8_File.txt');
CE = @FILE('C:\Files_\8_File.txt') ;
FC = @FILE('C:\Files_\8_File.txt');
VC = @FILE('C:\Files_\8_File.txt');
AF = @FILE('C:\Files_\8_File.txt');
PD = @FILE('C:\Files_\8_File.txt');
ENDDATA
!8.12; MIN = @SUM(Bars(p): @SUM(Units(i):@SUM(Periods(t): VC(i,t)*y
(i,p,t))))
+ @SUM(Units(i):@SUM(Periods(t): FC(i,t)*x(i,t)));
!8.8;
@FOR(Bars(p):@FOR(Periods(t): @SUM(Units(i):y(i,p,t))>=PD(p,t)));
!8.9; @FOR(Units(i):@FOR(Bars(p):@FOR(Periods(t): y(i,p,t)<=(AF(i,
t)/100)*(CE(i,t)+x(i,t)))));
!8.10;@FOr(Units(i):@FOr(Periods(t)|t#LT#5:x(i,t)<=x(i,t+1)));
!8.11;
@FOR(Units(i):x(i,5)>= (W(i)/100) * @SUM(Units(ii): x(ii,5)));
END
```

Solution

The solution is shown in Fig. 4.3.

4.3 Problem 9. Purchase of Processing Machines

Given a set of n production tasks, each with a given process time. The cost of the machines that process the tasks is C $/machine. It is about obtaining the lowest possible cost in the purchase of machines so that all tasks are processed in less than 8 h (the machines start working at the same time and there are no tasks with a duration greater than 8 h). On the other hand, a job is processed on a single machine, and a machine cannot process more than five tasks.

```
!Units; 5~

!Bars; 4~

!Periods; 5~

!W; 20 50 10 10 10~

!TD; 2920 3650 1280 910~

!CE; 15000 18000 20000 21000 21000
     40000 350000 300000 280000 280000
     50000 400000 400000 400000 300000
     10000 100000 120000 130000 150000
      1000  11000 12000  14000  16000 ~

!FC; 100 110 110 120 120
     200 200 200 200 200
     300 290 290 285 285
     200 190 190 180 180
     50 50 50 50 50  ~

!VC;  4 4 4 4 4
     5 5 5 5 5
     3 3 3 3 3
     3 3 3 3 3
     1 1 1 1 1 ~

!AF; 80 80 85 85 85
     85 85 85 85 85
     90 90 90 90 90
     90 90 95 95 95
     85 85 90 90 90~

!PD;  6000 6000 6000 6000 6000
     9000 10000 110000 12000 13000
     25000 27000 27000 27000 27000
     50000 58000 58000 58000 57000~
```

Fig. 4.2 Problem 8: input file

```
Solution Report - 8_Lingo
Global optimal solution found.
Objective value:                                    1460300.
Infeasibilities:                                    0.000000
Total solver iterations:                                  50
Elapsed runtime seconds:                                0.03

Model Class:                                              LP

Total variables:                     130
Nonlinear variables:                   0
Integer variables:                     0

Total constraints:                   146
Nonlinear constraints:                 0

Total nonzeros:                      490
Nonlinear nonzeros:                    0
```

Fig. 4.3 Problem 8: LINGO solution report

Modeling

Generally, in job scheduling problems (Pinedo 1995), machines are treated individually due to their individual identification in system decisions or even because there are explicit specifications regarding each machine, such as that they cannot process any more of a job at the same time. There are problems where this last specification can be ignored because the scheduling in time of the jobs is obtained after the solution of the model, through an assignment procedure (e.g., the case of the problem 'Pm | | Cmax').

There is a rare scenario where individual treatment of machines can be ignored. They are very specific cases where preemption situations are imposed and there are no specifications that treat the machines individually. An example would be the "P_m | prmp | Cmax" problem. The assignment of machines to jobs is carried out a posteriori by an assignment procedure.

In our example, although the machines are identical, we are in a general case that requires individual treatment of the machines, and there are also direct specifications on each machine. However, the concrete scheduling in the time of the tasks can be obtained later.

Well, as reflected in Garcia (2021b), if the a priori items of an element with an indeterminate quantity have an individual treatment, it is necessary to establish an upper bound on the number of individual elements to be established in the problem. For simplicity, we can use the number of tasks n as the upper bound of the number of machines.

List of elements in Table 4.3.

Table 4.3 Problem 9: table of elements

			Data				
Elements	Set	QN	Name	Parameter	Type	Belonging	Value
Tasks	$i = 1...n$	I_U	Processing time	t_i	C(h)	W	...
Machines	$j = 1...n$	I_U					
System	–	I_U	Machine cost	C	C($)	W	C$
			Max tasks/machine	M	I	W	5
			Work limit	TT	C(h)	W	8 h

Decision Activities

Assign (Schedule) Tasks (*i*) to Machines (*j*)

$$\forall i,j : \alpha_{ij} = 1 \text{ if task } i \text{ is scheduled on machine } j; 0 \text{ otherwise.} \tag{9.1}$$

Specifications

– Quantitative selection specifications:

Activity	Elements selecting	Selectable elements	Selection	Constraints
α_{ij}	Jobs ($i = 1...n$)	Machines ($j = 1...n$)	$=1$	$\forall i : \sum_{j=1}^{n} \alpha_{ij} = 1$ (9.2)
	Machines ($j = 1...n$)	Jobs ($j = 1...n$)	$\leq M$	$\forall j : \sum_{i=1}^{n} \alpha_{ij} \leq M$ (9.3)

– Machine work limit (capacity specification):

$$\forall j : \sum_{i=1}^{n} t_i \alpha_{ij} \leq \text{TT} \tag{9.4}$$

Objective Criterion

Minimize machine purchase cost:

We must buy the machines that we have used. We need a calculation to know which machines we have used, since this information is not directly collected by any attribute or decision variable.

Binary Logical Calculation

$$\forall j : \beta_j = 1 \text{ if machine } j \text{ has been used; 0 otherwise.} \tag{9.5}$$

$$\forall j : \beta_j = 1 \text{ IF AND ONLY IF } \sum_{i=1}^{n} \alpha_{ij} > 0$$

$$\Rightarrow f_{\mathrm{Sv}} \Rightarrow \forall j : \text{IF } \sum_{i=1}^{n} \alpha_{ij} > 0 \text{ THEN } \beta_j = 1$$

$$\Rightarrow f_3 \Rightarrow \forall j : \text{IF } \beta_j = 0 \text{ THEN } \sum_{i=1}^{n} \alpha_{ij} \leq 0$$

$$\Rightarrow f_7 \Rightarrow \forall j : \text{IF } 1 - \beta_j = 1 \text{ THEN } \sum_{i=1}^{n} \alpha_{ij} \leq 0$$

$$\Rightarrow f_{14}\left(\mathrm{UB}_{\sum_{i=1}^{n} \alpha_{ij}} = M\right) \Rightarrow \forall j : \sum_{i=1}^{n} \alpha_{ij} \leq M\left(1 - \left(1 - \beta_j\right)\right)$$

$$\Rightarrow \forall j : \sum_{i=1}^{n} \alpha_{ij} \leq M\beta_j \tag{9.6}$$

This constraint makes it unnecessary to consider constraint (9.3).
Objective Function
Minimize Costs:

$$\text{Min } \sum_{j=1}^{n} C\beta_j \tag{9.7}$$

Complete Model

$$\text{Min } \sum_{j=1}^{n} C\beta_j \tag{9.7}$$

s.t.

$$\forall i : \sum_{j=1}^{n} \alpha_{ij} = 1 \tag{9.2}$$

$$\forall j : \sum_{i=1}^{n} t_i \alpha_{ij} \leq \mathrm{TT} \tag{9.4}$$

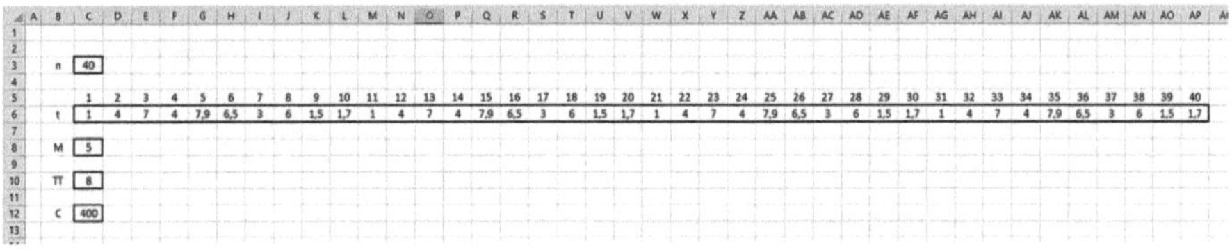

Fig. 4.4 File "c: \ Files_\ 9_Excel.xlsx"

$$\forall j : \sum_{i=1}^{n} \alpha_{ij} \leq M\beta_j \tag{9.6}$$

$$\forall i, j : \alpha_{ij} \text{ binary} \tag{9.1}$$

$$\forall j : \beta_j \text{ binary} \tag{9.5}$$

Implementation in LINGO

We use the data collected in the file "c: \ Files_\ 9_Excel.xlsx" (Fig. 4.4):

```
MODEL:
DATA:
 n= @OLE("c:\Files_\9_Excel.xlsx",'n');
ENDDATA
SETS:
Jobs/1..n/:t;
Machines/1..n/:Beta;
JM(Jobs,Machines):Alpha;
ENDSETS
DATA:
 t= @OLE("c:\Files_\9_Excel.xlsx",'t');
 M= @OLE("c:\Files_\9_Excel.xlsx",'M');
 TT = @OLE("c:\Files_\9_Excel.xlsx",'TT');
 C = @OLE("c:\Files_\9_Excel.xlsx",'Cost');
ENDDATA
!9.7; Min = @SUM(Machines(j): C*BEta(j));
!9.2; @FOR(Jobs(i):@SUM(Machines(j):Alpha(i,j))=1);
!9.4;
@FOR(Machines(j):@SUM(Jobs(i):t(i)*Alpha(i,j))<=TT);
!9.6;
@FOR(Machines(j):@SUM(Jobs(i):Alpha(i,j))<=M*Beta(j));
!9.1; @FOr(JM(i,j):@BIN(Alpha(i,j)));
!9.5; @FOr(Machines(j):@BIN(Beta(j)));
END
```

Solution

The solution is shown in Fig. 4.5.

```
Solution Report - 9_LingoE
Global optimal solution found.
Objective value:                              8800.000
Objective bound:                              8800.000
Infeasibilities:                              0.000000
Extended solver steps:                           17813
Total solver iterations:                        752252
Elapsed runtime seconds:                         42.13

Model Class:                                      PILP

Total variables:                  1640
Nonlinear variables:                 0
Integer variables:                1640

Total constraints:                 121
Nonlinear constraints:               0

Total nonzeros:                   4880
Nonlinear nonzeros:                  0
```

Fig. 4.5 Problem 9: LINGO solution report

4.4 Problem 10. Project Selection (Castro et al. 2011)

A company is considering five projects. Each approved project will be executed over a 3-year period. The expected returns and annual expenses for each project, together with the annual funds available in thousands of euros, are shown in Table 4.4.

Table 4.4 Problem data

	Expenses			
Project	Year 1	Year 2	Year 3	Yields
1	5	1	8	20
2	4	7	10	40
3	3	9	2	20
4	7	4	1	15
5	8	6	10	30
Funds	25	25	25	

The company, taking into account its available capital, must choose the projects with the aim of maximizing total yields. In addition, the following information is available:

Project 3 cannot be selected if it is done on 5.

Projects 1 and 2 are completed jointly only if neither Project 4 nor Project 5 is completed.

The company must reduce its available funds by $5000 in one of the 3 years and must decide in which year to do so.

Modeling

List of elements in Table 4.5.

Table 4.5 Problem 10: table of elements

			Data				
Elements	Set	QN	Name	Parameter	Type	Belonging	Value
Projects	$i = 1\ldots5$	I_U	Yields	L_i	C($)	W	…
			Expenses	g_{it}	C($)	S	…
Years	$t = 1\ldots3$	I_U	Funds	f_t	C($)	W	…
				g_{it}			
System	–	I_U	Funds reduction	R	C($)		5000

Decision Activities
Select projects (*i*)

$$\forall i : \alpha_i = 1 \text{ if Project } i \text{ is selected; 0 otherwise.} \tag{10.1}$$

Select years(*t*) *for funds reduction* (information for specifications).

$$\forall t : \beta_t = 1 \text{ if we select year } t\text{; 0 otherwise.} \tag{10.2}$$

Taking as an activity reducing funds in years, of a measurable nature, is feasible but not very efficient, since it entails a specification of propositional logic to impose that only 1 year be done. For the methodology, the funds should also be made explicit as an individual measurable element.

Setting the selection of a year as an activity, it will only be necessary to include a selection quantitative rule to know how many years to select, one in this case, and use such activity as an activity for the consumption of the available funds in the specification of capacity generated for funds.

Specifications

– Quantitative selection specification:

Activity	Elements selecting	Selectable elements	Selection	Constraints
β_t	System	Years $(t = 1\ldots3)$	=1	$\sum_{t=1}^{3} \beta_t = 1$ (10.3)

– Availability of funds (capacity specification):

$$\forall t : \sum_{t=1}^{3} g_{it}\alpha_i + R\beta_t \le F_t \tag{10.4}$$

– Project 3 is not selected if Project 5 is selected (propositional logic specification):

Statement: If Project 5 is selected, then Project 3 is not selected.
Mathematically:

$$
\begin{aligned}
&\text{IF } \alpha_5 = 1 \text{ THEN NOT}(\alpha_3 = 1)\\
&\Rightarrow f_{12} \Rightarrow \text{IF } \alpha_5 = 1 \text{ THEN } \alpha_3 = 0\\
&\Rightarrow f_7 \Rightarrow \text{IF } \alpha_5 = 1 \text{ THEN } 1 - \alpha_3 = 1\\
&\Rightarrow f_{13} \Rightarrow \alpha_5 \le 1 - \alpha_3 \Rightarrow\\
&\qquad \Rightarrow \alpha_3 + \alpha_5 \le 1
\end{aligned} \tag{10.5}
$$

- Projects 1 and 2 are done jointly only if neither Project 4 nor Project 5 is done (propositional logic specification).

Statement: If you select Project 1 and Project 2, then you do not select Project 4 AND you do not select Project 5.

$$\text{IF } \alpha_1 = 1 \text{ AND } \alpha_2 = 1 \text{ THEN NOT}(\alpha_4 = 1) \text{ AND NOT}(\alpha_5 = 1)$$

$$\Rightarrow f_{12} \Rightarrow \text{IF } \alpha_1 = 1 \text{ AND } \alpha_2 = 1 \text{ THEN } \alpha_4 = 0 \text{ AND } \alpha_5 = 0$$

$$\Rightarrow f_{41} \Rightarrow \text{IF } \alpha_1 = 1 \text{ AND } \alpha_2 = 1 \text{ THEN } \alpha_4 = 0$$

$$\Rightarrow f_{37} \Rightarrow \text{IF } \alpha_1 + \alpha_2 \ge 2 \text{ THEN } \alpha_4 = 0$$

$$\Rightarrow f_3 \Rightarrow \text{IF NOT}(\alpha_4 = 0) \text{ THEN NOT}(\alpha_1 + \alpha_2 \ge 2)$$

$$\Rightarrow f_{11}, f_2 \Rightarrow \text{IF } \alpha_4 = 1 \text{ THEN } \alpha_1 + \alpha_2 \le 1$$

$$\Rightarrow f_{14}(\text{UB}_{\alpha_1 + \alpha_2} = 2) \Rightarrow \alpha_1 + \alpha_2 \le 2 - \alpha_4 \tag{10.6}$$

$$\Rightarrow f_{41} \Rightarrow \text{IF } \alpha_1 = 1 \text{ AND } \alpha_2 = 1 \text{ THEN } \alpha_5 = 0$$

$$\Rightarrow f_{37}, f_3, f_{11}, f_{12}, f_{14}(\text{UB}_{\alpha_1 + \alpha_2} = 2) \Rightarrow \alpha_1 + \alpha_2 \le 2 - \alpha_5 \tag{10.7}$$

Objective Function
Maximize returns:

$$\text{Max} \sum_{i=1}^{5} L_i \alpha_i \tag{10.8}$$

Complete Model

$$\text{Max} \sum_{i=1}^{5} L_i \alpha_i \tag{10.8}$$

s.t.

$$\sum_{t=1}^{3} \beta_t = 1 \tag{10.3}$$

$$\forall t : \sum_{t=1}^{3} g_{it}\alpha_i + R\beta_t \leq F_t \tag{10.4}$$

$$\alpha_3 + \alpha_5 \leq 1 \tag{10.5}$$

$$\alpha_1 + \alpha_2 \leq 2 - \alpha_4 \tag{10.6}$$

$$\alpha_1 + \alpha_2 \leq 2 - \alpha_5 \tag{10.7}$$

$$\forall i : \alpha_i \text{ binaries} \tag{10.1}$$

$$\forall t : \beta_t \text{ binaries} \tag{10.2}$$

Implementation in LINGO

```
MODEL:
SETS:
Projects/1..5/:L,ALpha;
Years/1..3/:f,Beta;
PY(Projects,Years):g;
ENDSETS
DATA:
 L = 20 40 20 15 30;
 g =  5 1 8 4 7
    10 3 9 2 7
    4 1 8 6 10;
 f = 25 25 25;
 R=5;
ENDDATA
!10.8; MAX = @SUM(Projects(i): L(i)*Alpha(i));
!10.3; @SUM(Years(j):Beta(j))=1;
!10.4; @FOR(Years(j):@SUM(Projects(i):g(i,j)*Alpha(i))+ R*Beta(j)
<=f(j));
!10.5; Alpha(3)+Alpha(5)<= 1;
!10.6; Alpha(1)+Alpha(2)<= 2 - Alpha(4);
!10.7; Alpha(1)+Alpha(2)<= 2 - Alpha(5);
!10.1; @FOR(Projects(i): @BIN(Alpha(i)));
!10.2; @FOR(Years(j): @BIN(Beta(j)));
END
```

Solution

The solution is shown in Fig. 4.6.

4.5 Problem 11. Assignment of Workers to Centers

Let there be a set of n workers. Each worker lives in a specific city in the province. There are m work centers in the province, each located in a specific city in the province. We consider L cities. Each center has a minimum and a maximum number of workers needed.

```
Solution Report - 10_LingoE
Global optimal solution found.
Objective value:                              85.00000
Objective bound:                              85.00000
Infeasibilities:                              0.000000
Extended solver steps:                               0
Total solver iterations:                             0
Elapsed runtime seconds:                          0.03

Model Class:                                      PILP

Total variables:                  8
Nonlinear variables:              0
Integer variables:                8

Total constraints:                8
Nonlinear constraints:            0
```

Fig. 4.6 Problem 10: LINGO solution report

The company wants to reassign workers to centers in such a way as to minimize the total distance that workers travel. Obviously, the distances between each pair of cities are known.

Modeling

Unitary Version (inefficient version)

In this first version, we consider the workers as individual elements (Table 4.6).

Table 4.6 Problem 11 (unitary version): table of elements

			Data				
Elements	Set	QN	Name	Parameter	Type	Belonging	Value
Workers	$i = 1\ldots n$	I_U	City	CT_i	O	W	...
Cities	$k = 1\ldots L$	I_U	Distance between cities	$D_{ii'}$	C (Km)	S	...
Work centers	$j = 1\ldots m$	I_U	Minimum of workers	Mini_j	I	W	...
			Maximum of workers	Maxi_j			
			City	C_j	O	W	...

Decision Activities

Reassign workers (*i*) to centers (*j*)

$$\forall i,j : \alpha_{ij} = 1 \text{ if worker } i \text{ is reassigned to center } j$$

Specifications

– Quantitative selection specification:

Activity	Elements selecting	Selectable elements	Selection	Constraints
α_{ij}	Workers ($i = 1\ldots n$)	Centers ($j = 1\ldots m$)	$=1$	$\forall i : \sum_{j=1}^{m} \alpha_{ij} = 1$
	Centers ($j = 1\ldots m$)	Workers ($i = 1\ldots n$)	$\geq \text{Mini}_j$	$\forall j : \sum_{i=1}^{n} \alpha_{ij} \geq \text{Mini}_j$
			$\leq \text{Maxi}_j$	$\forall j : \sum_{i=1}^{n} \alpha_{ij} \leq \text{Maxi}_j$

Objective Function
Minimize the total distance:

$$\text{Min} \sum_{i=1}^{n} \sum_{j=1}^{m} D_{\text{CT}_i C_j} \alpha_{ij}$$

Collective Version (Efficient Version)

Workers can be grouped into groups from workers in each city, which is the data that differentiate them (Table 4.7). We can do this since workers are not referred to individually. Cities as such are represented in each of these groups, so distances between cities can be integrated as a shared attribute between groups, without the need to create an element for each city.

An example of a data instance could be the following (distance data are not displayed):

	Workers	Centers	City to which it belongs	Maximum of workers	Minimum of workers
City 1	135	Center 1	City 2	10	8
City 2	23	Center 2	City1	6	5
City 3	12	Center 3	City 2	8	7
...	...	...	...	...	...

Decision Activities
Reassign workers of cities (*i*) to centers (*j*)

$$\forall i,j : x_{ij} = \text{number of workers of city } i \text{ reassigned to center } j \tag{11.1}$$

Specifications

– Minimum and maximum of workers (bound impositions):

$$\forall j : \sum_{i=1}^{L} x_{ij} \geq \text{Mini}_j \tag{11.2}$$

Table 4.7 Problem 11 (collective version): table of elements

Elements	Set	QN	Data				
			Name	Parameter	Type	Belonging	Value
Workers of cities	$i = 1 \ldots L$	C_D	Number of workers	n_i	I	W	...
			Distance between cities	$D_{ii'}$	C	S	...
Centers	$j = 1 \ldots m$	I_U	Minimum of workers	Mini_j	I	W	...
			Maximum of workers	Maxi_j	I	W	...
			City	C_j	O	W	...

$$\forall j : \sum_{i=1}^{L} x_{ij} \leq \text{Maxi}_j \tag{11.3}$$

– Allocation of workers (allocation specification):

$$\forall i : \sum_{j=1}^{m} x_{ij} = n_i \tag{11.4}$$

Objective Function

$$\text{Min} \sum_{i=1}^{L} \sum_{j=1}^{m} D_{iC_j} x_{ij} \tag{11.5}$$

Complete Model

$$\text{Min} \sum_{i=1}^{L} \sum_{j=1}^{m} D_{iC_j} x_{ij} \tag{11.5}$$

s.t.

$$\forall j : \sum_{i=1}^{L} x_{ij} \geq \text{Mini}_j \tag{11.2}$$

$$\forall j : \sum_{i=1}^{L} x_{ij} \leq \text{Maxi}_j \quad (11.3)$$

$$\forall i : \sum_{j=1}^{m} x_{ij} = n_i \quad (11.4)$$

$$\forall i,j : x_{ij} \text{ integers} \quad (11.1)$$

Implementation in LINGO

For resolution, we are going to use a set of data collected in the Excel sheet "11_Excel.xlsx":

```
MODEL:
DATA:
 L= @OLE("c:\Files_\11_Excel.xlsx",'L');
 m= @OLE("c:\Files_\11_Excel.xlsx",'m');
ENDDATA
SETS:
Cities/1..L/:n;
Centers/1..m/:Mini,Maxi,C;
DD(Cities,Cities):D;
CC(Cities,Centers):x;
ENDSETS
DATA:
 n= @OLE("c:\Files_\11_Excel.xlsx",'nt');
 m= @OLE("c:\Files_\11_Excel.xlsx",'m');
 D = @OLE("c:\Files_\11_Excel.xlsx",'Distance');
 Mini = @OLE("c:\Files_\11_Excel.xlsx",'Mini');
 Maxi = @OLE("c:\Files_\11_Excel.xlsx",'Maxi');
 C= @OLE("c:\Files_\11_Excel.xlsx",'City');
ENDDATA
CALC:
@FOR(Cities(i):@FOR(Cities(ii)|i#LT#ii: D(ii,i)=D(i,ii)));
ENDCALC
!11.5; MIN = @SUM(Cities(i):@SUM(Centers(j):D(i,C(j))*x(i,j)));
!11.2; @FOR(Centers(j):@SUM(Cities(i):x(i,j))>=Mini(j));
!11.3; @FOR(Centers(j):@SUM(Cities(i):x(i,j))<=Maxi(j));

!11.4; @FOR(Cities(i):@SUM(Centers(j):x(i,j))=n(i));
!11.1; @FOR(Cities(i):@FOR(Centers(j):@GIN(x(i,j))));
DATA:
@OLE("c:\Files_\11_Excel.xlsx",'Solution') = x ;
ENDDATA
END
```

Solution

The solution is shown in Fig. 4.7 and it is also printed in file "11_Excel.xlsx."

```
Solution Report - 11_LingoE
Global optimal solution found.
Objective value:                               280.0000
Objective bound:                               280.0000
Infeasibilities:                               0.000000
Extended solver steps:                                0
Total solver iterations:                             64
Elapsed runtime seconds:                           1.03


Export Summary Report
---------------------
Transfer Method:         OLE BASED
Workbook:                c:\Files_\11_Excel.xlsx
Ranges Specified:               1
    Solution
Ranges Found:                   1
Range Size Mismatches:          0
Values Transferred:          1000

Model Class:                                       PILP

Total variables:                     1000
Nonlinear variables:                    0
Integer variables:                   1000

Total constraints:                    106
Nonlinear constraints:                  0

Total nonzeros:                      3961
Nonlinear nonzeros:                     0
```

Fig. 4.7 Problem 11: LINGO solution report

4.6 Problem 12. Warehouse Rental

An entrepreneur wants to rent a series of industrial buildings for the next year for a cement distribution business. In the industrial estate where he is going to establish the business, there are six warehouses ($i = 1\ldots6$) available for lease.

The businessman starts the business with 50 tons of cement and five vehicles for transportation. The warehouses he rents have to accommodate both cement and vehicles.

Industrial buildings offer the storage capacities for cement and vehicles collected in Table 4.8.

The annual rent price of the warehouses ($i = 1\ldots6$) is $\$R_i$.

The entrepreneur imposes as requirements in the business that each of the rented warehouses stores only one of the two concepts (cement or vehicles). In other words, a warehouse cannot be used to store cement and vehicles at the same time.

The objective is to minimize the total cost of renting for 1 year.

Modeling

There are several interpretations here:

Rent initially seems like a primary decision of the system. However, we cannot work with it alone as we need to know what is stored in each warehouse. That, in a

Table 4.8 Problem data

	Wh 1	Wh 2	Wh 3	Wh 4	Wh 5	Wh 6
Cement tons	25	30	20	20	15	30
No vehicles	3	3	2	3	2	4

context that takes into account costs, will mean that renting a warehouse will only occur if I introduce a concept into it. From that point of view, the action of renting will proceed from a logical calculation based on whether or not we put cement or vehicles in the warehouses.

On the other hand, there are two postulates regarding the activity of introducing concepts in the ships. The first, more general and precise, measures the amount of cement that we store in each warehouse and the number of vehicles (version 1). This first version considers the two concepts measurable and focuses on measuring their content in distribution by the warehouses.

It is also possible to propose a second postulate (version 2) that does not measure in the decisions the cement and vehicles to be introduced in the warehouses. It does not focus on putting quantities of the two concepts in the warehouses, but on deciding which warehouses to rent for each concept and making sure in the specifications that with the selection of warehouses made for each concept, we will have enough capacity for the content of each concept. We will achieve this with a specification of the variable capacity contribution on a capacity demand. The decisions in this second version are logical decisions about whether each concept is assigned to each warehouse. Decisions work with the vehicle concept and the cement concept as indirect objects, not direct objects; therefore, they are not measured in decisions.

We will finally see how from version 1 we can simplify the model and arrive at the version 2 model.

Version 1

List of elements in Table 4.9.

Table 4.9 Problem 12: table of elements

			Data				
Elements	Set	QN	Name	Parameter	Type	Belonging	Value
Warehouses	$i = 1...6$	I_U	Rent price	R_i	C($)	W	...
			Vehicles Capacity	vK_i	I	S	...
			Cement Capacity	cK_i	C(m^3)	S	...
Vehicles	–	C_D	Stock	EV	I	W	5
				vK_i			
Cement	–	I_M	Stock	EC	C(m^3)	W	50 tn
				cK_i			

Decision Activities

Store cement in warehouses (*i*)

$$x_i = \text{Cement amount stored in warehouse } i.$$

Store vehicles in warehouses (*i*)

$$y_i = \text{Number of vehicles stored in warehouse } i.$$

Specifications

- Capacity of the warehouses (capacity specification)

$$\forall i : x_i \leq cK_i \tag{12.1}$$

$$\forall i : y_i \leq vK_i \tag{12.2}$$

- Availability of cement and vehicles (allocation specification):

  Availability generates an allocation specification in this system:

$$\sum_{i=1}^{6} x_i = \text{EC} \tag{12.3}$$

$$\sum_{i=1}^{6} y_i = \text{EV} \tag{12.4}$$

- Store a single concept at most in a warehouse (propositional logic specification):

$$\text{NOT}(x_i > 0 \ \text{ AND } \ y_i > 0) \tag{12.5}$$

Modeling:
Binary Logical Calculation:

$$\forall i : \omega_i = 1 \ \text{IF AND ONLY IF} \ x_i > 0$$
$$\Rightarrow f_5 \Rightarrow \forall i : \omega_i = 1 \ \text{IF AND ONLY IF} \ x_i \geq \xi$$

Binary Logical Calculation:

$$\forall i : \pi_i = 1 \text{ IF AND ONLY IF } y_i > 0$$
$$\Rightarrow f_5 \Rightarrow \forall i : \omega_i = 1 \text{ IF AND ONLY IF } y_i \geq 1$$

Later ω_i and π_i will determine the rental calculation, which involves a cost, so in this case these binary logic calculations can be simplified; it is not necessary to define the value 0 of these calculations (the system will take it as long as it has freedom).

$$\forall i : \omega_i = 1 \text{ IF AND ONLY IF } x_i \geq \xi$$
$$\Rightarrow f_{S_V} \Rightarrow \forall i : \text{IF } x_i \geq \xi \text{ THEN } \omega_i = 1$$
$$\Rightarrow f_3, f_7 \Rightarrow \forall i : \text{IF } 1 - \omega_i = 1 \text{ THEN } x_i \leq 0$$
$$\forall i : \pi_i = 1 \text{ IF AND ONLY IF } y_i \geq 1$$
$$\Rightarrow f_{S_V} \Rightarrow \forall i : \text{IF } y_i \geq 1 \text{ THEN } \pi_i = 1$$
$$\Rightarrow f_3, f_7 \Rightarrow \forall i : \text{IF } 1 - \pi_i = 1 \text{ THEN } y_i \leq 0$$
$$\Rightarrow f_{14}(\text{UB}_{x_i} = cK_i) \Rightarrow \forall i : x_i \leq cK_i\omega_i \tag{12.6}$$
$$\Rightarrow f_{14}\left(\text{UB}_{y_i} = cV_i\right) \Rightarrow \forall i : y_i \leq vK_i\pi_i \tag{12.7}$$

(12.6) and (12.7) allow us to ignore constraints (12.1) and (12.2), respectively.

$$\forall i : \text{NOT}(x_i > 0 \text{ AND } y_i > 0)$$
$$\Rightarrow \forall i : \text{NOT}(w_i = 1 \text{ AND } \pi_i = 1) \Rightarrow f_{37} \Rightarrow \forall i : \text{NOT}(w_i + \pi_i \geq 2)$$
$$\Rightarrow f_{11} \Rightarrow \forall i : w_i + \pi_i \leq 1 \tag{12.8}$$

Objective Criterion
Minimize the cost of rented warehouses:
Binary Logical Calculation: Rented warehouses (i).

$$\forall i : \alpha_i = 1 \text{ if warehouse } i \text{ has been rented; 0 otherwise} \tag{12.9}$$

For the definition of the calculation and to avoid repeating the already defined logical calculations, we will take advantage of the auxiliary logical calculations used in the specification (12.5).

$$\forall i : \alpha_i = 1 \text{ IF AND ONLY IF } (\omega_i = 1 \text{ OR } \pi_i = 1)$$
$$\forall i : \alpha_i = 1 \text{ IF AND ONLY IF } (\omega_i + \pi_i \geq 1)$$
$$\Rightarrow f_{S_V} \Rightarrow \forall i : \text{IF } \omega_i + \pi_i \geq 1 \text{ THEN } \alpha_i = 1$$
$$\Rightarrow f_3, f_7 \Rightarrow \forall i : \text{IF } 1 - \alpha_i = 1 \text{ THEN } \omega_i + \pi_i \leq 0$$

$$\Rightarrow f_{14}(\mathrm{UB}_{\omega_i+\ \pi_i}=1) \Rightarrow \forall i: \omega_i + \ \pi_i \le \alpha_i \tag{12.10}$$

Constraint (12.10) allows us to ignore Constraint (12.8).
Objective Function:

$$\text{Min} \sum_{i=1}^{6} p_i \alpha_i \tag{12.11}$$

Version 2
Decision Activities
Rent warehouses (*i*) for cement

$$w_i = 1 \text{ if we rent warehouse } i \text{ for cement; 0 otherwise.} \tag{12.12}$$

Rent warehouses (*i*) for vehicles

$$\pi_i = 1 \text{ if we rent warehouse } i \text{ for vehicles; 0 otherwise.} \tag{12.13}$$

Specifications

- Capacity variable contribution for a capacity demand (capacity specification).

$$\sum_{i=1}^{6} cK_i w_i \ge \mathrm{EC} \tag{12.14}$$

$$\sum_{i=1}^{6} vK_i \pi_i \ge \mathrm{EV} \tag{12.15}$$

- Rent a warehouse for a single concept at most

$$\forall i: \mathrm{NOT}(\omega_i = 1 \ \text{AND} \ \pi_i = 1) \Rightarrow f_{37} \Rightarrow \forall i: \mathrm{NOT}(\omega_i + \ \pi_i \ge 2)$$
$$\Rightarrow f_{11} \Rightarrow \forall i: \omega_i + \ \pi_i \le 1 \tag{12.16}$$

Objective Criterion
Identical modeling to version 1.
Complete Model

$$\text{Min} \sum_{i=1}^{6} p_i \alpha_i \tag{12.11}$$

s.t.

$$\sum_{i=1}^{6} cK_i w_i \geq \text{EC} \tag{12.14}$$

$$\sum_{i=1}^{6} vK_i \pi_i \geq \text{EV} \tag{12.15}$$

$$\forall i : \omega_i + \ \pi_i \leq \alpha_i \tag{12.10}$$

$$\forall i : \alpha_i, \omega_i, \ \pi_i \text{ binaries} \qquad (12.9,\ 12.12,\ 12.13)$$

Implementation in LINGO

```
MODEL:
SETS:
Warehouses/1..6/: p, vK, cK, Alpha,Omega,Pi;
ENDSETS
DATA:
p= 35 33 26 23 30 29;
vK= 2 2 2 1 3 3;
cK= 20 18 13 19 22 22;
EV= 5;
EC= 50;
ENDDATA
!12.11;
MIN = @SUM(Warehouses(i):P(i)*Alpha(i));
!12.14;
@SUM(Warehouses(i):cK(i)*Omega(i))>=EC;
!12.15;
@SUM(Warehouses(i):vK(i)*Pi(i))>=EV;
!12.10;
@FOR(Warehouses(i): Omega(i) + Pi(i)<= Alpha(i));
!12.9;
@FOR(Warehouses(i):@BIN(Alpha(i)));
!12.12;
@FOR(Warehouses(i):@BIN(Omega(i)));
!12.13;
@FOR(Warehouses(i):@BIN(Pi(i)));
END
```

Solution

It is shown in Fig. 4.8.

Step from Version 1 to Version 2

Looking at the version 1 model:

It is true that:

$$\text{EC} = \sum_{i=1}^{6} x_i \leq \text{(adding 12.6)} \leq \sum_{i=1}^{6} cK_i \omega_i \text{ We get constraint } (12.14)$$

```
Solution Report - 12_LingoE
Global optimal solution found.
Objective value:                                    141.0000
Objective bound:                                    141.0000
Infeasibilities:                                    0.000000
Extended solver steps:                                     0
Total solver iterations:                                   0
Elapsed runtime seconds:                                0.03

Model Class:                                            PILP

Total variables:                      18
Nonlinear variables:                   0
Integer variables:                    18

Total constraints:                     9
Nonlinear constraints:                 0

Total nonzeros:                       36
Nonlinear nonzeros:                    0
```

Fig. 4.8 Problem 12: LINGO solution report

$$\text{EV} = \sum_{i=1}^{6} y_i \leq (\text{adding } 12.7) \leq \sum_{i=1}^{6} vK_i\pi_i \text{ We get constraint } (12.15)$$

And we managed to eliminate the variables x_i and *yi*.

4.7 Problem 13. Article and Position Planning

Plan the placement of a set of articles in the set of positions in a warehouse.

The data available for this management are summarized below:

Each article has a weight. The compatibility between them is known. Two incompatible articles cannot be in the same position. The positions have a supported weight capacity, and we also know the distance between them. The distance between two occupied positions must be at least 3 m.

The goal is to place the largest number of articles.

Modeling
List of elements in Table 4.10.

Table 4.10 Problem 13: table of elements

Elements	Set	QN	Data				
			Name	Parameter	Type	Belonging	Value
Articles	$i, j = 1 \ldots n$	I_U	Weight	p_i	C(Kg)	W	...
			Compatibility	A_{ij}	B	S	...
Positions	$k, k' = 1 \ldots m$	I_U	Supported weight	SW_k	C(Kg)	W	...
			Distance	$D_{kk'}$	C (m)	S	...

Decision Activities
Place articles (*i*) in positions (*k*).

$$\forall i, k : \alpha_{ik} = 1 \text{ if article } i \text{ is placed in position } k; 0 \text{ otherwise.} \tag{13.1}$$

Specifications

- Quantitative selection specifications:

  An article is at most in one position.

Activity	Elements selecting	Selectable elements	Selection	Constraints
α_{ik}	Articles ($i = 1 \ldots n$)	Positions ($k = 1 \ldots m$)	≤ 1	$\forall i : \sum_{k=1}^{m} \alpha_{ik} \leq 1$ (13.2)

- Supported weight (capacity specification based on attribute):

$$\forall k : \sum_{i=1}^{n} p_i \alpha_{ik} \leq SW_k \tag{13.3}$$

- Articles incompatibility: Two articles i and j that are not compatible ($A_{ij} = 0$) cannot go in the same position (any position) (propositional logic specification).

$$\forall k, i, j / i < j \& A_{ij} = 0 : \text{NOT}(\alpha_{ik} = 1 \text{ AND } \alpha_{jk} = 1) \Rightarrow$$
$$\Rightarrow f_{37}, f_{11} \Rightarrow \forall k, i, j / i < j \& A_{ij} = 0 : \alpha_{ik} + \alpha_{jk} \leq 1 \tag{13.4}$$

- Minimum distance between occupied positions (logical calculation + propositional logic specification): We must calculate for each position whether or not it is occupied, and between each two occupied positions, impose that its distance be at least 3 m.

  Binary Logical Calculation: Occupied Position

$$\forall k : \beta_k = 1 \text{ if position } k \text{ has been occupied; } 0 \text{ otherwise.} \tag{13.5}$$

$$\forall k : \beta_k = 1 \text{ IF AND ONLY IF } \sum_{i=1}^{n} \alpha_{ik} > 0$$

The system does not want a position that is not occupied to be considered occupied, so we can ignore the definition of the value $\beta_k = 0$.

$$\begin{aligned}
&\forall k : \beta_k = 1 \text{ IF AND ONLY IF } \sum_{i=1}^{n} \alpha_{ik} > 0\\
&\Rightarrow f_{S_V} \Rightarrow \forall k : \text{IF } \sum_{i=1}^{n} \alpha_{ik} > 0 \text{ THEN } \beta_k = 1\\
&\Rightarrow f_3, f_7 \Rightarrow \forall k : \text{IF } 1 - \beta_k = 1 \text{ THEN } \sum_{i=1}^{n} \alpha_{ik} \le 0\\
&\Rightarrow f_{14}\left(\text{UB}_{\sum_{i=1}^{n} \alpha_{ik}} = n\right) \Rightarrow \forall k : \sum_{i=1}^{n} \alpha_{ik} \le n\beta_k
\end{aligned} \tag{13.6}$$

Propositional logic specification:

$$\begin{aligned}
&\forall k, k'/k < k' : IF \ \beta_k = 1 \text{ AND } \beta_{k'} = 1 \text{ THEN } D_{kk'} > 3\\
&\Rightarrow f_3 \Rightarrow \forall k, k'/k < k' :\\
&\text{IF NOT}(D_{kk'} > 3) \text{ THEN NOT}(\beta_k = 1 \text{ AND } \beta_{k'} = 1)\\
&\Rightarrow f_{45} \Rightarrow \forall k, k'/k < k' \& D_{kk'} \le 3 : \ \text{NOT}(\beta_k = 1 \text{ AND } \beta_{k'} = 1)\\
&\Rightarrow f_{37}, f_{11} \Rightarrow \forall k, k'/k < k' \& D_{kk'} \le 3 : \ \beta_k + \beta_{k'} \le 1
\end{aligned} \tag{13.7}$$

Objective Function

Maximize Placed Articles:

It is not necessary to perform the logical calculation to know if each object has been placed or not because an object has a quantitative norm of one position at most. Therefore, we can include the decision variables directly in the objective function:

$$\text{Max} \sum_{i=1}^{n} \sum_{k=1}^{m} \alpha_{ik} \tag{13.8}$$

Complete Model

$$\text{Max} \sum_{i=1}^{n} \sum_{k=1}^{m} \alpha_{ik} \tag{13.8}$$

s.t.

$$\forall i : \sum_{k=1}^{m} \alpha_{ik} \leq 1 \tag{13.2}$$

$$\forall k : \sum_{i=1}^{n} p_i \alpha_{ik} \leq \mathrm{SW}_k \tag{13.3}$$

$$\forall k, i, j / i < j \& A_{ij} = 0 : \alpha_{ik} + \alpha_{jk} \leq 1 \tag{13.4}$$

$$\forall k : \sum_{i=1}^{n} \alpha_{ik} \leq n\beta_k \tag{13.6}$$

$$\forall k, k' / k < k' \& D_{kk'} \leq 3 : \beta_k + \beta_{k'} \leq 1 \tag{13.7}$$

$$\forall i, k : \alpha_{ik} \text{ binaries} \tag{13.1}$$

$$\forall k : \beta_k \text{ binaries} \tag{13.5}$$

Implementation in LINGO

We implement an instance collected in the Excel sheet "13_Model.xlsx" (Fig. 4.9):

```
MODEL:
DATA:
 n =@OLE('c:\Files_\13_Excel.xlsx');
 m =@OLE('c:\Files_\13_Excel.xlsx');
ENDDATA
SETS:
Articles/1..n/:Weight;
Positions/1..m/:SW,Beta;
Pr_Pos(Articles,Positions):Alpha;
Pr_Pr(Articles,Articles):A;
Pos_Pos(Positions,Positions):D;
ENDSETS
DATA:
 Weight =@OLE('c:\Files_\13_Excel.xlsx','Weight');
 SW =@OLE('c:\Files_\13_Excel.xlsx','SW');
 A =@OLE('c:\Files_\13_Excel.xlsx');
 D =@OLE('c:\Files_\13_Excel.xlsx');
ENDDATA

!13.8; MAX = @SUM(Pr_Pos(i,k):Alpha(i,k));
!13.2; @FOR(Articles(i):@SUM(Positions(k):Alpha(i,k))<=1);
!13.3; @FOR(Positions(k):@SUM(Articles(i):Weight(i)*Alpha(i,k))
<=SW(k));
!13.4; @FOR(Positions(k):@FOR(Articles(i):@FOR(Articles(j)|i#LT#j
#AND# A(i,j)#EQ#0: Alpha(i,k)+Alpha(j,k)<=1))) ;
!13.6; @FOR(Positions(k):@SUM(Articles(i): Alpha(i,k))<= n*Beta
(k));
!13.7; @FOR(Positions(k):@FOR(Positions(kk)|k#LT#kk #AND# D(k,kk)
```

n	25
m	15

	1	2	3	4	5	6	7	8	9	10	11	12	13	14	15	16	17	18	19	20	21	22	23	24	25
Weight	12	14	15	16	4	7	5	9	12	15	14	13	12	17	3	5	7	6	3	4	5	5	3	2	2

A	1	2	3	4	5	6	7	8	9	10	11	12	13	14	15	16	17	18	19	20	21	22	23	24	25
1	1	1	0	1	0	1	1	1	1	0	0	1	0	0	0	1	0	0	1	0	0	0	0	1	1
2	0	1	1	0	1	0	0	1	1	0	0	0	0	0	0	0	0	1	1	0	0	0	1	0	0
3	0	0	1	0	0	1	1	0	1	1	1	0	1	0	0	1	0	0	0	0	0	0	0	1	0
4	0	0	0	1	0	0	0	1	0	1	1	0	0	0	0	0	0	0	0	0	0	1	1	0	0
5	0	0	0	0	1	1	0	0	1	0	1	0	0	0	1	0	1	0	1	0	0	0	0	0	1
6	0	0	0	0	0	1	0	1	0	0	1	0	1	0	0	0	0	0	0	0	1	0	1	0	0
7	0	0	0	0	0	0	1	0	0	0	0	0	0	1	1	1	0	0	0	0	0	0	0	0	0
8	0	0	0	0	0	0	0	1	0	1	0	1	1	0	0	0	0	0	0	0	0	0	0	1	1
9	0	0	0	0	0	0	0	0	1	1	1	1	0	1	0	0	1	0	1	0	1	0	0	0	0
10	0	0	0	0	0	0	0	0	0	1	0	0	1	1	0	1	0	0	0	0	0	0	0	0	0
11	0	0	0	0	0	0	0	0	0	0	1	1	0	0	0	0	0	0	0	0	1	0	0	0	1
12	0	0	0	0	0	0	0	0	0	0	0	1	0	0	1	0	1	0	1	1	0	0	0	1	0
13	0	0	0	0	0	0	0	0	0	0	0	0	1	0	0	0	0	0	0	0	0	0	0	0	0
14	0	0	0	0	0	0	0	0	0	0	0	0	0	1	0	0	0	0	0	0	0	0	1	0	1
15	0	0	0	0	0	0	0	0	0	0	0	0	0	0	1	0	1	0	0	1	0	0	0	0	0
16	0	0	0	0	0	0	0	0	0	0	0	0	0	0	0	1	0	0	0	0	1	0	0	0	0
17	0	0	0	0	0	0	0	0	0	0	0	0	0	0	0	0	1	0	0	0	0	0	1	1	0
18	0	0	0	0	0	0	0	0	0	0	0	0	0	0	0	0	0	1	0	0	0	0	0	0	0
19	0	0	0	0	0	0	0	0	0	0	0	0	0	0	0	0	0	0	1	0	0	0	0	0	0
20	0	0	0	0	0	0	0	0	0	0	0	0	0	0	0	0	0	0	0	1	0	0	1	0	0
21	0	0	0	0	0	0	0	0	0	0	0	0	0	0	0	0	0	0	0	0	1	0	0	1	0
22	0	0	0	0	0	0	0	0	0	0	0	0	0	0	0	0	0	0	0	0	0	1	0	0	0
23	0	0	0	0	0	0	0	0	0	0	0	0	0	0	0	0	0	0	0	0	0	0	1	0	0
24	0	0	0	0	0	0	0	0	0	0	0	0	0	0	0	0	0	0	0	0	0	0	0	1	0
25	0	0	0	0	0	0	0	0	0	0	0	0	0	0	0	0	0	0	0	0	0	0	0	0	1

	1	2	3	4	5	6	7	8	9	10	11	12	13	14	15
supported weight	15	25	25	30	30	40	40	30	30	25	25	15	15	10	10

Distance	1	2	3	4	5	6	7	8	9	10	11	12	13	14	15
1	0	10	2	1	1,5	2,5	3,5	4	5	5,5	6	6,5	7	7,7	7
2	0	0	1,5	2	1,5	3	4	5	2	3	2	3	4	5	5
3	0	0	0	1,5	4	1,5	1,5	4	2	2	2	2	5	3	2
4	0	0	0	0	2	4	4	1,5	4	2	2	1	1	1	1
5	0	0	0	0	0	3	4	2	1	2	2	2	2	3	2,5
6	0	0	0	0	0	0	3,5	2,5	1	2	1	4,6	1	2,7	4
7	0	0	0	0	0	0	0	1	2	1	2,6	1	2,7	1,5	1,5
8	0	0	0	0	0	0	0	0	1	1	3	4	6	3,5	2,5
9	0	0	0	0	0	0	0	0	0	1	2	1	2,6	1	2,7
10	0	0	0	0	0	0	0	0	0	0	1	2	1	1,5	2,5
11	0	0	0	0	0	0	0	0	0	0	0	1	3,4	3,5	2,7
12	0	0	0	0	0	0	0	0	0	0	0	0	4	2,5	1
13	0	0	0	0	0	0	0	0	0	0	0	0	0	2,5	2
14	0	0	0	0	0	0	0	0	0	0	0	0	0	0	3
15	0	0	0	0	0	0	0	0	0	0	0	0	0	0	0

Fig. 4.9 Excel sheet "13_Model.xlsx"

```
#LE#3: Beta(k)+Beta(kk) <=1));
!13.1; @FOR(Articles(i):@FOR(Positions(k): @BIN(Alpha(i,k))));
!13.5; @FOR(Positions(k): @BIN(Beta(k)));
DATA:
@OLE('c:\Files_\13_Excel.xlsx','Positions')=@WRITEFOR(Articles
(i):@SUM(Positions(k):k*Alpha(i,k)));
ENDDATA
END
```

Solution

The solution is shown in Fig. 4.10.

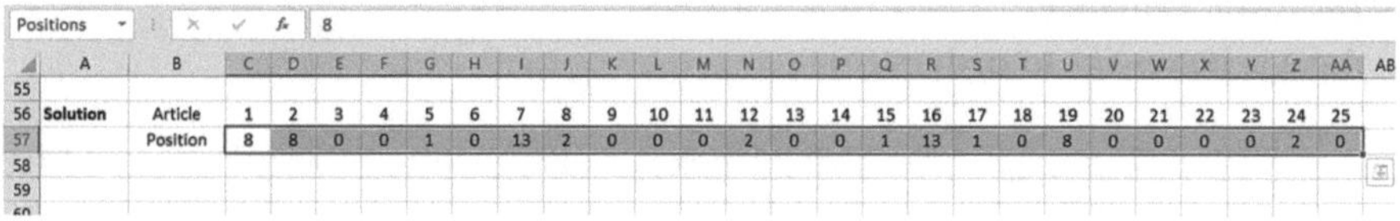

	A	B	C	D	E	F	G	H	I	J	K	L	M	N	O	P	Q	R	S	T	U	V	W	X	Y	Z	AA
55																											
56	**Solution**	Article	1	2	3	4	5	6	7	8	9	10	11	12	13	14	15	16	17	18	19	20	21	22	23	24	25
57		Position	8	8	0	0	1	0	13	2	0	0	0	2	0	0	1	13	1	0	8	0	0	0	0	2	0

Fig. 4.10 Printed solution in Excel sheet "13_Model.xlsx"

4.8 Problem 14: Online Sale

A greengrocer sells 45 different products online, including fruits, vegetables, and derivatives. 25 of them are sold by kilograms and the remaining 20 by units.

Each order is prepared in boxes to be determined for shipment. It has four models of boxes, each with a specific volume. It has a sufficient stock of boxes of any model. The requested quantities are entered into the boxes. The entrepreneur has determined the volume that each kilo of his products occupies and the volume of products that are sold by units.

The stocks of each product are known and may not be sufficient. Therefore, we need to determine which orders are shipped. A dealt order is served in full.

We model the problem of preparing 1-day orders, considering the minimization of the total volume of shipped boxes plus the volume of unattended products multiplied by a factor, so that all possible orders are made. We could choose to force orders whenever there is stock, but it would be more complex. Assume that we have 100 orders, each order with a specified quantity of products.

Modeling

List of elements in Table 4.11.

Table 4.11 Problem 14: table of elements

			Data				
Elements	Set	QN	Name	Parameter	Type	Belonging	Value
K_Products	$i = 1\ldots25$	I_U	Product in order	KP_{ip}	C (Kg)	S	–
			Stock	KS_i	C (Kg)	W	–
				KV_i			
U_Products	$j = 1\ldots20$	C_D	Product in order	UP_{ip}	I	S	–
			Stock	US_j	I	W	–
				UV_j			
Orders	$p = 1\ldots100$	I_U		KP_{ip}, UP_{ip}			
Boxes	$k = 1\ldots4$	C_I	Volume	V_k			
System	–	I_U	Factor	F	C	W	–

Decision Activities
Assign boxes (*k*) to orders (*p*)

$$\forall k, p : x_{kp} = \text{Number of boxes } k \text{ assigned to order } p. \tag{14.1}$$

Ship Orders (*p*)

$$\forall p : \alpha_p = 1 \text{ if order } p \text{ is shipped; 0 otherwise.} \tag{14.2}$$

It can be defined as a logical calculation. In the specifications, we will indicate the relationship between both activities that would be equivalent to the definition as a logical calculation.

Specifications

- Stock availability (capacity specifications):

$$\forall i : \sum_{p=1}^{100} \mathrm{KP}_{ip}\alpha_p \le \mathrm{KS}_i \tag{14.3}$$

$$\forall j : \sum_{p=1}^{100} \mathrm{UP}_{jp}\alpha_p \le \mathrm{US}_j \tag{14.4}$$

- Packaging served orders (propositional logic specification)

$$\begin{aligned} &\forall p : \text{IF } \alpha_p = 1 \text{ THEN } \sum_{k=1}^{4} V_k x_{kp} \ge \sum_{i=1}^{25} \mathrm{KP}_{ip}\mathrm{KV}_i + \sum_{j=1}^{20} \mathrm{UP}_{jp}\mathrm{UV}_j \\ &\Rightarrow f_4\left(\mathrm{LB}_{\sum_{k=1}^{4} V_k x_{kp}} = 0\right) \\ &\Rightarrow \forall p : \sum_{k=1}^{4} V_k x_{kp} \ge \left(\sum_{i=1}^{25} \mathrm{KP}_{ip}\mathrm{KV}_i + \sum_{j=1}^{20} \mathrm{UP}_{jp}\mathrm{UV}_j\right)\alpha_p \end{aligned} \tag{14.5}$$

This specification would already be modeled if α_p had been defined as a logical calculation:

$$\forall p : \alpha_p = 1 \text{ IF AND ONLY IF } \sum_{k=1}^{4} V_k x_{kp} \geq \sum_{i=1}^{25} \text{KP}_{ip}\text{KV}_i + \sum_{j=1}^{20} \text{UP}_{jp}\text{UV}_j$$

$$\text{IF } \sum_{k=1}^{4} V_k x_{kp} \geq \sum_{i=1}^{25} \text{KP}_{ip}\text{KV}_i + \sum_{j=1}^{20} \text{UP}_{jp}\text{UV}_j \text{ THEN } \alpha_p = 1$$

$$\text{IF } \sum_{k=1}^{4} V_k x_{kp} < \sum_{i=1}^{25} \text{KP}_{ip}\text{KV}_i + \sum_{j=1}^{20} \text{UP}_{jp}\text{UV}_j \text{ THEN } \alpha_p = 0 \Rightarrow$$

$$\Rightarrow f_3 \Rightarrow \text{IF } \alpha_p = 1 \text{ THEN } \sum_{k=1}^{4} V_k x_{kp} \geq \sum_{i=1}^{25} \text{KP}_{ip}\text{KV}_i + \sum_{j=1}^{20} \text{UP}_{jp}\text{UV}_j$$

For the objective function, it is only necessary to define the value 0.

Objective Function

Minimize the total volume of shipped boxes plus the volume of unattended products:

$$\text{Min} \quad \sum_{p=1}^{100} \sum_{k=1}^{4} V_k x_{kp} + F \sum_{p=1}^{100} \left(\sum_{i=1}^{25} \text{KP}_{ip}\text{KV}_i + \sum_{j=1}^{20} \text{UP}_{jp}\text{UV}_j \right) (1 - \alpha_p) \quad (14.6)$$

Complete Model

$$\text{Min} \quad \sum_{p=1}^{100} \sum_{k=1}^{4} V_k x_{kp} + F \sum_{p=1}^{100} \left(\sum_{i=1}^{25} \text{KP}_{ip}\text{KV}_i + \sum_{j=1}^{20} \text{UP}_{jp}\text{UV}_j \right) (1 - \alpha_p) \quad (14.6)$$

s.t.

$$\forall i : \sum_{p=1}^{100} \text{KP}_{ip}\alpha_p \leq \text{KS}_i \quad (14.3)$$

$$\forall j : \sum_{p=1}^{100} \text{UP}_{jp}\alpha_p \leq \text{US}_j \quad (14.4)$$

$$\forall p : \quad \sum_{k=1}^{4} V_k x_{kp} \geq \left(\sum_{i=1}^{25} \text{KP}_{ip}\text{KV}_i + \sum_{j=1}^{20} \text{UP}_{jp}\text{UV}_j \right) \alpha_p \quad (14.5)$$

$$\forall k, p : x_{kp} \text{ integers} \quad (14.1)$$

$$\forall p : \alpha_p \text{ binaries} \quad (14.2)$$

Implementation in LINGO

We implement an instance collected in the Excel sheet “14_Model.xlsx.” The excel file is not displayed because it is too large.

```
MODEL:
SETS:
KProducts/1..25/:KS,KV;
UProducts/1..20/:US,UV;
Orders/1..100/:Alpha;
Boxes/1..4/:V;
OK(KProducts,Orders):KP;
OU(UProducts,Orders):UP;
OB(Boxes,Orders):x;
ENDSETS
DATA:
 KS =@OLE('C:\Files_\14_Excel.xlsx','KS');
 US =@OLE('C:\Files_\14_Excel.xlsx','US');
 KV =@OLE('C:\Files_\14_Excel.xlsx','KS');
 UV =@OLE('C:\Files_\14_Excel.xlsx','US');
 KP =@OLE('C:\Files_\14_Excel.xlsx','KP');
 UP=@OLE('C:\Files_\14_Excel.xlsx','UP');
 V=@OLE('C:\Files_\14_Excel.xlsx','V');
 F=100;
ENDDATA
!14.6; MIN = @SUM(Boxes(k):@SUM(Orders(p):V(k)*x(k,p)))+ F*@SUM
(Orders(p):@SUM(KProducts(i):KP(i,p)*KV(i))+@SUM(UProducts(j):UP
(j,p)*UV(j))*(1-Alpha(p)));
!14.3; @FOR(KProducts(i):@SUM(Orders(p):KP(i,p)*Alpha(p))<=KS(i));
!14.4; @FOR(UProducts(j):@SUM(Orders(p):UP(j,p)*Alpha(p))<=US(j));
!14.5; @FOR(Orders(p):@SUM(Boxes(k):V(k)*x(k,p))>=(@SUM(KProducts
(i):KP(i,p)*KV(i))+@SUM(UProducts(j):UP(j,p)*UV(j)))*Alpha(p));
!14.1; @FOR(Boxes(k):@FOR(Orders(p):@GIN(x(k,p))));
!14.2; @FOR(Orders(p):@BIN(Alpha(p)));
END
```

Solution

The solution is shown in Fig. 4.11.

4.9 Problem 15. Plot Cultivation

An owner is going to study the planting of crops on his plots. He owns N plots and the crops he can plant are 4 (wheat, corn, oats, and olive). Each plot has an area. The boundaries between the plots are known, and we express them as a binary attribute between the pairs of plots (= 1 delimiting; = 0 otherwise).

The considerations in the cultivation process are the following:

Each plot can be cultivated with a maximum of two crops and can be left uncultivated.

It is necessary to plant at least D_i m^2 of each crop i.

Wheat cannot be grown with any other crop on the same plot.

You cannot grow corn and wheat or corn and oats in adjoining plots.

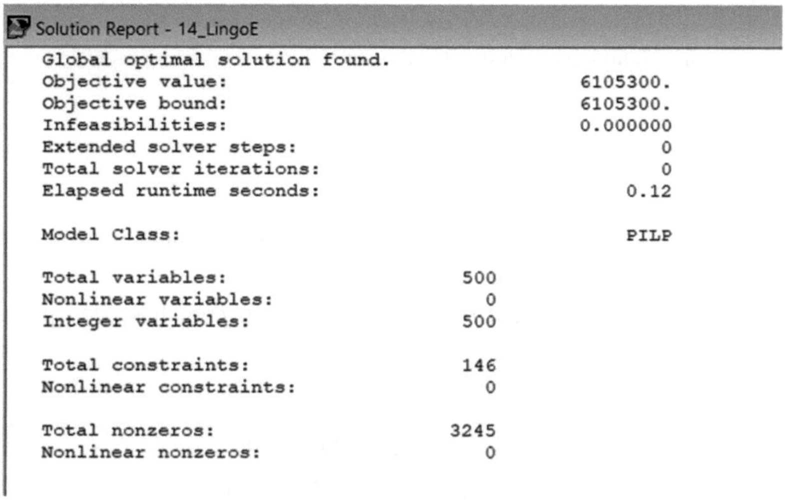

```
Solution Report - 14_LingoE
Global optimal solution found.
Objective value:                          6105300.
Objective bound:                          6105300.
Infeasibilities:                          0.000000
Extended solver steps:                           0
Total solver iterations:                         0
Elapsed runtime seconds:                      0.12

Model Class:                                  PILP

Total variables:                    500
Nonlinear variables:                  0
Integer variables:                  500

Total constraints:                  146
Nonlinear constraints:                0

Total nonzeros:                    3245
Nonlinear nonzeros:                   0
```

Fig. 4.11 Problem 14: LINGO solution report

For each crop *i,* a subsidy is received S_i \$/m^2, and for cultivating, we have a fixed cost F_j and a variable cost per m^2 of each crop in each plot, c_{ij} (\$/m^2). We seek to maximize profit.

Modeling

List of elements in Table 4.12.

Table 4.12 Problem 15: table of elements

			Data				
Elements	Set	QN	Name	Parameter	Type	Belonging	Value
Plot	$j, j' = 1 \ldots N$	I_U	Extension	E_j	C(m^2)	W	...
			Delimit	$L_{jj'}$	B	S	...
			Variable cost	c_{ij}	C (\$/m^2)	S	...
			Fixed cost	F_j	C(\$)	S	...
Crops	$i = 1 \ldots 4$ { wheat, corn, oats, olive }	I_M	Subsidy	S_i	C (\$/m^2)	W	...
			Minimum	D_i	C(m^2)	W	...
				c_{ij}			

Decision Activities
Plant crops (*i*) in plots (*j*)

$$\forall i, j : x_{ij} = \quad \text{Amount of m}^2 \text{ of crop } i \text{ planted on plot } j.$$

Specifications

- Plot extension (capacity specification):

$$\forall j: \sum_{i=1}^{4} x_{ij} \leq E_j \tag{15.1}$$

- Maximum of 2 crops per plot (logical calculation + quantitative selection specification):

We propose a logical calculation to know if each crop has been planted in each plot:

Binary Logical Calculation

$$\forall i,j: \alpha_{ij} = \quad 1 \text{ if crop } i \text{ has been planted in plot } j; 0 \text{ otherwise} \tag{15.2}$$

$$\forall i,j: \alpha_{ij} = 1 \text{ IF AND ONLY IF } x_{ij} > 0$$

We can ignore the definition of the value 0, since α_{ij} will be used to assign the fixed costs in the objective function.

$$\Rightarrow f_{S_V} \Rightarrow \forall i,j: \text{IF } x_{ij} > 0 \text{ THEN } \alpha_{ij} = 1$$

$$\Rightarrow f_3, f_7 \Rightarrow \forall i,j: \text{IF } 1 - \alpha_{ij} = 1 \text{ THEN } x_{ij} \leq 0$$

$$\Rightarrow f_{14}\left(\text{UB}_{x_{ij}} = E_j\right) \Rightarrow \forall i,j: x_{ij} \leq E_j \alpha_{ij} \tag{15.3}$$

Quantitative selection specification:

$$\forall j: \sum_{i=1}^{4} \alpha_{ij} \leq 2 \tag{15.4}$$

- Minimum to be cultivated of each crop (lower bound imposition):

$$\forall i: \sum_{j=1}^{N} x_{ij} \geq D_i \tag{15.5}$$

– Wheat cannot be grown with any other crop on the same plot (propositional logic specification): We put it with the following statement: If we plant wheat in a plot, we cannot plant any other crop.

$$\forall j : \text{IF } \alpha_{1j} = 1 \text{ THEN } \sum_{i=2}^{4} \alpha_{ij} = 0$$

$$\Rightarrow f_{\mathrm{LB}} \Rightarrow \forall j : \text{IF } \alpha_{1j} = 1 \text{ THEN } \sum_{i=2}^{4} \alpha_{ij} \leq 0$$

$$\Rightarrow f_{14}\left(UB_{\sum_{i=2}^{4} \alpha_{ij}} = 2\right) \Rightarrow \forall j : \sum_{i=2}^{4} \alpha_{ij} \leq 2(1 - \alpha_{1j}) \tag{15.6}$$

– We cannot grow corn and wheat or corn and oats in adjoining plots (propositional logic specification):

$$\forall j, j'/j < j' \& L_{jj'} = 1 : \text{NOT}(\alpha_{1j} = 1 \text{ AND } \alpha_{2j'} = 1)$$
$$\forall j, j'/j < j' \& L_{jj'} = 1 : \text{NOT}(\alpha_{2j} = 1 \text{ AND } \alpha_{3j'} = 1)$$

$$\Rightarrow f_{37}, f_{11} \Rightarrow \forall j, j'/j < j' \& L_{jj'} = 1 : \alpha_{1j} + \alpha_{2j'} \leq 1 \tag{15.7}$$

$$\Rightarrow f_{37}, f_{11} \Rightarrow \forall j, j'/j < j' \& L_{jj'} = 1 : \alpha_{2j} + \alpha_{3j'} \leq 1 \tag{15.8}$$

Objective Function

$$\text{Maximize profit} = (\text{Incomes} - \text{Costs})$$

$$\text{Max} \sum_{i=1}^{4} \sum_{j=1}^{N} S_i x_{ij} - \left(\sum_{i=1}^{4} \sum_{j=1}^{N} c_{ij} x_{ij} + \sum_{i=1}^{4} \sum_{j=1}^{N} F_j \alpha_{ij} \right) \tag{15.9}$$

Complete Model

$$\text{Max} \sum_{i=1}^{4} \sum_{j=1}^{N} S_i x_{ij} - \left(\sum_{i=1}^{4} \sum_{j=1}^{N} c_{ij} x_{ij} + \sum_{i=1}^{4} \sum_{j=1}^{N} F_j \alpha_{ij} \right) \tag{15.9}$$

s.t.

$$\forall j : \sum_{i=1}^{4} x_{ij} \leq E_j \tag{15.1}$$

$$\forall i,j : x_{ij} \leq E_j\alpha_{ij} \tag{15.3}$$

$$\forall j : \sum_{i=1}^{4} \alpha_{ij} \leq 2 \tag{15.4}$$

$$\forall i : \sum_{j=1}^{N} x_{ij} \geq D_i \tag{15.5}$$

$$\forall j : \sum_{i=2}^{4} \alpha_{ij} \leq 2(1 - \alpha_{1j}) \tag{15.6}$$

$$\forall j, j'/j < j' \& L_{jj'} = 1 : \alpha_{1j} + \alpha_{2j'} \leq 1 \tag{15.7}$$

$$\forall j, j'/j < j' \& L_{jj'} = 1 : \alpha_{2j} + \alpha_{3j'} \leq 1 \tag{15.8}$$

$$\forall i,j : x_{ij} \geq 0$$

$$\forall i,j : \alpha_{ij} \geq 0 \text{ binaries} \tag{15.2}$$

Implementation in LINGO

We collect the data from the file "15_Text.txt." The image of the file is in Fig. 4.12.

```
MODEL:
SETS:
Plots/1..@FILE('C:\Files_\15_Text.txt')/:E;
Crops/1..4/:S,F,D;
CP(Crops,Plots):C,x,Alpha;
Delimit(Plots,Plots):L;
ENDSETS
DATA:
 E =@FILE('C:\Files_\15_Text.txt');
 L =@FILE('C:\Files_\15_Text.txt');
 C =@FILE('C:\Files_\15_Text.txt');
 F =@FILE('C:\Files_\15_Text.txt');
 S =@FILE('C:\Files_\15_Text.txt');
 D =@FILE('C:\Files_\15_Text.txt');
 Eps= 0.00001;
ENDDATA
!15.9; MAX = @SUM(Crops(i):@SUM(Plots(j):S(i)*x(i,j)))-(@SUM(Crops
(i):@SUM(Plots(j):c(i,j)*x(i,j)))+@SUM(Crops(i):@SUM(Plots(j):F
(i)*Alpha(i,j))));
!15.1; @FOR(Plots(j):@SUM(Crops(i):x(i,j))<=E(j));
!15.3; @FOR(Crops(i):@FOR(Plots(j):x(i,j)<=E(j)*Alpha(i,j)));
!15.4; @FOR(Plots(j):@SUM(Crops(i):Alpha(i,j))<=2);
!15.5; @FOR(Crops(i):@SUM(Plots(j):x(i,j))>=D(i));
!15.6; @FOR(Plots(j):@SUM(Crops(i)|i#GE#2:Alpha(i,j))<=2*(1-Alpha
(1,j)));
!15.7; @FOR(Plots(j): @FOR(Plots(jj)|j#LT#jj#AND#L(j,jj)#EQ#1:
Alpha(1,j)+Alpha(2,jj)<=1));
```

```
15_Text.txt - Notepad
File Edit Format View Help
!N;
20 ~

!Extension;
80 70 80 100 500 1000 400 2500 4000 1000 500 1000 400 2500 4000 500 1000 400 2500 4000 ~

!Delimit;
1 0 0 0 0 0 0 0 0 0 0 0 1 1 0 0 0 0 0 1
0 1 0 0 0 0 0 0 0 0 0 0 0 0 0 0 0 0 0 0
0 0 1 0 1 0 0 1 0 0 1 0 0 1 0 0 0 0 0 0
0 0 0 1 0 1 0 0 0 0 0 0 0 0 0 0 0 0 0 0
0 0 0 0 1 0 1 0 0 0 0 0 0 1 0 0 0 0 0 0
0 0 0 0 0 1 0 0 0 1 0 1 0 1 0 0 1 0 0 1
0 0 0 0 0 0 1 0 0 0 0 0 0 0 0 0 0 0 0 0
0 0 0 0 0 0 0 1 0 0 0 0 1 0 0 0 1 0 0 0
0 0 0 0 0 0 0 0 1 0 0 0 0 0 0 1 0 0 1 0
0 0 0 0 0 0 0 0 0 1 0 0 0 1 0 0 0 0 0 0
0 0 0 0 0 0 0 0 0 0 1 0 0 0 0 1 1 0 0 0
0 0 0 0 0 0 0 0 0 0 0 1 0 0 0 0 0 1 0 0
0 0 0 0 0 0 0 0 0 0 0 0 1 0 0 1 0 1 0 0
0 0 0 0 0 0 0 0 0 0 0 0 0 1 0 0 1 0 1 0
0 0 0 0 0 0 0 0 0 0 0 0 0 0 1 0 0 1 0 0
0 0 0 0 0 0 0 0 0 0 0 0 0 0 0 1 0 0 0 0
0 0 0 0 0 0 0 0 0 0 0 0 0 0 0 0 1 0 0 0
0 0 0 0 0 0 0 0 0 0 0 0 0 0 0 0 0 1 0 0
0 0 0 0 0 0 0 0 0 0 0 0 0 0 0 0 0 0 1 0
0 0 0 0 0 0 0 0 0 0 0 0 0 0 0 0 0 0 0 1 ~

!Variable costs;
10 12 14 13 14 15 17 13 12 10 10 12 11 9 8 7 9 5 6 7
7 7 7 7 8 8 9 9 10 5 5 5 5 9 8 7 9 5 6 7
12 11 9 8 7   9 5 6 7 8  12 14 13 14 15   17 13 12 10 10
6 12 14 13 14  15 17 13 12 10  10 12 11 9 8  7 9 5 6 7  ~

!Fixed costs;
40 10 15 20 ~

!Subsidy:;
100 200 150 200 ~

!Minimum:;
5000 2000 3000 2000 ~
```

Fig. 4.12 “15_Text.txt”

```
!15.8; @FOR(Plots(j): @FOR(Plots(jj)|j#LT#jj#AND#L(j,jj)#EQ#1:
Alpha(2,j)+Alpha(3,jj)<=1));
!15.2; @FOR(Crops(i):@FOR(Plots(j):@BIN(Alpha(i,j))));
END
```

Solution

The solution is shown in Fig. 4.13.

4.10 Problem 16. Production and Distribution

A company has two factories where units of its product are produced. Each factory i has a daily production capacity of K_i units. Assume L working days/week for production.

```
Solution Report - 15_LingoE
Global optimal solution found.
Objective value:                              4457875.
Objective bound:                              4457878.
Infeasibilities:                              0.000000
Extended solver steps:                               1
Total solver iterations:                           625
Elapsed runtime seconds:                          0.06

Model Class:                                      MILP

Total variables:                    160
Nonlinear variables:                  0
Integer variables:                   80

Total constraints:                  201
Nonlinear constraints:                0

Total nonzeros:                     752
Nonlinear nonzeros:                   0
```

Fig. 4.13 Problem 15: LINGO solution report

There are two hubs where manufactured products can be sent later to customers. Manufactured products can also remain in the factory as each factory has a small storage capacity FA_i. Each hub j has a storage capacity of A_j units (the capacity is considered with respect to the products that remain daily in the hub, that is, those that are more arrive and those that are less leave, daily). The hubs can ship units between them, but due to the distance between them, the units do not arrive until the next day. There is a known distance between hubs and factories, just like between hubs. The maximum number of units that each hub j can ship each day is known, E_j.

There is a set of n customers with a known weekly demand for units of the product, which can be sent to them from the hubs or directly from the factory, during the working days of the week. There is a known distance between the hubs and the customers and between the factories and the customers.

Model the problem of minimizing total shipping distances in a week (the distance from one point to another is imputed each day product units are shipped, regardless of the quantity shipped).

Modeling

List of elements in Table 4.13.

The ideal is to represent in a graph all the actions related to the product that flow between the elements of the system (Fig. 4.14).

Variables

$$\forall i, t : x_{it} = \text{Units of product produced at factory } i \text{ on day } t; \tag{16.1}$$

$$\forall i, t : F_{it} = \text{Units of product stored at factory } i \text{ on day } t; \tag{16.2}$$

Table 4.13 Problem 16: table of elements

			Data				
Elements	Set	QN	Name	Parameter	Type	Belonging	Value
Factories	$i = 1, 2$	I_U	Production capacity	K_i	I	S	...
			Storage capacity	FA_i	I	S	...
			Distance to hub	DH_{ij}	C (Km)	S	...
			Distance to customer	DC_{ik}	C (Km)	S	...
Hubs	j, $j' = 1 \ldots 2$	I_U	Capacity	A_j	I	S	...
			Distance to hubs	$D_{jj'}$	C	S	...
			No maximum of units	E_j	I	S	...
			Distance to customer	DHC_{jk}	C (Km)	S	...
Customers	$k = 1 \ldots n$	I_U	Demand	M_k	I	S	...
				DC_{ik}; DHC_{jk}			
Days	$t = 1 \ldots L$	I_U					
Product	–	C_D		K_i; FA_i; A_j; E_j; M_k			

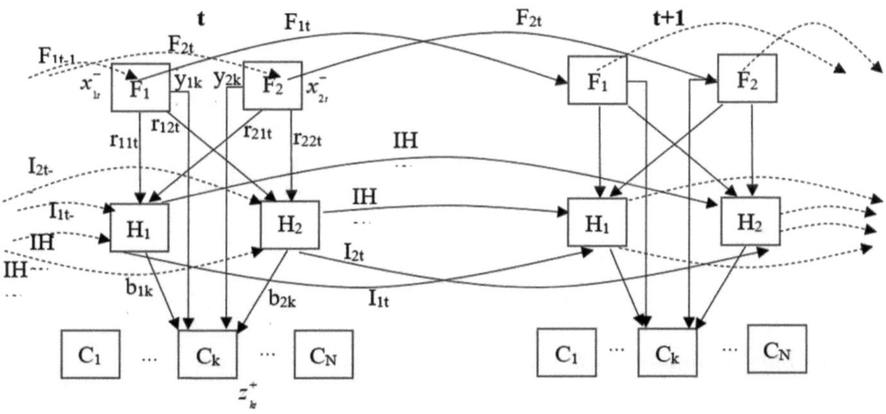

Fig. 4.14 Graph representing activity

$$\forall i, j, t : r_{ijt} = \text{Units of product shipped from factory } i \text{ to hub } j \text{ on day } t; \quad (16.3)$$

$$\forall i, k, t : y_{ikt} = \text{Units of product shipped from factory } i \text{ to customer } k \text{ on day } t; \quad (16.4)$$

$\forall j, k, t : b_{jkt}$
$=$ Units of product shipped from hub j to customer k on day t; (16.5)

$\forall j, t : I_{jt} =$ Units of product stored in hub j on day t; (16.6)

$\forall j, j' / j \neq j', t : \mathrm{IH}_{jj't}$
$=$ Units of product shipped from hub j to hub j' on day t; (16.7)

$\forall k, t : z_{kt} =$ Units of product delivered to customer k on day t; (16.8)

Specifications

- A specification of flow balance in each node of the graph.

  Each factory i each day $t > 1$:

$$\forall i, t/t > 1 : x_{it} + F_{it-1} = F_{it} + \sum_{k=1}^{n} y_{ikt} + \sum_{j=1}^{2} r_{ijt} \tag{16.9}$$

  Each factory i day $t = 1$:

$$\forall i, t = 1 : x_{it} = F_{it} + \sum_{k=1}^{n} y_{ikt} + \sum_{j=1}^{2} r_{ijt} \tag{16.10}$$

  Each hub j each day $t > 1$, $t < L$:

$$\forall j, t/t > 1 \& t < L : \sum_{i=1}^{2} r_{ijt} + I_{jt-1} + \sum_{j'/j' \neq j} \mathrm{IH}_{j'jt-1} = I_{jt} + \sum_{j'/j' \neq j} \mathrm{IH}_{jj't} + \sum_{k=1}^{n} b_{jkt} \tag{16.11}$$

  Each hub j each day $t = L$:

$$\forall j, t/t = L : \sum_{i=1}^{2} r_{ijt} + I_{jt-1} + \sum_{j'/j' \neq j} \mathrm{IH}_{j'jt-1} = I_{jt} + \sum_{k=1}^{n} b_{jkt} \tag{17.12}$$

  Each hub j each day $t = 1$:

$$\forall j, t/t = 1 : \sum_{i=1}^{2} r_{ijt} = I_{jt} + \sum_{j'/j' \neq j} \mathrm{IH}_{jj't} + \sum_{k=1}^{n} b_{jkt} \tag{16.13}$$

  Each customer k each day t:

$$\forall k,t: \sum_{i=1}^{2} y_{ikt} + \sum_{j=1}^{2} b_{jkt} = z_{kt} \tag{16.14}$$

– Production capacity of each factory (capacity specification):

$$\forall i,t: x_{it} \leq K_i \tag{16.15}$$

– Storage capacity of each factory (capacity specification):

$$\forall i,t: F_{it} \leq \mathrm{FA}_i \tag{16.16}$$

– Storage capacity of each hub (capacity specification):

$$\forall j,t: I_{jt} \leq A_j \tag{16.17}$$

– Shipping capacity of each hub (capacity specification):

$$\forall j,t: \sum_{k=1}^{n} b_{jkt} + \sum_{j'/j'\neq j} \mathrm{IH}_{jj't} \leq E_j \tag{16.18}$$

– Customer demand (demand specification):

$$\forall k: \sum_{t=1}^{L} z_{kt} = M_k \tag{16.19}$$

Objective Criterion

Minimize distances, taking into account each shipment.

Logical calculations are required with each activity to know whether or not units were shipped.

Binary Logical Calculations

$$\forall i,k,t: \beta_{ikt} = 1 \text{ if units are shipped from factory } i \text{ to customer } k; \text{ 0 otherwise.} \tag{16.20}$$

$$\forall i,j,t: \pi_{ijt} = 1 \text{ if units are shipped from factory } i \text{ to hub } j; \text{ 0 otherwise} \tag{16.21}$$

$$\forall j,k,t: \omega_{jkt} = 1 \text{ if units are shipped from hub } j \text{ to client } k; \text{0 otherwise.} \tag{16.22}$$

$$\forall j, j'/j \neq j', t : \delta_{jj't} = 1 \quad \text{if units are shipped from hub } j \text{ to hub } j'; 0 \text{ otherwise.} \tag{16.23}$$

$\forall i, k, t : \beta_{ikt} = 1$ IF AND ONLY IF $y_{ikt} > 0$

$\Rightarrow f_5 \Rightarrow \forall i, k, t : \beta_{ikt} = 1$ IF AND ONLY IF $y_{ikt} \geq 1$

$\forall i, j, t : \pi_{ijt} = 1$ IF AND ONLY IF $r_{ijt} > 0$

$\Rightarrow f_5 \Rightarrow \forall i, j, t : \pi_{ijt} = 1$ IF AND ONLY IF $r_{ijt} \geq 1$

$\forall j, k, t : \omega_{jkt} = 1$ IF AND ONLY IF $b_{jkt} > 0$

$\Rightarrow f_5 \Rightarrow \forall j, k, t : \omega_{jkt} = 1$ IF AND ONLY IF $b_{jkt} \geq 1$

$\forall j, j'/j \neq j', t : \delta_{jj't} = 1$ IF AND ONLY IF $\mathrm{IH}_{jj't} > 0 \Rightarrow$

$\Rightarrow f_5 \Rightarrow \forall j, j'/j \neq j', t : \delta_{jj't} = 1$ IF AND ONLY IF $\mathrm{IH}_{jj't} \geq 1$

$\Rightarrow f_{S_V} \Rightarrow \forall i, k, t :$ IF $y_{ikt} \geq 1$ THEN $\beta_{ikt} = 1$

$\Rightarrow f_{S_V} \Rightarrow \forall i, j, t :$ IF $r_{ijt} \geq 1$ THEN $\pi_{ijt} = 1$

$\Rightarrow f_{S_V} \Rightarrow \forall j, k, t :$ IF $b_{jkt} \geq 1$ THEN $\omega_{jkt} = 1$

$\Rightarrow f_{S_V} \Rightarrow \forall j, j'/j \neq j', t :$ IF $\mathrm{IH}_{jj't} \geq 1$ THEN $\delta_{jj't} = 1$

$$\begin{aligned} &\Rightarrow f_3, f_7, f_{14}\left(\mathrm{UB}_{y_{ikt}} = \min\{K_i, M_k\}\right) \\ &\Rightarrow \forall i, k, t : y_{ikt} \leq \min\{K_i, M_k\}\beta_{ikt} \end{aligned} \tag{16.24}$$

$$\begin{aligned} &\Rightarrow f_3, f_7, f_{14}\left(\mathrm{UB}_{r_{ijt}} = \min\{K_i, A_j\}\right) \\ &\Rightarrow \forall i, j, t : \quad r_{ijt} \leq \min\{K_i, A_j\}\pi_{ijt} \end{aligned} \tag{16.25}$$

$$\begin{aligned} &\Rightarrow f_3, f_7, f_{14}\left(\mathrm{UB}_{b_{jkt}} = \min\{A_j, M_k\}\right) \\ &\Rightarrow \forall j, k, t : \quad b_{jkt} \leq \min\{A_j, M_k\}\omega_{jkt} \end{aligned} \tag{16.26}$$

$$\begin{aligned} &\Rightarrow f_3, f_7, f_{14}\left(\mathrm{UB}_{\mathrm{IH}_{jj't}} = \min\{A_j, A_{j'}\}\right) \\ &\Rightarrow \forall j, j', t : \quad \mathrm{IH}_{jj't} \leq \min\{A_j, A_{j'}\}\delta_{jj't} \end{aligned} \tag{16.27}$$

Objective Function

$$\begin{aligned} \text{Min} \ &\sum_{t=1}^{L}\sum_{i=1}^{2}\sum_{j=1}^{2}\mathrm{DH}_{ij}\pi_{ijt} + \sum_{t=1}^{L}\sum_{i=1}^{2}\sum_{k=1}^{n}\mathrm{DC}_{ik}\beta_{ikt} + \\ &\sum_{t=1}^{L}\sum_{j=1}^{2}\sum_{k=1}^{n}\mathrm{DHC}_{jk}\omega_{jkt} + \sum_{t=1}^{L}\sum_{j=1}^{2}\sum_{j'/j'\neq j} D_{jj'}\delta_{jj't} \end{aligned} \tag{16.28}$$

Complete Model

$$\text{Min} \sum_{t=1}^{L}\sum_{i=1}^{2}\sum_{j=1}^{2}\text{DH}_{ij}\pi_{ijt} + \sum_{t=1}^{L}\sum_{i=1}^{2}\sum_{k=1}^{n}\text{DC}_{ik}\beta_{ikt} + \sum_{t=1}^{L}\sum_{j=1}^{2}\sum_{k=1}^{n}\text{DHC}_{jk}\omega_{jkt} + \sum_{t=1}^{L}\sum_{j=1}^{2}\sum_{j'/j'\neq j} D_{jj'}\delta_{jj't} \tag{16.28}$$

s.t.

$$\forall i, t/t > 1 : x_{it} + F_{it-1} = F_{it} + \sum_{k=1}^{n} y_{ikt} + \sum_{j=1}^{2} r_{ijt} \tag{16.9}$$

$$\forall i, t = 1 : x_{it} = F_{it} + \sum_{k=1}^{n} y_{ikt} + \sum_{j=1}^{2} r_{ijt} \tag{16.10}$$

$$\forall j, t/t > 1 \& t < L : \sum_{i=1}^{2} r_{ijt} + I_{jt-1} + \sum_{j'/j'\neq j} \text{IH}_{j'jt-1} = I_{jt} + \sum_{j'/j'\neq j} \text{IH}_{jj't} + \sum_{k=1}^{n} b_{jkt} \tag{16.11}$$

$$\forall j, t/t = L : \sum_{i=1}^{2} r_{ijt} + I_{jt-1} + \sum_{j'/j'\neq j} \text{IH}_{j'jt-1} = I_{jt} + \sum_{k=1}^{n} b_{jkt} \tag{16.12}$$

$$\forall j, t/t = 1 : \sum_{i=1}^{2} r_{ijt} = I_{jt} + \sum_{j'/j'\neq j} \text{IH}_{jj't} + \sum_{k=1}^{n} b_{jkt} \tag{16.13}$$

$$\forall k, t : \sum_{i=1}^{2} y_{ikt} + \sum_{j=1}^{2} b_{jkt} = z_{kt} \tag{16.14}$$

$$\forall i, t : x_{it} \leq K_i \tag{16.15}$$

$$\forall i, t : F_{it} \leq \text{FA}_i \tag{16.16}$$

$$\forall j, t : I_{jt} \leq A_j \tag{16.17}$$

$$\forall j, t : \sum_{k=1}^{n} b_{jkt} + \sum_{j'/j'\neq j} \text{IH}_{jj't} \leq E_j \tag{16.18}$$

$$\forall k : \sum_{t=1}^{L} z_{kt} \geq M_k \tag{16.19}$$

$$\forall i, k, t : y_{ikt} \leq \min\{K_i, M_k\}\beta_{ikt} \tag{16.24}$$

$$\forall i,j,t: \quad r_{ijt} \leq \min\left\{K_i, A_j\right\}\pi_{ijt} \tag{16.25}$$

$$\forall j,k,t: \quad b_{jkt} \leq \min\left\{A_j, M_k\right\}\omega_{jkt} \tag{16.26}$$

$$\forall j,j'/j \neq j',t: \quad \text{IH}_{jj't} \leq \quad \min\left\{A_j, A_{j'}\right\}\delta_{jj't} \tag{16.27}$$

$$\forall i,t: x_{it} \geq 0 \ \text{ integers} \tag{16.1}$$

$$\forall i,t: F_{it} \geq 0 \ \text{ integers} \tag{16.2}$$

$$\forall i,j,t: r_{ijt} \geq 0 \ \text{ integers} \tag{16.3}$$

$$\forall i,k,t: y_{ikt} \geq 0 \ \text{ integers} \tag{16.4}$$

$$\forall j,k,t: b_{jkt} \geq 0 \ \text{ integers} \tag{16.5}$$

$$\forall j,t: I_{jt} \geq 0 \ \text{ integers} \tag{16.6}$$

$$\forall j,j'/j \neq j',t: \text{IH}_{jj't} \geq 0 \ \text{ integers} \tag{16.7}$$

$$\forall k,t: z_{kt} \geq 0 \ \text{ integers} \tag{16.8}$$

$$\forall i,k,t: \beta_{ikt} \ \text{ binaries} \tag{16.20}$$

$$\forall i,j,t: \quad \pi_{ijt} \ \text{ binaries} \tag{16.21}$$

$$\forall j,k,t: \quad \omega_{jkt} \ \text{ binaries} \tag{16.22}$$

$$\forall j,j'/j \neq j',t: \quad \delta_{jj't} \ \text{ binaries} \tag{16.23}$$

Implementation in LINGO

We consider the data collected in the database "16_DB.accdb." We take 5 days and 10 customers.

```
MODEL:
SETS:
 Factories/1..2/:Ka,FA;
 Hubs/1..2/:A,E;
 Customers/1..10/:M;
 Days/1..5/;
 FC(Factories,Customers):DC;
 FH(Factories,Hubs):DH;
 HH(Hubs,Hubs)/ 1 2, 2 1 /:D;
 HC(Hubs,Customers):DHC;
 FD(Factories, Days):x,F;
 HD(Hubs, Days):I;
 FHD(Factories, Hubs, Days):r, Pi;
 HHD(Hubs,Hubs,Days):IH,Delta;
 HCD(Hubs, Customers,Days):b,Omega;
 FCD(Factories,Customers, Days):y,Beta;
 CD(Customers, Days):z;
ENDSETS
DATA:
 Ka= @ODBC('16_ODBC','Factories','Dayly_Prod_Cap');
```

```
 FA= @ODBC('16_ODBC','Factories','St_Cap');
 A= @ODBC('16_ODBC','Hubs','St_Cap');
 E= @ODBC('16_ODBC','Hubs','Maximum');
 M= @ODBC('16_ODBC','Customers','Demand');
 DC= @ODBC('16_ODBC','Distance_F_C','Distance');
 DH= @ODBC('16_ODBC','Distance_F_H','Distance');
 D= @ODBC('16_ODBC','Distance_H_H','Distance');
 DHC= @ODBC('16_ODBC','Distance_H_C','Distance');
ENDDATA
!16.28;
MIN = @SUM(Days(t):@SUM(Factories(ii):@SUM(Hubs(j):DH(ii,j)*Pi(ii,
j,t)))) + @SUM(Days(t):@SUM(Factories(ii):@SUM(Customers(k):DC(ii,
k)*Beta(ii,k,t))))
+ @SUM(Days(t):@SUM(Hubs(j):@SUM(Customers(k):DHC(j,k)*Omega(j,k,
t)))) + @SUM(Days(t):@SUM(Hubs(j):@SUM(Hubs(jj)|j#NE#jj:D(j,jj)
*Delta(j,jj,t)))) ;
!16.9;
@FOR(Factories(ii):@FOR(Days(t)|t#GT#1:x(ii,t)+F(ii,t-1)=F(ii,t)
+@SUM(Customers(k):y(ii,k,t))+@SUM(Hubs(j):r(ii,j,t))));
!16.10;
@FOR(Factories(ii):@FOR(Days(t)|t#EQ#1:x(ii,t)=F(ii,t)+@SUM
(Customers(k):y(ii,k,t))+@SUM(Hubs(j):r(ii,j,t))));
!16.11;
@FOR(Hubs(j):@FOR(Days(t)|t#GT#1#AND#t#LT#25:@SUM(Factories(ii):r
(ii,j,t))+I(j,t-1)+@SUM(Hubs(jj)|jj#NE#j:IH(jj,j,t-1))=I(j,t)
+@SUM(Hubs(jj)|jj#NE#j:IH(j,jj,t))+@SUm(Customers(k):b(j,k,t))));
!16.12;
@FOR(Hubs(j):@FOR(Days(t)|t#EQ#25:@SUM(Factories(ii):r(ii,j,t))+I
(j,t-1)+@SUM(Hubs(jj)|jj#NE#j:IH(jj,j,t-1))=I(j,t)+@SUm(Customers
(k):b(j,k,t))));
!16.13;
@FOR(Hubs(j):@FOR(Days(t)|t#EQ#1:@SUM(Factories(ii):r(ii,j,t))=I
(j,t)+@SUM(Hubs(jj)|jj#NE#j:IH(j,jj,t))+@SUm(Customers(k):b(j,k,
t))));
!16.14;
@FOR(Customers(k):@FOR(Days(t):@SUM(Factories(ii):y(ii,k,t))+@SUM
(Hubs(j):b(j,k,t))=z(k,t)));
!16.15;
@FOR(Factories(ii):@FOR(Days(t):x(ii,t)<=Ka(ii)));
!16.16;
@FOR(Factories(ii):@FOR(Days(t):F(ii,t)<=FA(ii)));
!16.17;
@FOR(Hubs(j):@FOR(Days(t):I(j,t)<=A(j)));
!16.18;
@FOR(Hubs(j):@FOR(Days(t):@SUM(Customers(k):b(j,k,t))+@SUM(Hubs
(jj)|jj#NE#j:IH(j,jj,t)) <=E(j)));
!16.19;
@FOR(Customers(k):@SUM(Days(t):z(k,t))>=M(k));
!16.24;
@FOR(Factories(ii):@FOR(Customers(k):@FOR(Days(t):y(ii,k,t)
<=@SMIN(Ka(ii),M(k))*Beta(ii,k,t))));
!16.25;
@FOR(Factories(ii):@FOR(Hubs(j):@FOR(Days(t):r(ii,j,t)<=@SMIN(Ka
```

```
(ii),A(j))*Pi(ii,j,t))));
!16.26;
@FOR(Hubs(j):@FOR(Customers(k):@FOR(Days(t):b(j,k,t)<=@SMIN(A(j),
M(k))*Omega(j,k,t))));
!16.27;
@FOR(Hubs(j):@FOR(hubs(jj)|jj#NE#j:@FOR(Days(t):IH(j,jj,t)<=@SMIN
(A(j),A(jj))*Delta(j,jj,t))));
!16.1;
@FOR(Factories(ii):@FOR(Days(t):@GIN(x(ii,t))));
!16.2;
@FOR(Factories(ii):@FOR(Days(t):@GIN(F(ii,t))));
!16.3;
@FOR(Factories(ii):@FOR(Hubs(j):@FOR(Days(t):@GIN(r(ii,j,t)))));
!16.4;
@FOR(Factories(ii):@FOR(Customers(k):@FOR(Days(t):@GIN(y(ii,k,
t)))));
!16.5;
@FOR(Hubs(j):@FOR(Customers(k):@FOR(Days(t):@GIN(b(j,k,t)))));
!16.6;
@FOR(Hubs(j):@FOR(Days(t):@GIN(I(j,t))));
!16.7;
@FOR(Hubs(j):@FOR(hubs(jj)|jj#NE#j:@FOR(Days(t):@GIN(IH(j,jj,
t)))));
!16.8;
@FOR(Customers(k):@FOR(Days(t):@GIN(z(k,t))));
!16.20;
@FOR(Factories(ii):@FOR(Customers(k):@FOR(Days(t):@BIN(Beta(ii,k,
t)))));
!16.21;
@FOR(Factories(ii):@FOR(Hubs(j):@FOR(Days(t):@BIN(Pi(ii,j,t)))));
!16.22;
@FOR(Hubs(j):@FOR(Customers(k):@FOR(Days(t):@BIN(Omega(j,k,
t)))));
!16.23;
@FOR(Hubs(j):@FOR(hubs(jj)|jj#NE#j:@FOR(Days(t):@BIN(Delta(j,jj,
t)))));
END
```

Solution

The solution is shown in Fig. 4.15.

4.11 Problem 17. Production of Divisible Tasks

Given a set of n production tasks, each with a given process time. There is a set of 5 machines to process the tasks. All tasks need to be processed. A task is considered processed when the sum of the processing times on each machine is equal to the task time. A job can be partially processed on up to three machines, but always one machine must process at least one-third of the task time. Each task that is processed on any machine causes the machine to incur a setup time TT plus the time that the

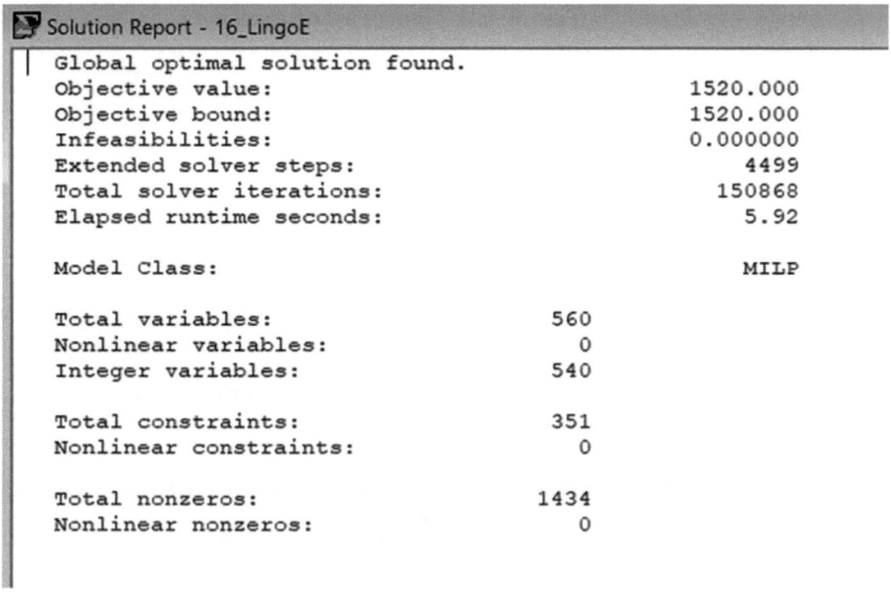

```
Solution Report - 16_LingoE
Global optimal solution found.
Objective value:                              1520.000
Objective bound:                              1520.000
Infeasibilities:                              0.000000
Extended solver steps:                            4499
Total solver iterations:                        150868
Elapsed runtime seconds:                          5.92

Model Class:                                      MILP

Total variables:                    560
Nonlinear variables:                  0
Integer variables:                  540

Total constraints:                  351
Nonlinear constraints:                0

Total nonzeros:                    1434
Nonlinear nonzeros:                   0
```

Fig. 4.15 Problem 16: LINGO solution report

machine processes the task. It is about balancing the working time of the machines, minimizing the time of the machine that works the most.

For simplicity: There is no overlap to take into account: that is, there is no need to control or decide when a task is processed on a machine.

Modelling

List of elements in Table 4.14.

Table 4.14 Problem 17: table of elements

			Data				
Elements	Set	QN	Name	Parameter	Type	Belonging	Value
Tasks	$i = 1 \ldots n$	I_M	Process time	t_i	C(m)	W	...
Machines	$j = 1 \ldots 5$	I_U					
System	–	I_U	Setup time	TT	C(m)	W	...

Decision Activities

Process tasks (i) on machines (j).

$$\forall i,j : x_{ij} = \text{Part of task } i \text{ processed on machine } j$$

Specifications

- Process all tasks (allocation specification)

$$\forall i : \sum_{j=1}^{10} x_{ij} = t_i \tag{17.1}$$

- Processing of each task in a maximum of three machines (logical calculation + quantitative selection specification).

  We collect in a logical calculation if a task has been processed on a machine: *Binary Logical Calculation*

$$\forall i,j : \beta_{ij} = 1 \text{ if task } i \text{ has been processed on machine } j; 0 \text{ otherwise.} \tag{17.2}$$

$$\forall i,j : \beta_{ij} = 1 \ \text{ IF AND ONLY IF } \ x_{ij} > 0$$

We can ignore the definition of the value 0. In no case, the definition of the value 1:

$$\Rightarrow f_{S_V} \Rightarrow \forall i,j : \text{IF } \ x_{ij} > 0 \ \text{ THEN } \ \beta_{ij} = 1$$

$$\Rightarrow f_3, f_7 \Rightarrow \forall i,j : \text{IF } \ 1 - \beta_{ij} = 1 \ \text{ THEN } \ x_{ij} \leq 0$$

$$\Rightarrow f_{14}\left(\text{UB}_{x_{ij}} = t_i\right) \Rightarrow \forall i,j : x_{ij} \leq t_i \beta_{ij} \tag{17.3}$$

Quantitative selection specification:

$$\forall i : \sum_{j=1}^{10} \beta_{ij} \leq 3 \tag{17.4}$$

- Processing on a machine for at least one-third of the task time (propositional logic specification).

$$\forall i,j : \textit{IF } \ \beta_{ij} = 1 \ \textit{ THEN } \ x_{ij} \geq \frac{1}{3} t_i$$

$$\Rightarrow f_{15}\left(LB_{x_{ij}} = 0\right) \Rightarrow \forall i,j : x_{ij} \geq \frac{1}{3} t_i \beta_{ij} \tag{17.5}$$

Objective Criterion
Balance machines workload (minimize maximum workload):
Auxiliary Calculation

$$\forall j : y_j = \text{Machine workload } j$$

$$\forall j : y_j = \sum_{i=1}^{n} x_{ij} + \sum_{i=1}^{n} TT\beta_{ij} \tag{17.6}$$

Upper Bound Calculation

$$y_{\text{MAX}} : \text{upper bound of the workload}$$

$$\forall j : y_{\text{MAX}} \geq y_j \tag{17.7}$$

Objective Function

$$\text{Min } y_{\text{MAX}} \tag{17.8}$$

Complete Model

$$\text{Min } y_{\text{MAX}} \tag{17.8}$$

$$\text{s.t.}$$

$$\forall i : \sum_{j=1}^{10} x_{ij} = t_i \tag{17.1}$$

$$\forall i,j : x_{ij} \leq t_i \beta_{ij} \tag{17.3}$$

$$\forall i,j : x_{ij} \geq \frac{1}{3} t_i \beta_{ij} \tag{17.4}$$

$$\forall i : \sum_{j=1}^{10} \beta_{ij} \leq 3 \tag{17.5}$$

$$\forall j : y_j = \sum_{i=1}^{n} x_{ij} + \sum_{i=1}^{n} \text{TT}\beta_{ij} \tag{17.6}$$

$$\forall j : y_{\text{MAX}} \geq y_j \tag{17.7}$$

$$\forall i,j : \beta_{ij} \text{ binaries} \tag{17.2}$$

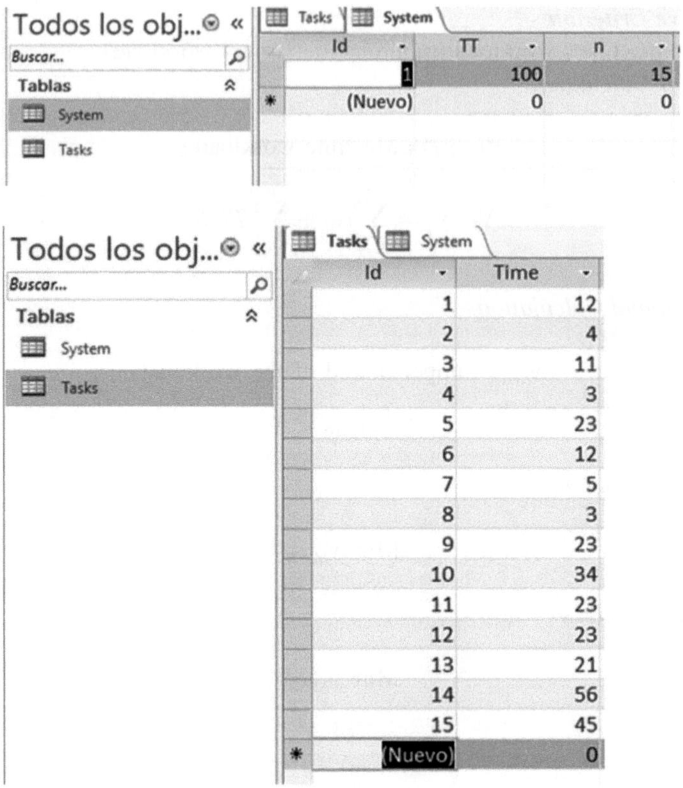

Id	TT	n
1	100	15
(Nuevo)	0	0

Id	Time
1	12
2	4
3	11
4	3
5	23
6	12
7	5
8	3
9	23
10	34
11	23
12	23
13	21
14	56
15	45
(Nuevo)	0

Fig. 4.16 Database "17_DB.accdb"

$$\forall i,j : x_{ij} \geq 0$$
$$\forall j : y_j \geq 0$$
$$y_{\text{MAX}} \geq 0$$

Implementation in LINGO

We store the information in the database "17_DB.accdb" (Fig. 4.16).

```
MODEL:
DATA:
 n= @ODBC('17_ODBC','System','n');
ENDDATA
SETS:
 Tasks/1..n/:t;
 Machines/1..5/:y;
 TM(Tasks,Machines):x,Beta;
ENDSETS
DATA:
 t= @ODBC('17_ODBC','Tasks','time');
```

```
Solution Report - 17_LingoE
Global optimal solution found.
Objective value:                              362.0000
Objective bound:                              362.0000
Infeasibilities:                              0.000000
Extended solver steps:                          108482
Total solver iterations:                       1240820
Elapsed runtime seconds:                         34.18

Model Class:                                      MILP

Total variables:                    156
Nonlinear variables:                  0
Integer variables:                   75

Total constraints:                  191
Nonlinear constraints:                0

Total nonzeros:                     616
Nonlinear nonzeros:                   0
```

Fig. 4.17 Problem 17: LINGO solution report

```
 TT= @ODBC('17_ODBC','System','TT');
 Epsilon= 0.01;
ENDDATA
!17.9; MIN= yMAX;
!17.1; @FOR(Tasks(i):@SUM(Machines(j):x(i,j))=t(i));
!17.3;@FOR(Tasks(i):@FOR(Machines(j):x(i,j)<=t(i)*Beta(i,j)));
!17.4;@FOR(Tasks(i):@FOR(Machines(j):x(i,j)>=(1/3)*t(i)*Beta(i,
j)));
!17.5;@FOR(Tasks(i):@SUM(Machines(j):Beta(i,j))<=3);
!17.6;@FOR(Machines(j):y(j)=@SUM(Tasks(i):x(i,j))+@SUM(Tasks(i):
TT*Beta(i,j)));
!17.7;@FOR(Machines(j):y(j)<=yMAX);
!17.2;@FOR(Tasks(i):@FOR(Machines(j):@BIN(Beta(i,j))));
END
```

Solution

It is shown in Fig. 4.17.

4.12 Problem 18. Selection of Nodes and Edges

Given an undirected graph G(*N*, *E*), where each edge has a cost, we have to select five nodes not connected to each other by any edge and select two edges adjacent to each of those nodes, so that the sum of costs of those edges is minimal.

Modeling

We use the format that uses two indices to reference the set of edges, the format described in Table 3.41 in the chapter "The elements of a system" (Garcia 2021a) (Table 4.15).

Table 4.15 Problem 18: table of elements

Elements	Set	QN	Data				
			Name	Parameter	Type	Belonging	Value
Nodes	$i = 1 \ldots n$	I_U					
Edges	$(i,j)/i,j \in$ Nodes	I_U	Cost	c_{ij}	C($)	W	...

Decision Activities

$$\textbf{Select nodes}(i): \forall i: \alpha_i = 1 \text{ if we select node } i; 0 \text{ otherwise.} \quad (18.1)$$

Select edges (*i*, *j*)

$$\forall (i,j): \beta_{ij} = 1 \text{ if we select edge } (i,j); 0 \text{ otherwise.} \quad (18.2)$$

Specifications

– Quantitative selection specifications:

Activity	Elements selecting	Selectable elements	Selection	Constraints
α_i	–	Nodes ($i = 1 \ldots n$)	=5	$\sum_{i=1}^{n} \alpha_i = 5$ (18.3)

Select 10 edges: It is not necessary because the selection by pairs on the selected nodes will be imposed as a propositional logic specification.

– Two selected nodes cannot be connected by any edge (propositional logic specification):

$$\forall (i,j): \ \text{NOT}(\alpha_i = 1 \ \text{AND} \ \alpha_j = 1)$$
$$\Rightarrow f_{37}, f_{11} \Rightarrow \forall (i,j) \ : \alpha_i + \alpha_j \leq 1 \quad (18.4)$$

– Select two edges of each selected node (propositional logic specification):

$$\forall i: \text{IF} \ \alpha_i = 1 \ \text{THEN} \sum_{(k,j)/k=i \vee j=i} \beta_{kj} = 2$$

We can apply f_{UB} reduction since the objective function minimizes costs and, therefore, will never select more than two edges.

$$\Rightarrow f_{\text{UB}} \Rightarrow \forall i: \text{IF} \ \alpha_i = 1 \ \text{THEN} \sum_{(k,j)/k=i \vee j=i} \beta_{kj} \geq 2$$

$$\Rightarrow f_{15}\left(\text{LB} \sum_{(k,j)/k=i\vee j=i} \beta_{kj} = 0\right) \Rightarrow \forall i : \sum_{(k,j)/k=i\vee j=i} \beta_{kj} \geq 2\alpha_i \tag{18.5}$$

Objective Function
Minimize costs

$$\text{Min} \sum_{(i,j)\in Edges} c_{ij}\beta_{ij} \tag{18.6}$$

Complete Model

$$\text{Min} \sum_{(i,j)\in \text{Aristas}} c_{ij}\beta_{ij} \tag{18.6}$$

s.t.

$$\sum_{i=1}^{n} \alpha_i = 5 \tag{18.3}$$

$$\forall(i,j) : \alpha_i + \alpha_j \leq 1 \tag{18.4}$$

$$\forall i : \sum_{(k,j)/k=i\vee j=i} \beta_{kj} \geq 2\alpha_i \tag{18.5}$$

$$\forall i : \alpha_i \text{ binaries} \tag{18.1}$$

$$\forall(i,j) : \beta_{ij} \text{ binaries} \tag{18.2}$$

Implementation in LINGO

We collect the information from file "18_Text.txt" and print the results on file "out.18.txt."

```
MODEL:
DATA:
 n=@FILE('C:\Files_\18_Text.txt');
ENDDATA
SETS:
Nodes/1..n/:Alfa;
Edges(Nodes,Nodes)/@FILE('C:\Files_\18_Text.txt')/:c,Beta;
ENDSETS
DATA:
 c = @FILE('C:\Files_\18_Text.txt');
ENDDATA
!18.6; MIN= @SUM(Edges(i,j):c(i,j)*Beta(i,j));
!18.3; @SUM(Nodes(i):Alfa(i))=5;
!18.4; @FOR(Edges(i,j):Alfa(i)+Alfa(j)<=1);
!18.5; @FOR(Nodes(i):@SUM(Edges(k,j)|k#EQ#i #OR# j#EQ#i: Beta(k,j))
>=2*Alfa(i));
!18.1; @FOR(Nodes(i):@BIN(Alfa(i)));
```

```
Solution Report - 18_LingoE
Global optimal solution found.
Objective value:                              14.00000
Objective bound:                              14.00000
Infeasibilities:                              0.000000
Extended solver steps:                               1
Total solver iterations:                           870
Elapsed runtime seconds:                          0.08

Model Class:                                      PILP

Total variables:                 113
Nonlinear variables:               0
Integer variables:               113

Total constraints:               115
Nonlinear constraints:             0

Total nonzeros:                  415
Nonlinear nonzeros:                0
```

Fig. 4.18 Problem 18: LINGO solution report

```
*Out18.txt - Notepad
File Edit Format View Help
Selected Nodes:5 13 21 37 48
Selected Edges:(4,5)(4,48)(9,21)(10,13)(10,48)(13,34)(21,12)(29,37)(31,5)(37,31)
```

Fig. 4.19 Output file "Out18.txt"

```
!18.2; @FOR(Edges(i,j):@BIN(Beta(i,j)));
DATA:
@TEXT("c:\Files_\Out18.txt")= @WRITE('Selected Nodes:');
 @TEXT("c:\Files_\Out18.txt")= @WRITEFOR(Nodes(i): @IF(Alfa(i)
#EQ#1,i,""),@IF(Alfa(i)#EQ#1," ",""));
 @TEXT("c:\Files_\Out18.txt")= @WRITE(@NEWLINE(1));
 @TEXT("c:\Files_\Out18.txt")= @WRITE('Selected Edges:');
 @TEXT("c:\Files_\Out18.txt")= @WRITEFOR(Edges(i,j): @IF(Beta(i,j)
#EQ#1,'(',""),@IF(Beta(i,j)#EQ#1, i,""),@IF(Beta(i,j)
#EQ#1,',',""),@IF(Beta(i,j)#EQ#1,j,""),@IF(Beta(i,j)#EQ#1,
')',""));
ENDDATA
END
```

Solution

It is shown in Figs. 4.18 and 4.19.

References

Castro B, Diez H, Urrutia A M (2011) Ejercicios resueltos de investigación o p e r a t i v a. Servicio Editorial de la Universidad del País Vasco, Bilbao. isbn:978-84-694-5889-1

Garcia JM (2021a) The elements of a system. In: Modelling in mathematical programming. Springer, Cham. https://doi.org/10.1007/978-3-030-57250-1_3

Garcia JM (2021b) The quantitative nature of the elements. In: Modelling in mathematical programming. Springer, Cham. https://doi.org/10.1007/978-3-030-57250-1_7

Pinedo M (1995) Scheduling. theory, algorithms, and systems. Prentice Hall international series in industrial and systems engineering. isbn:0-13-706757-7

Sarker RA, Newton CS (2007) Optimization modelling. A practical approach. CRC Press. isbn:978-1-4200-4310-5

Chapter 5
Medium-Level Problems

5.1 Introduction

Input files with a lot of data are not displayed due to their size. They are available for download.

The characteristics of the problems are shown as in previous chapters below.

Problems:

Number	Problem
19	Management of polluting products
20	Assignment of tasks to operators
21	Assignment of students to groups
22	Home building
23	Assignment of extracurricular activities
24	Assignment of subjects to classrooms
25	Disciplines management
26	Installation of agencies

Versions and elements:

		Elements				
Problem	Versions	Collective elements	Measurable individual elements	Grouping of items	Upper bound in the number of elements	Data handling
19	2	✓	✓			Medium
20	1					Medium
21	1	✓		✓		Medium
22	1	✓				Medium
23	1					High
24	1				✓	Medium

(continued)

J. M. García Sánchez, *Building and Solving Mathematical Programming Models*, International Series in Operations Research & Management Science 329, https://doi.org/10.1007/978-3-030-97626-2_5

Problem	Versions	Elements				
		Collective elements	Measurable individual elements	Grouping of items	Upper bound in the number of elements	Data handling
25	1					Medium
26	2	✓	✓			Medium

Calculations:

Problem	Binary logical calculations	Not-binary logical calculations	Bounds calculations	Conditioned bounds calculations
19	✓			
20	✓			
21	✓		✓	
22				
23				
24	✓			
25	✓			
26				

Specifications:

Problem	QSS[a]	Capacity	Demand	Bound imposition	Allocation/assignment
19	✓	✓			
20	✓	✓			✓
21		✓		✓	✓
22		✓		✓	
23	✓	✓		✓	
24	✓			✓	
25	✓				
26	✓	✓		✓	

[a]*QSS* quantitative selection specifications

Problem	Flow balance	PLS[a]—One operator	PLS—Several operators	PLS—Attributes	Variable costs cases
19		✓			
20		✓			
21			✓	✓	
22		✓			✓
23		✓		✓	
24					
25		✓	✓		
26		✓		✓	

[a]*PLS* propositional logic specifications

Objective criterion, input data and output data:

	Objective criterion		Input data				
Problem	Calculations	Particular cases	Integrated data	Text file	Spreadsheet	Database	Output data
19		✓			✓		
20					✓		
21	✓		✓				
22		✓	✓				
23			✓				
24	✓					✓	✓
25	✓			✓			
26					✓		

5.2 Problem 19. Management of Polluting Products (Castro et al. 2011)

A company is considering producing its three products P1, P2, and P3 at just one of the locations U1, U2, and U3. The production of each product generates a volume of contamination of 0.5, 2, and 1 cm^3, respectively, per unit produced, regardless of the location. Table 5.1 shows for each of the locations: the unit income ($) of each product, the daily production capacity (units), the maximum volumes of pollution allowed (cm^3), and the penalty for excess pollution volume ($/$cm^3$).

The company, aware of environmental problems, proposes goals with an order of priorities given by the following:

Priority 1. Maximize daily income.
Priority 2. Do not exceed the maximum level of contamination of the location.
Priority 3. The company does not want to spend more than $9000 per day due to excess pollution.

Let us formulate a model that allows us to determine how many daily units of each product should be produced and at what location.

Table 5.1 Problem data

	U1	U2	U3
Unit income P1 ($)	2	4	3
Unit income P2 ($)	5	3	6
Unit income P3 ($)	3	4	2
Daily production capacity (units)	200	400	300
Daily maximum volume of pollution allowed (cm^3)	150	250	200
Penalty for excess pollution volume ($/$cm^3$)	20	15	10

Modeling

This system allows us to raise the problem from a first version with a methodological modeling and a second version that models the elements and activities taking into account some of the specifications.

Version 1

List of elements in Table 5.2.

Table 5.2 Problem 19: table of elements

Elements	Set	QN	Data				
			Name	Parameter	Type	Belonging	Value
Products	$i = 1\ldots3$	C_I	Income	PR_{ij}	C ($/ud))	S	...
			Volume of contamination	CC_i	C(cm^3/ud)	S	...
Locations	$j = 1\ldots3$	I_U	Production capacity	K_j	I (uds)	W	...
			Maximum volume of pollution allowed	V_j	C C (cm^3)	S	...
			Penalty	p_j	C ($/cm^3)	W	...
Pollution	–	I_M		CC_i, V_j			
System	–	I_U	Maximum expenditure	G	C($)	W	9000 €

Decision Activities

This first version introduces two decision activities, producing the products at the locations and increasing the allowable contamination at the locations. For this reason, this first version declares contamination as a measurable element.

Produce products (*i*) at locations (*j*)

$$\forall i,j : x_{ij} = \text{units of products } i \text{ produced at location } j. \tag{19.1}$$

Allow additional pollution at locations (*j*)

$$\forall j : z_j = \text{Additional contamination allowed at location } j$$

This second activity also motivates the use of a second version for the formulation since, although it is a decision activity, the specifications and the objective function determine that its value can be calculated from the value of x_{ij}, but that is something we will analyze in the second version.

Specifications

– Produce in a single location (binary logical calculation + quantitative selection specification): To model this specification, we need to introduce a logical calculation that returns us whether or not it has produced in each location. Next, we model the specification as a quantitative selection specification of the logical variables created.

Binary Logical Calculation

$$\forall j : \alpha_j = 1 \text{ if we produce at location } j; 0 \text{ otherwise.} \tag{19.2}$$

$$\text{Logical proposition} : \forall j : \alpha_j = 1 \text{ IF AND ONLY IF } \sum_{i=1}^{3} x_{ij} > 0$$

Since we are going to use a quantitative rule with respect to the variables by imposing that we only select one, we can reduce the definition of the calculation to use fewer restrictions, although the previous definition is also correct.

$$\forall j : \alpha_j = 1 \text{ IF AND ONLY IF } \sum_{i=1}^{3} x_{ij} > 0$$

$$\Rightarrow f_{S_v} \Rightarrow \forall j : \text{IF } \sum_{i=1}^{3} x_{ij} > 0 \text{ THEN } \alpha_j = 1$$

$$\Rightarrow f_3 \Rightarrow \forall j : \text{IF } \alpha_j = 0 \text{ THEN } \sum_{i=1}^{3} x_{ij} \leq 0$$

$$\Rightarrow f_7 \Rightarrow \forall j : \text{IF } 1 - \alpha_j = 1 \text{ THEN } \sum_{i=1}^{3} x_{ij} \leq 0$$

$$\Rightarrow f_{14}\left(\text{UB}_{\sum_{i=1}^{3} x_{ij}} = K_j\right) \Rightarrow \forall j : \sum_{i=1}^{3} x_{ij} \leq K_j \alpha_j \tag{19.3}$$

Quantitative selection rule with respect to the number of locations used:

$$\sum_{j=1}^{3} \alpha_j = 1 \tag{19.4}$$

– Production capacity (capacity specification):

$$\forall j : \sum_{i=1}^{3} x_{ij} \leq K_j \tag{19.5}$$

Specification not necessary, already covered by constraint (19.3).

– Volume of pollution (capacity specifications): We are in case 2 within the most common capacity specification (Table 6.2 in García 2021). There is a fixed contribution of the pollution capacity (V_j) plus a variable contribution motivated by the decision variables z_j:

$$\forall j : \sum_{i=1}^{3} \mathrm{CC}_i x_{ij} \leq V_j + z_j \tag{19.6}$$

In this type of capacity specifications, it is not necessary to impose that z_j take values if the capacity V_j has been exhausted, since as the increase in capacity supposes a cost, z_j will only take positive values when we have exhausted the fixed capacity V_j.

- Maximum allowable spending on additional pollution (capacity specifications): The maximum spenditure G functions as capacity data, of the one that the pollution variable z_j consumes, and p_j is the unit consumption:

$$\sum_{j=1}^{3} p_j z_j \leq G \tag{19.7}$$

This specification can also be proposed individually for each location since there is a specification that forces the use of only one location:

$$\forall j : p_j z_j \leq G \tag{19.8}$$

Objective Function

$$\text{Maximize Profits} = \text{Income} - \text{Costs}$$

$$\text{Income} = \sum_{j=1}^{3} \sum_{i=1}^{3} \mathrm{PR}_{ij} x_{ij}$$

$$\text{Costs} = \sum_{j=1}^{3} p_j z_j$$

$$\text{Max} \sum_{j=1}^{3} \sum_{i=1}^{3} \mathrm{PR}_{ij} x_{ij} - \sum_{j=1}^{3} p_j z_j \tag{19.9}$$

Version 2

In version 2, we are going to simplify the constraint of allowed contamination capacity, eliminating the direct decision associated with allowing more contamination in a location (this also means that it is not necessary to explicitly consider the contamination element).

From the first version, we have:

$$\forall j: \sum_{i=1}^{3} \mathrm{CC}_i x_{ij} \le V_j + z_j \tag{19.6}$$

$$\forall j: p_j z_j \le G \tag{19.8}$$

Therefore:

$$\forall j: z_j \le \frac{G}{p_j}$$

$$\forall j: \sum_{i=1}^{3} \mathrm{CC}_i x_{ij} \le V_j + z_j \Rightarrow \forall j: \sum_{i=1}^{3} \mathrm{CC}_i x_{ij} \le V_j + \frac{G}{p_j} \tag{19.10}$$

In this way, we no longer need the variable z_j in the specifications.

However, this will mean that for the cost calculation we will need a particular case of deviation over a threshold.

Objective Criterion

$$\text{Maximize Profits} = \text{Income} - \text{Costs}$$

$$\text{Income} = \sum_{j=1}^{3} \sum_{i=1}^{3} PR_{ij} x_{ij}$$

Costs:

We need to collect the excess produced over the initial volume V_j allowed. The excess supposes a penalty in the objective function. We are obviously talking about penalizing a deviation for excess with respect to the volume threshold V_j. (Sect. 1.6.3.1). Data and variables are shown in Fig. 5.1.

We define an auxiliary calculation to collect the total contamination at each location:

Auxiliary Calculation:

$$\forall j: y_j = \text{Pollution in location } j$$

$$\forall j: y_j = \sum_{i=1}^{3} \mathrm{CC}_i x_{ij} \tag{19.11}$$

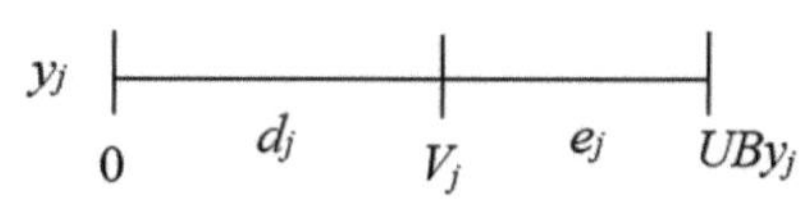

Fig. 5.1 Data and variables of penalization for excess

$$\forall j : \mathrm{UB}_{y_j} = V_j + \frac{G}{p_j}$$

Auxiliary Calculations:

$\forall j : d_j$: Pollution defect with respect to the threshold V_j

$\forall j : e_j$: Pollution excess with respect to the threshold V_j

$$\forall j : y_j + d_j - e_j = V_j \tag{19.12}$$

$$\text{Cost Term} = \sum_{j=1}^{3} p_j e_j$$

Objective Function:

$$\text{Max} \sum_{j=1}^{3} \sum_{i=1}^{3} \mathrm{PR}_{ij} x_{ij} - \sum_{j=1}^{3} p_j e_j \tag{19.13}$$

Complete Model

$$\text{Max} \sum_{j=1}^{3} \sum_{i=1}^{3} \mathrm{PR}_{ij} x_{ij} - \sum_{j=1}^{3} p_j e_j \tag{19.13}$$

s.t.

$$\sum_{i=1}^{3} x_{ij} \leq K_j \alpha_j \tag{19.3}$$

$$\sum_{j=1}^{3} \alpha_j = 1 \tag{19.4}$$

$$\forall j : \sum_{i=1}^{3} \mathrm{CC}_i x_{ij} \leq V_j + \frac{G}{p_j} \tag{19.10}$$

$$\forall j : y_j = \sum_{i=1}^{3} \mathrm{CC}_i x_{ij} \tag{19.11}$$

$$\forall j : y_j + d_j - e_j = V_j \tag{19.12}$$

$$\forall i,j : x_{ij} \geq 0 \text{ Integers} \tag{19.1}$$

$$\forall j : \alpha_j \text{ binaries} \tag{19.2}$$

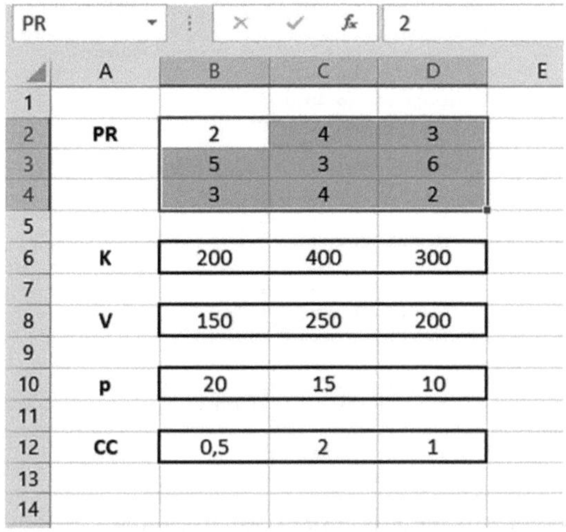

PR | 2

	A	B	C	D	E
1					
2	PR	2	4	3	
3		5	3	6	
4		3	4	2	
5					
6	K	200	400	300	
7					
8	V	150	250	200	
9					
10	p	20	15	10	
11					
12	CC	0,5	2	1	
13					
14					

Fig. 5.2 "19_Excel.xlsx"

$$\forall j : y_j \ \geq 0$$

$$\forall j : d_j \geq 0, e_j \geq 0$$

Implementation in LINGO

In this case, we are going to implement version 2. The data in this example are found in an Excel file, "19_Excel.xlsx," where the ranges that contain each data have been labeled with the data name (Fig. 5.2).

```
MODEL:
SETS:
Products/1..3/:CC;
Locations/1..3/:K,V,p,Alpha,y,e,d;
P_U(Products,Locations):PR,x;
ENDSETS
DATA:
 PR,K,V,p,CC =@OLE("c:\Files_\19_Excel.xlsx");
 G=9000;
ENDDATA
!19.13;
Max = @SUM(P_U(i,j): PR(i,j)*x(i,j))-@SUM(Locations(j):p(j)*e(j));
!19.3; @FOR(Locations(j):@SUM(Products(i):x(i,j))<=K(j)*Alpha(j));
!19.4; @SUM(Locations(j):Alpha(j))=1;
!19.10;
@FOR(Locations(j):@SUM(Products(i):CC(i)*x(i,j))<=V(j)+(G/p(j)));
!19.11; @FOR(Locations(j): y(j)=@SUM(Products(i):CC(i)*x(i,j)));
!19.12; @FOR(Locations(j): y(j)+d(j)-e(j)=V(j));
!19.1; @FOR(P_U(i,j):@GIN(x(i,j)));
!19.2; @FOR(Locations(j):@BIN(Alpha(j)));
END
```

```
Solution Report - 19_LingoE
Global optimal solution found.
Objective value:                              1600.000
Objective bound:                              1600.000
Infeasibilities:                              0.000000
Extended solver steps:                               0
Total solver iterations:                            24
Elapsed runtime seconds:                          0.19

Model Class:                                      MILP

Total variables:                  21
Nonlinear variables:               0
Integer variables:                12

Total constraints:                14
Nonlinear constraints:             0

Total nonzeros:                   57
Nonlinear nonzeros:                0
```

Fig. 5.3 Problem 19: LINGO solution report

Solution
The solution is shown in Fig. 5.3.

5.3 Problem 20. Assignment of Tasks to Operators

Let there be a system for assigning tasks to operators. The system has *N* tasks and *M* operators. Each task has the following characteristics:

- Processing time of the task by each operator.
- Task priority: Value between 0 and 10.
- Each operator has the following characteristics:
- Compatibility with tasks: There are tasks that they can perform and others that they cannot.
- Maximum working time.
- Cost: Each operator has a fixed cost.

Given a completed task, the number of tasks with a higher priority than it and not performed cannot exceed 2.

The company has a spending limit that must not be exceeded. The goal is to maximize the number of tasks performed.

Modeling
List of elements in Table 5.3.

Table 5.3 Problem 20: table of elements

			Data				
Elements	Set	QN	Name	Parameter	Type	Belonging	Value
Tasks	$i = 1 \ldots N$	I_U	Processing time	T_{ij}	B	S	…
			Priority	P_i	I	W	…
			Compatibility	A_{ij}	B	S	…
Operators	$j = 1 \ldots M$	I_U	Cost	C_j	C ($)	W	…
			Max. working time	K_j	C(h)	W	…
				T_{ij}; C_{ij}			
System	–		Spending limit	G	C(h)	W	

Decision Activities

Assign tasks (*i*) to operators (*j*)

$$\forall i,j : \alpha_{ij} = 1 \text{ if assign task } i \text{is assigned to operator } j; 0 \text{ otherwise.} \quad (20.1)$$

Specifications

– Quantitative selection specifications:

Activity	Elements selecting	Selectable elements	Selection	Constraints
α_{ij}	Tasks ($i = 1 \ldots N$)	Operators ($j = 1 \ldots M$)	≤ 1	$\forall i : \sum_{j=1}^{M} \alpha_{ij} \leq 1$ (20.2)

– Maximum working time (capacity specification):

$$\forall j : \sum_{i=1}^{N} T_{ij}\alpha_{ij} \leq K_j \quad (20.3)$$

– Task operator incompatibility (assignment specification):

$$\forall i,j / A_{ij} = 0 : \alpha_{ij} = 0 \quad (20.4)$$

This specification could have been included defining only the decision-making events in the decision activity:

$$\forall i,j / A_{ij} = 1 : \alpha_{ij} = 1 \quad \text{if assign task } i \text{ is assigned to operator } j; 0 \text{ otherwise}$$

– Spending limit (binary logical calculation + capacity specification):

To allocate the cost of each operator, we need to know if the operator has processed any task.

Binary Logical Calculation

$\forall j : \beta_j$

$= 1$ if the operator j has been used (has processed any tasks); 0 otherwise.

(20.5)

$$\forall j : \beta_j = 1 \quad \text{IF AND ONLY IF} \sum_{i=1}^{N} \alpha_{ij} > 0$$

$$\Rightarrow f_{S_V} \Rightarrow \forall j : \text{IF} \sum_{i=1}^{N} \alpha_{ij} > 0 \quad \text{THEN} \quad \beta_j = 1$$

$$\Rightarrow f_3, f_7 \Rightarrow \forall j : \text{IF } 1 - \beta_j = 1 \text{ THEN } \sum_{i=1}^{N} \alpha_{ij} \leq 0$$

$$\Rightarrow f_{14}\left(\text{UB}_{\sum_{i=1}^{N} \alpha_{ij}} = N\right) \Rightarrow \forall j : \sum_{i=1}^{N} \alpha_{ij} \leq N\beta_j \tag{20.6}$$

Capacity specification:

$$\sum_{j=1}^{M} C_j \beta_j \leq G \tag{20.7}$$

- Given a processed task, the number of tasks with a higher priority than it and not carried out cannot exceed 2 (propositional logic specification):

If we perform a task, then the number of tasks not performed with a higher priority than it must be less than or equal to 2:

$$\forall i : \text{IF} \sum_{j=1}^{M} \alpha_{ij} = 1 \text{ THEN } \sum_{i'/P_{i'} > P_i} \left(1 - \sum_{j=1}^{M} \alpha_{i'j}\right) \leq 2$$

$$\sum_{j=1}^{M} \alpha_{ij} \quad \text{is a binary expression :}$$

$$\Rightarrow f_{15}\left(\text{LB}_{\sum_{i'/P_{i'} > P_i} \left(1 - \sum_{j=1}^{M} \alpha_{i'j}\right)} = 0\right) \forall i : \sum_{i'/P_{i'} > P_i} \left(1 - \sum_{j=1}^{M} \alpha_{i'j}\right) \leq 2 \sum_{j=1}^{M} \alpha_{ij} \tag{20.8}$$

Objective Function
Maximize the number of tasks

$$\text{Max} \sum_{i=1}^{N} \sum_{j=1}^{M} \alpha_{ij} \tag{20.9}$$

Complete Model

$$\text{Max} \sum_{i=1}^{N} \sum_{j=1}^{M} \alpha_{ij} \tag{20.9}$$

s.t.

$$\forall i : \sum_{j=1}^{M} \alpha_{ij} \leq 1 \tag{20.2}$$

$$\forall j : \sum_{i=1}^{N} T_{ij}\alpha_{ij} \leq K_j \tag{20.3}$$

$$\forall i,j/A_{ij} = 0 : \alpha_{ij} = 0 \tag{20.4}$$

$$\forall j : \sum_{i=1}^{N} \alpha_{ij} \leq N\beta_j \tag{20.6}$$

$$\sum_{j=1}^{M} C_j\beta_j \leq G \tag{20.7}$$

$$\forall i : \sum_{i'/P_{i'}>P_i} \left(1 - \sum_{j=1}^{M} \alpha_{i'j}\right) \leq 2\sum_{j=1}^{M} \alpha_{ij} \tag{20.8}$$

$$\forall i,j : \alpha_{ij} \text{ binaries} \tag{20.1}$$

$$\forall j : \beta_j \text{ binaries} \tag{20.5}$$

Implementation in LINGO

We use the data collected in the file "c: \ Files_\ 20_Excel.xlsx" (Fig. 5.4):

```
MODEL:
DATA:
 N =@OLE("c:\Files_\20_Excel.xlsx",'N');
 M= @OLE("c:\Files_\20_Excel.xlsx",'M');
ENDDATA
SETS:
Tasks/1..N/:P;
Operators/1..M/:C,K,Beta;
TO(Tasks,Operators):T,A,Alpha;
ENDSETS
DATA:
 P = @OLE("c:\Files_\20_Excel.xlsx",'Priority');
```

N	25
M	10

	1	2	3	4	5	6	7	8	9	10	11	12	13	14	15	16	17	18	19	20	21	22	23	24	25
Priority	4	8	9	9	10	2	3	4	7	7	7	5	5	3	2	2	1	1	1	2	1	1	2	3	4

Processing times	1	2	3	4	5	6	7	8	9	10
1	10	0	23	18	19	12	0	0	32	25
2	0	14	0	0	0	4	0	0	5	0
3	0	7	0	0	15	19	0	0	9	0
4	0	0	13	0	0	13	0	0	0	0
5	23	0	0	34	12	15	0	0	45	13
6	0	0	15	13	45	0	35	0	0	0
7	12	15	0	0	45	0	15	0	22	12
8	0	0	0	35	13	13	19	0	10	0
9	0	0	12	0	25	0	25	0	0	0
10	0	24	0	34	12	0	72	0	0	0
11	0	35	0	0	19	0	19	0	0	21
12	23	24	0	34	0	19	0	19	0	0
13	14	35	0	13	0	0	19	13	0	0
14	18	19	0	12	0	72	0	21	0	23
15	14	35	0	0	19	0	0	25	0	0
16	0	24	0	34	0	12	0	72	0	0
17	0	0	13	0	0	0	0	25	0	25
18	0	24	0	0	0	0	0	72	0	21
19	20	14	35	13	13	19	39	0	25	0
20	0	0	0	0	0	13	0	0	20	21
21	0	22	0	0	35	0	0	0	13	0
22	12	13	0	35	0	0	19	0	19	15
23	11	0	35	0	0	0	39	0	12	13
24	10	0	0	35	13	0	19	0	11	12
25	19	0	0	0	0	0	25	0	25	19

A	1	2	3	4	5	6	7	8	9	10
1	1	0	1	1	1	1	0	0	1	1
2	0	1	0	0	0	1	0	0	1	0
3	0	1	0	0	1	1	0	0	1	0
4	0	0	1	0	0	1	0	0	1	0
5	1	0	0	1	1	1	0	0	0	0
6	0	0	1	1	1	0	1	0	1	1
7	1	1	0	0	1	0	1	0	0	0
8	1	0	0	1	1	1	1	0	1	1
9	0	0	1	0	1	0	1	0	1	0
10	0	1	0	1	1	0	1	0	0	0
11	0	1	0	0	1	0	1	0	0	0
12	1	1	0	1	0	1	0	1	0	1
13	1	1	0	1	0	0	1	1	0	0
14	1	1	0	1	0	1	0	1	0	0
15	1	1	0	0	1	0	0	1	0	1
16	0	1	0	1	0	1	0	1	0	0
17	0	0	1	0	0	1	0	1	0	0
18	0	1	0	0	0	0	0	1	0	1
19	1	1	1	1	1	1	1	0	1	0
20	0	0	0	0	0	1	0	0	1	1
21	0	1	0	0	1	0	0	0	1	0
22	1	1	0	1	0	0	1	0	1	1
23	1	0	1	0	0	0	1	0	1	1
24	1	0	0	1	1	0	1	0	1	1
25	1	0	0	0	0	0	1	0	1	1

	1	2	3	4	5	6	7	8	9	10
Costs	35	34	38	41	42	50	39	31	25	29
Maximum Times	100	80	70	70	50	40	80	90	50	50

Fig. 5.4 "20_Excel.xlsx"

```
 C = @OLE("c:\Files_\20_Excel.xlsx",'Cost');
 K = @OLE("c:\Files_\20_Excel.xlsx",'TimeMax');
 T = @OLE("c:\Files_\20_Excel.xlsx",'Times');
 A = @OLE("c:\Files_\20_Excel.xlsx",'A');
 G=900;
ENDDATA
!20.9;
Max = @SUM(Tasks(i):@SUM(Operators(j):Alpha(i,j)));
!20.2;
@FOR(Tasks(i):@SUM(Operators(j):Alpha(i,j))<=1);
!20.3;
@FOR(Operators(j):@SUM(Tasks(i):T(i,j)*Alpha(i,j))<=K(j));
!20.4;
@FOR(Tasks(i):@FOR(Operators(j)|A(i,j)#EQ#0:Alpha(i,j)=0));
!20.6;
@FOR(Operators(j):@SUM(Tasks(i):Alpha(i,j))<=N*Beta(j));
!20.7;
@SUM(Operators(j):C(j)*Beta(j))<=G;
!20.8;
@FOR(Tasks(i):@SUM(Tasks(ii)|P(ii)#GT#P(i):1-@SUM(Operators(j):
Alpha(ii,j)))<=2*@SUM(Operators(j):Alpha(i,j)));

!20.1;
@FOR(Tasks(i):@FOR(Operators(j):@BIN(Alpha(i,j))));
!20.5;
```

```
Solution Report - 20_LingoE
Global optimal solution found.
Objective value:                                  25.00000
Objective bound:                                  25.00000
Infeasibilities:                                  0.000000
Extended solver steps:                                   0
Total solver iterations:                               134
Elapsed runtime seconds:                              0.28

Model Class:                                          PILP

Total variables:                    125
Nonlinear variables:                  0
Integer variables:                  125

Total constraints:                   72
Nonlinear constraints:                0

Total nonzeros:                    1772
Nonlinear nonzeros:                   0
```

Fig. 5.5 Problem 20: LINGO solution report

```
@FOR(Operators(j):@BIN(Beta(j)));
END
```

Solution

The solution is shown in Fig. 5.5.

5.4 Problem 21. Assignment of Students to Groups

A school considers, after the marks of the first evaluation, improving the performance of its 100 students in the subject of mathematics. The grade of each student for the subject is known, between 0 and 10. The school establishes grade categories (Insufficient [0–5), Sufficient [5,7), Notable [7,9), Excellent [9,10]) to work more comfortably.

The school has decided to create study groups with the aim of ensuring that students with the worst grades are related to students with the best grades. To do this, 15 study groups are created. Each group can have a maximum of 10 students. The school establishes that there are students with an Excellent grade in all groups if possible (if there are at least 15). If a group does not have Excellent's, it should have at least as many Notable's as any group with Excellent's. The objective is to balance the level of grades of the groups, assigning as level values, 1 to insufficient ... 4 to Excellent.

Modeling

This is a problem in which the 100 students do not have to be considered individual, since they are not referred to individually. Since they differ in the grade and the system establishes four categories of grades, we can create four groups of students, depending on the grade they have.

List of elements in Table 5.4.

Table 5.4 Problem 21: table of elements

Elements	Set	QN	Data				
			Name	Parameter	Type	Belonging	Value
Student groups	$i = 1\ldots4$ 1: Insufficient 2: Sufficient 3: Notable 4: Excellent	C_D	No of students	n_i	I	W	...
			Level Value	v_i	I	W	{1,2,3,4}
Study groups	$j = 1\ldots15$	I_U					
System	–	I_U	Maximum students per group	M	I	W	10

Decision Activities
Assign student groups (*i*) to study groups (*j*)

$$x_{ij} = \text{number of students from the student group } i \text{ assigned to the studygroup } j. \quad (21.1)$$

Specifications

- Assignment of the number of students of each group (balance specification based on attribute):

$$\forall i : \sum_{j=1}^{15} x_{ij} = n_i \quad (21.2)$$

- Maximum number of students in study groups (capacity specification):

$$\forall j : \sum_{i=1}^{4} x_{ij} \leq M \quad (21.3)$$

- If there are at least 15 students with Excellent grade, then every study group must have students with Excellent grade (propositional logic specification with attributes):

$$\forall j : \text{IF } n_4 \geq 15 \text{ THEN } x_{4j} \geq 1$$

$$\Rightarrow f_{45} \Rightarrow \forall j / n_4 \geq 15 : \; x_{4j} \geq 1 \tag{21.4}$$

– If a group does not have students with excellent grade, it has at least as many students with a notable grade as any group with students with excellent grade (propositional logic specification):

$$\forall j,k/j \neq k : \text{IF } x_{4j} = 0 \text{ AND } x_{4k} > 0 \text{ THEN } x_{3j} \geq x_{3k}$$

$$\Rightarrow f_5 \Rightarrow \forall j,k/j \neq k : \text{IF } x_{4j} = 0 \text{ AND } x_{4k} \geq 1 \text{ THEN } x_{3j} \geq x_{3k}$$

$$\Rightarrow \forall j,k/j \neq k : \text{IF } x_{4j} = 0 \text{ AND } x_{4k} \geq 1 \text{ THEN } x_{3j} - x_{3k} \geq 0$$

$$\Rightarrow f_{\text{LB}} \Rightarrow \forall j,k/j \neq k : \text{IF } x_{4j} \leq 0 \text{ AND } x_{4k} \geq 1 \text{ THEN } x_{3j} - x_{3k} \geq 0$$

$$\forall j : \omega_{1j} = 1 \text{ IF AND ONLY IF } x_{4j} \leq 0 \tag{21.5}$$

$$\Rightarrow f_{S_V} \Rightarrow \forall j : \; \text{IF } x_{4j} \leq 0 \text{ THEN } \omega_{1j} = 1$$

$$\Rightarrow f_3, f_7, f_{15}\left(\text{LB}_{x_{4j}} = 0\right) \Rightarrow \forall j : \; x_{4j} \geq 1 - \omega_{1j} \tag{21.6}$$

$$\forall j : \omega_{2j} = 1 \text{ IF AND ONLY IF } x_{4j} \geq 1 \tag{21.7}$$

$$\Rightarrow f_{S_V} \Rightarrow \forall j : \; \text{IF } x_{4j} \geq 1 \text{ THEN } \omega_{2j} = 1$$

$$\Rightarrow f_3, f_7, f_{14}\left(\text{UB}_{x_{4j}} = n_4\right) \Rightarrow \forall j : \; x_{4j} \leq n_4 \omega_{2j} \tag{21.8}$$

$$\Rightarrow \forall j,k/j \neq k : \text{IF } \omega_{1j} = 1 \text{ AND } \omega_{2k} = 1 \text{ THEN } x_{3j} - x_{3k} \geq 0$$

$$\Rightarrow f_{37} \Rightarrow \forall j,k/j \neq k : \text{IF } \omega_{1j} + \omega_{2k} \geq 2 \text{ THEN } x_{3j} - x_{3k} \geq 0$$

$$\Rightarrow f_{18} \Rightarrow \forall j,k/j \neq k : \omega_{1j} + \omega_{2k} \leq 1 + \pi_{jk} \tag{21.9}$$

$$\Rightarrow f_{21} \Rightarrow \forall j,k/j \neq k : x_{3j} - x_{3k} \geq -n_3\left(1 - \pi_{jk}\right) \tag{21.10}$$

Objective Criterion

Balance the level of notes of the groups, assigning as level values, 1 to insufficient ... 4 to excellent.

Auxiliary Calculation

$$\forall j : \; y_j : \text{Level of each study group} :$$

$$\forall j : \; y_j = \sum_{i=1}^{4} v_i x_{ij} \tag{21.11}$$

Bound Calculation

$$y_{\text{MIN}} : \text{Lower bound of the grade level}$$

$$\forall j: \quad y_{\text{MIN}} \leq y_j \tag{21.12}$$

y_{MAX} : Upper bound of the grade level

$$\forall j: \quad y_{\text{MAX}} \geq y_j \tag{21.13}$$

Objective Function

$$\text{Min } y_{\text{MAX}} - y_{\text{MIN}} \tag{21.14}$$

Complete Model

$$\text{Min } y_{\text{MAX}} - y_{\text{MIN}} \tag{21.14}$$

s.t.

$$\forall i: \sum_{j=1}^{15} x_{ij} = n_i \tag{21.2}$$

$$\forall j: \sum_{i=1}^{4} x_{ij} \leq M \tag{21.3}$$

$$\forall j/n_4 \geq 15: \quad x_{4j} \geq 1 \tag{21.4}$$

$$\forall j: \quad x_{4j} \geq 1 - \omega_{1j} \tag{21.6}$$

$$\forall j: \quad x_{4j} \leq n_4 \omega_{2j} \tag{21.8}$$

$$\forall j, k/j \neq k: \omega_{1j} + \omega_{2k} \leq 1 + \pi_{jk} \tag{21.9}$$

$$\forall j, k/j \neq k: x_{3j} - x_{3k} \geq -n_3(1 - \pi_{jk}) \tag{21.10}$$

$$\forall j: \quad y_j = \sum_{i=1}^{4} v_i x_{ij} \tag{21.11}$$

$$\forall j: \quad y_{\text{MIN}} \leq y_j \tag{21.12}$$

$$\forall j: \quad y_{\text{MAX}} \geq y_j \tag{21.13}$$

$$\forall i, j: \quad x_{ij} \geq 0 \text{ integer} \tag{21.1}$$

$$\forall j: \quad \omega_{1j}, \omega_{2j} \text{ binaries} \tag{21.5, 21.7}$$

$$\forall j, k/j \neq k: \pi_{jk} \text{ binaries} \tag{21.15}$$

$$\forall j: \quad y_j \geq 0$$

$$y_{\text{MAX}} \geq 0; y_{\text{MIN}} \geq 0$$

Implementation in LINGO

```
MODEL:
SETS:
 StudentGroups/1..4/:n,v;
 StudyGroups/1..15/:Omega1,Omega2,y;
 AE(StudentGroups,StudyGroups):x;
 EE(StudyGroups,StudyGroups):Pi;
ENDSETS
DATA:
 n=24,48,22,6;
 v=1,2,3,4;
 M= 10;
ENDDATA

!21.14;MIN= yMAX-yMIN;
!21.2;@FOR(StudentGroups(i):@SUM(StudyGroups(j):x(i,j))=n(i));
!21.3;@FOR(StudyGroups(j):@SUM(StudentGroups(i):x(i,j))<=M);
!21.4; @FOR(StudyGroups(j)|n(4)#GE#15:x(4,j)>=1);
!21.6; @FOR(StudyGroups(j): x(4,j)>=1-Omega1(j));
!21.8; @FOR(StudyGroups(j):x(4,j)<=n(4)*Omega2(j));
!21.9; @FOR(StudyGroups(j):@FOR(StudyGroups(k)|k#NE#j:Omega1(j)
+Omega2(k)<=1+Pi(j,k)));
!21.10; @FOR(StudyGroups(j):@FOR(StudyGroups(k)|k#NE#j:x(3,j)-x(3,
k)>=-n(3)*(1-Pi(j,k))));
!21.11; @FOR(StudyGroups(j):y(j)=@SUM(StudentGroups(i):v(i)*x(i,
j)));
!21.12; @FOR(StudyGroups(j):y(j)<=yMAX);
!21.13; @FOR(StudyGroups(j):y(j)>=yMIN);
!21.1;@FOR(StudentGroups(i):@FOR(StudyGroups(j):@GIN(x(i,j))));
!21.5;@FOR(StudyGroups(j):@BIN(Omega1(j)));!21.7;@FOR(StudyGroups
(j):@BIN(Omega2(j)));

!21.15; @FOR(StudyGroups(j):@FOR(StudyGroups(k)|k#NE#j:@BIN(Pi(j,
k))));
END
```

Solution

It is shown in Fig. 5.6.

5.5 Problem 22. Home Building

A developer wants to optimize the use of a rectangular block that he has acquired for housing construction. The block has a surface area of 25,000 m^2 (500 m long by 50 m wide). It is going to build houses in two rows (500 m long each row and 20 m wide). In the middle there is a 10m wide area for access. The developer must decide how many houses to build in each row within three models:

Independent houses on plots of 400 m^2 (20 m length $\times$ 20 m width).

Semi-detached houses on plots of 700 m^2 (35 m length $\times$ 20 m width) for the two houses.

Detached houses on plots of 200 m^2 (10 m long $\times$ 20 m wide) for each house.

```
Solution Report - 21_LingoE
Global optimal solution found.
Objective value:                              0.000000
Objective bound:                              0.000000
Infeasibilities:                              0.000000
Extended solver steps:                             115
Total solver iterations:                          2717
Elapsed runtime seconds:                          0.15

Model Class:                                      MILP

Total variables:                    332
Nonlinear variables:                  0
Integer variables:                  300

Total constraints:                  515
Nonlinear constraints:                0

Total nonzeros:                    1577
Nonlinear nonzeros:                   0
```

Fig. 5.6 Problem 21: LINGO solution report

The construction cost of each house is C_i ($i = 1, 2, 3$). The sale price for each type of house is B_i. The established investment ceiling is M \$. The investment money has a cost of 3%.

The city council imposes a minimum number of houses of each type, *mi*.

Detached houses can only be in a single row.

The promoter is not obliged to build the entire row. Propose a model that maximizes benefits taking into account that if the number of semi-detached houses is greater than the number of independent houses, it carries an additional cost of *CA* \$ for each additional semi-detached dwelling.

Modeling

List of elements in Table 5.5.

Table 5.5 Problem 22: table of elements

Elements	Set	QN	Data				
			Name	Parameter	Type	Belonging	Value
Rows	$j = 1, 2$	I_U	Length	LF_j	C (m)	W	500
House models	$i = 1\ldots3$	C_I	Length	LP_i	C (m)	W	…
			Cost	c_i	C(\$)	W	…
			Sale price	b_i	C(\$)	W	…
			Minimum number	m_i			…
Investment Money	–	I_M	Inv. cost percent	CI	C (%)	W	3%
Developer	–	I_U	Inv. ceiling	TI	C(\$)	W	M\$
			Additional cost	CA	C(\$)	W	–

Decision Activities
Build house models (*i*) in rows (*j*)

$$x_{ij} = \text{number of houses of model } i \text{ built in row } j \tag{22.1}$$

Auxiliary Calculation
Plots for semi-detached houses in rows

$$z_{2j} : \text{Number of plots for semi-detached houses in row } j. \tag{22.2}$$

$$\forall j : z_{2j} = \frac{x_{2j}}{2} \tag{22.3}$$

Specifications

- Row length (capacity specification):

$$\forall j : \text{LP}_1 x_{1j} + \text{LP}_2 z_{2j} + \text{LP}_3 x_{3j} \leq \text{LF}_j \tag{22.4}$$

- Minimum number of houses (lower bound imposition):

$$\forall i : \sum_{j=1}^{2} x_{ij} \geq m_i \tag{22.5}$$

- Investment ceiling (auxiliary calculation + capacity specification):

The consumption made is the following:

$$\text{Construction cost} + \text{Additional cost (CA)} + \text{Investment cost (CI)}$$

Construction cost:
Auxiliary Calculations

$$y = \text{construction cost}$$

$$y = \sum_{i=1}^{3} \sum_{j=1}^{2} c_i x_{ij} \tag{22.6}$$

Additional cost:
We obtain the total number of semi-detached and independent houses:
Auxiliary Calculations
P = Total of semi-detached houses
In = Total of independent houses

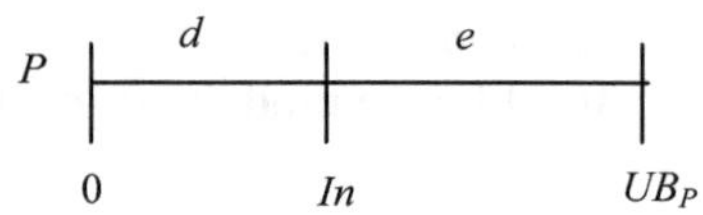

Fig. 5.7 Data and variables for the calculation of additional cost

$$P = \sum_{j=1}^{2} x_{2j} \tag{22.7}$$

$$\text{In} = \sum_{j=1}^{2} x_{1j} \tag{22.8}$$

The additional cost is applied on the excess of semi-detached over the number of independent (Fig. 5.7).

$$P + d - e = In \tag{22.9}$$

$$\text{Additional cost} = \text{CA}e$$

Investment cost:

$$\frac{\text{CI}}{100}(y + \text{CA}e)$$

Final constraint:

$$y + \text{CA}e + \frac{\text{CI}}{100}(y + \text{CA}e) \leq \text{TI} \tag{22.10}$$

– Detached houses in a single row (propositional logic specification):

$$\text{IF } x_{31} > 0 \text{ THEN } x_{32} = 0$$

$$\Rightarrow f_5, f_{\text{LB}} \Rightarrow \text{IF } x_{31} \geq 1 \text{ THEN } x_{32} \leq 0$$

$$\Rightarrow f_{18}\left(\text{UB}_{x_{31}} = \frac{\text{LF}_1}{\text{LP}_3}\right) \Rightarrow x_{31} \leq \frac{\text{LF}_1}{\text{LP}_3}\omega \tag{22.11}$$

$$\Rightarrow f_{20}\left(\text{UB}_{x_{32}} = \frac{\text{LF}_2}{\text{LP}_3}\right) \Rightarrow x_{32} \leq \frac{\text{LF}_2}{\text{LP}_3}(1 - \omega) \tag{22.12}$$

Objective Function

$$\text{Maximize incomes} - \text{Costs}$$

$$\text{Income} = \sum_{i=1}^{3}\sum_{j=1}^{2} b_i x_{ij}$$

$$\text{Costs} = y + \text{CA}e + \frac{\text{CI}}{100}(y + \text{CA}e)$$

$$\text{Max} \quad \sum_{i=1}^{3}\sum_{j=1}^{2} b_i x_{ij} - \left(y + \text{CA}e + \frac{\text{CI}}{100}(y + \text{CA}e)\right) \tag{22.13}$$

Complete Model

$$\text{Max} \quad \sum_{i=1}^{3}\sum_{j=1}^{2} b_i x_{ij} - \left(y + \text{CA}e + \frac{\text{CI}}{100}(y + \text{CA}e)\right) \tag{22.13}$$

s.t.

$$\forall j : z_{2j} = \frac{x_{2j}}{2} \tag{22.3}$$

$$\forall j : \text{LP}_1 x_{1j} + \text{LP}_2 z_{2j} + \text{LP}_3 x_{3j} \le \text{LF}_j \tag{22.4}$$

$$\forall i : \sum_{j=1}^{2} x_{ij} \ge m_i \tag{22.5}$$

$$y = \sum_{i=1}^{3}\sum_{j=1}^{2} c_i x_{ij} \tag{22.6}$$

$$P = \sum_{j=1}^{2} x_{2j} \tag{22.7}$$

$$In = \sum_{j=1}^{2} x_{1j} \tag{22.8}$$

$$P + d - e = \text{In} \tag{22.9}$$

$$y + \text{CA}e + \frac{\text{CI}}{100}(y + \text{CA}e) \le \text{TI} \tag{22.10}$$

$$x_{31} \le \frac{\text{LF}_1}{\text{LP}_3}\omega \tag{22.11}$$

$$x_{32} \le \frac{\text{LF}_2}{\text{LP}_3}(1 - \omega) \tag{22.12}$$

$$\forall i, j : x_{ij} \geq 0 \text{ integers} \tag{22.1}$$

$$\forall j : z_{2j} \geq 0 \text{ integers} \tag{22.2}$$

$$\omega \text{ binarie} \tag{22.14}$$

$$y, P, \text{In}, d, e \geq 0$$

Implementation in LINGO

```
MODEL:
SETS:
 Rows/1,2/:LF,z2;
 HModels/1..3/:LP,c,b,m;
 HR(HModels,Rows):x;
ENDSETS
DATA:
 LF= 500 500;
 LP= 20 35 10;
 c= 190000 150000 100000;
 b= 250000 200000 120000;
 m= 4 4 6;
 TI= 3000000;
 CI= 3;
 CA= 15000;
ENDDATA
!22.13;
MAX=@SUM(HModels(i):@SUM(Rows(j):b(i)*x(i,j)))-(y+CA*e+(CI/100)*
(y+CA*e));
!22.3;
@FOR(Rows(j): z2(j)=x(2,j)/2);
!22.4;
@FOR(Rows(j): (LP(1)*x(1,j))+(LP(2)*z2(j))+(LP(3)*x(3,j))<=LF(j));
!22.5;
@FOR(HModels(i):@SUM(Rows(j):x(i,j))>=m(i));
!22.6;
y=@SUM(HModels(i):@SUM(Rows(j):c(i)*x(i,j)));
!22.7;
P=@SUM(Rows(j):x(2,j));
!22.8;
In=@SUM(Rows(j):x(1,j));
!22.9;
P+d-e=In;
!22.10;
y+ A*e+ (CI/100)*(y+CA*e) <=TI;
!22.11;
x(3,1)<=(LF(1)/LP(3))*Omega;
!22.12;
x(3,2)<=(LF(2)/LP(3))*(1-Omega);
!22.1;
@FOR(HModels(i):@FOR(Rows(j):@GIN(x(i,j))));
!22.2;
@FOR(Rows(j):@GIN(z2(j)));
```

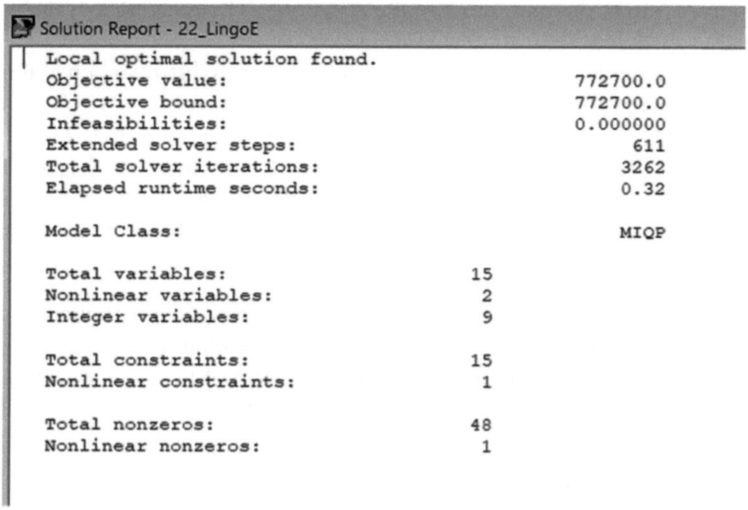

```
Solution Report - 22_LingoE
Local optimal solution found.
Objective value:                              772700.0
Objective bound:                              772700.0
Infeasibilities:                              0.000000
Extended solver steps:                             611
Total solver iterations:                          3262
Elapsed runtime seconds:                          0.32

Model Class:                                      MIQP

Total variables:                    15
Nonlinear variables:                 2
Integer variables:                   9

Total constraints:                  15
Nonlinear constraints:               1

Total nonzeros:                     48
Nonlinear nonzeros:                  1
```

Fig. 5.8 Problem 22: LINGO solution report

```
!22.14;
@BIN(Omega);

END
```

Solution

The solution is shown in Fig. 5.8.

5.6 Problem 23. Assignment of Extracurricular Activities

A school wants to plan the assignments of extracurricular activities to the students. The activities offered are the following:

- Gymnastics (two groups)
- Music (one group)
- Basketball (two groups)
- Crafts (one group)
- Drawing (one group)
- English (two groups)
- French (one group)
- Soccer (two groups)

The activities are carried out in 1-h classes (a number of weekly classes that we do not need to consider) between Monday, Tuesday, Wednesday, and Thursday, from

4 to 5 h, from 5 to 6, or from 6 to 7 in the afternoon. The schedules of the activities are already scheduled.

There are activities with two groups. Each group has its own schedule and its maximum number of students.

The 150 students have already requested the different activities (each student requests the activities they want to do), but they did so when they formalized their enrollment and were unaware of the schedule. We are going to assign groups of activities to the students with the objective of maximizing the sum of activities that are assigned to the total number of students, taking into account that:

- A student cannot be assigned to two groups of overlapping activities.
- A student can only be assigned activities they have requested.
- We cannot assign a student to more than one group of the same activity.
- A student cannot do more than two activities in one day.

Modelling

List of elements in Table 5.6.

Table 5.6 Problem 23: table of elements

			Data				
Elements	Set	QN	Name	Parameter	Type	Belonging	Value
Activities	$i = 1\ldots8$	I_U	Requested	S_{ji}	B	S	...
Groups	$k = 1\ldots12$	I_U	Activity	A_k	O	W	...
			Scheduling	P_{kh}	B	S	...
			Maximum of students	M_k	I	W	...
Students	j=1...150	I_U		S_{ji}			
Schedules	$h = 1..12$	I_U	Day	D_h	O	W	...
				P_{kh}			
Days	$t = 1, 2, 3, 4$	I_U					

Decision Activities

Assign groups (*k*) to students (*j*)

$$\forall j, k : \alpha_{jk} = 1 \text{ if student } j \text{ is assigned to group } k; 0 \text{ otherwise} \quad (23.1)$$

Specifications

- A student can only be assigned groups of activities that he has requested:

It is presented in a negative way, as described in Sect. 6.8.6 (García 2021): If the student has not requested an activity, that activity cannot be assigned to him (propositional logic specification with attributes).

$$\forall j, i : \text{IF } S_{ji} = 0 \text{ THEN } \sum_{k/A_k=i} \alpha_{jk} = 0$$

$$\Rightarrow f_{45} \Rightarrow \forall j, i/S_{ji} = 0 : \sum_{k/A_k=i} \alpha_{jk} = 0 \tag{23.2}$$

- A student cannot be assigned two groups of overlapping activities (propositional logic specification):

$$\forall j, h, k/S_{jA_k} = 1, k'/k' > k \& S_{jA_{k'}} = 1 \& P_{kh} = P_{k'h} = 1 :$$
$$\text{NOT}\left(\alpha_{jk} = 1 \ \ Y \ \ \alpha_{jk'} = 1\right)$$
$$\Rightarrow f_{37}, f_{11} \Rightarrow \forall j, h, k/S_{jA_k} = 1, k'/k' > k \& S_{jA_{k'}} = 1 \& P_{kh} = P_{k'h} = 1 :$$

$$\alpha_{jk} + \alpha_{jk'} \leq 1 \tag{23.3}$$

- We cannot assign more than one group of the same activity to a student (quantitative selection specification):

$$\forall i, j/S_{ji} = 1 : \sum_{k/A_k=i} \alpha_{jk} \leq 1 \tag{23.4}$$

- A student cannot do more than two activities in 1 day (upper bound imposition).

$$\forall j, i/S_{ji} = 1, \forall t : \sum_{h/D_h=t} \sum_{k/A_k=i \& P_{kh}=1} \alpha_{jk} \leq 2 \tag{23.5}$$

- Maximum of students (capacity specification).

$$\forall k : \sum_{j/S_{jA_k}=1} \alpha_{jk} \leq M_k \tag{23.6}$$

Objective Function

Maximize the sum of activities that all students could do.

$$\text{Max} \sum_{j=1}^{150} \sum_{k=1}^{12} \alpha_{jk} \tag{23.7}$$

Complete Model

$$\text{Max} \sum_{j=1}^{150} \sum_{k=1}^{12} \alpha_{jk} \tag{23.7}$$

s.t.

$$\forall j, i / S_{ji} = 0 : \sum_{k/A_k = i} \alpha_{jk} = 0 \tag{23.2}$$

$$\forall j, h, k / S_{jA_k} = 1, k' / k' > k \& S_{jA_{k'}} = 1 \& P_{kh} = P_{k'h} = 1 : \alpha_{jk} + \alpha_{jk'} \leq 1 \tag{23.3}$$

$$\forall i, j / S_{ji} = 1 : \sum_{k/A_k = i} \alpha_{jk} \leq 1 \tag{23.4}$$

$$\forall j, i / S_{ji} = 1, \forall t : \sum_{h/D_h = t} \sum_{k/A_k = i \& P_{kh} = 1} \alpha_{jk} \leq 2 \tag{23.5}$$

$$\forall k : \sum_{j/S_{jA_k} = 1} \alpha_{jk} \leq M_k \tag{23.6}$$

$$\forall j, k : \alpha_{jk} \ \text{binaries} \tag{23.1}$$

Implementation in LINGO

```
MODEL:

SETS:
Activities/1..8/;
Groups/1..12/:A,M;
Students/1..100/;
Schedules/1..12/:D;
Days/1..4/;
GS(Groups,Schedules):P;
SA(Students,Activities): S;
SG(Students,Groups):Alpha;
ENDSETS

DATA:
A= 1 1 2 3 3 4 5 6 6 7 8 8 ;
M= 10 10 10 10 10 10 10 10 10 10 12 12;
P = 1 0 0 0 0 0 1 0 0 0 0 0 0 1 0 0 0 0 0 1 0 0 0 0
   0 0 1 0 0 0 0 0 1 0 0 0 0 0 0 1 0 0 0 0 0 1 0 0
   0 0 0 0 1 0 0 0 0 0 1 0 0 0 0 0 0 1 0 0 0 0 0 1
   1 0 0 0 1 0 0 0 0 1 0 0 0 1 0 0 0 1 0 0 0 0 1 0
   0 0 1 0 0 0 1 0 0 0 0 1 0 0 1 0 0 0 1 0 0 0 0 0
   0 0 1 0 0 0 1 0 0 1 0 0 0 0 0 0 1 0 0 0 0 0 1 0;
S = 1 1 0 0 0 0 0 0 1 1 0 0 0 0 1 0 1 1 0 0 0 1 1 0 0 0 0 0 0 0 1 0 0 1 1 0 1 1 0 0 1 1 0 1 1 0
0 0 1 1 0 1 0 1 0 0 1 1 0 0 1 1 0 0 0 1 0 0 1 0 0 0 1 1 0 0 0 0 1 0 1 1 0 0 0 0 0 0 1 1 0 0 1 1
1 0 1 1 0 0 1 0 0 1 1 1 0 0 0 0 0 0 0 0 0 0 1 1 0 1 1 1 1 0 1 0 0 0 1 1 0 1 0 0 0 0 1 1 0 0
0 0 1 0 1 0 1 0 0 0 1 0 1 1 1 0 0 0 0 0 1 1 0 0 0 1 0 0 1 1 0 0 1 0 0 0 1 1 0 0 1 1 0 0 0 1 0 0
1 0 0 0 0 1 0 0 1 0 1 0 1 1 0 0 1 0 0 0 0 1 1 0 0 0 0 0 1 1 0 0 1 0 1 0 0 0 1 0 1 0 0 0 1 0 1 0
0 0 0 1 1 1 0 1 1 0 0 0 1 1 0 1 1 0 0 0 1 1 0 0 1 0 0 0 0 0 0 0 0 0 0 0 0 1 0 0 1 0 0 0 1 1 0 0
1 1 0 0 1 0 0 0 1 0 0 0 1 1 0 0 0 1 0 1 0 0 1 0 0 0 1 0 1 1 0 0 1 0 0 0 1 1 0 0 1 0 0 0 1 1 0 0
1 0 0 0 1 1 0 0 1 0 0 0 1 1 1 0 1 0 0 0 1 0 1 0 0 0 0 0 0 0 1 0 0 0 1 0 1 1 0 0 1 0 0 0 1 1 0 0
```

```
0 0 0 0 1 1 0 1 1 0 0 0 0 1 0 0 1 0 0 1 0 1 0 0 1 0 0 0 1 1 0 1 1 0 0 0 1 1 0 0 0 1 0 1 0 1 0 0
0 0 1 0 1 1 0 0 1 0 1 0 1 0 1 0 0 1 1 0 1 1 0 0 1 0 0 0 1 1 0 0 1 0 0 0 1 1 1 0 0 1 0 0 0 0 1 0
0 1 1 0 1 0 0 0 1 0 0 0 1 1 0 1 1 0 0 0 1 1 0 0 1 1 0 0 1 1 0 0 1 0 1 0 1 1 0 0 1 0 0 0 0 0 0 0
1 0 0 0 1 1 0 0 1 0 0 0 1 1 0 0 0 0 0 0 0 1 1 0 0 0 1 0 1 0 1 0 1 0 0 0 1 1 1 0 0 1 0 0 1 1 0 0
1 0 0 0 1 1 0 0 1 0 0 1 0 1 0 0 1 0 1 0 0 1 0 0 1 0 0 0 0 1 0 0 1 0 1 0 1 1 0 0 1 0 0 0 1 1 0 0
1 0 0 0 1 1 0 0 1 0 0 0 0 0 0 1 0 0 0 0 1 1 0 0 1 0 0 0 1 1 0 0 1 0 0 0 1 1 0 0 0 1 0 0 1 1 1 0
0 0 0 0 1 1 1 0 1 0 0 0 0 1 1 0 0 0 0 0 1 1 0 0 1 0 0 0 1 1 0 0 0 0 0 1 0 1 0 1 1 0 1 0 0 0 0 0
0 1 0 0 1 0 0 0 1 0 1 0 1 1 0 0 1 0 0 0 1 1 0 0 0 0 0 0 1 1 0 0 1 0 1 0 1 1 1 0 1 0 1 0 1 1 1 0
1 0 0 0 1 1 0 0 1 0 0 0 1 1 0 0 0 1 0 0 1 1 1 0 1 0 1 0 0 0 0 0 1 0 0 1;
D= 1 1 1 2 2 2 3 3 3 4 4 4;
ENDDATA
!23.7;
MAX= @SUM(Students(j):@SUM(Groups(k):Alpha(j,k)));
!23.2;
@FOR(Students(j):@FOR(Activities(i)|S(j,i)#EQ#0: @SUM(Groups(k)|A
(k)#EQ#i:Alpha(j,k))=0));
!23.3;
@FOR(Students(j):@FOR(Schedules(h):@FOR(Groups(k)|S(j,A(k))#EQ#1:
@FOR(Groups(kk)|S(j,A(kk))#EQ#1 #AND# kk#GT#k #AND# P(k,h)#EQ#1 #AND#
P(kk,h)#EQ#1 :
 Alpha(j,k)+Alpha(j,kk)<=1))));
!23.4;
@FOR(Students(j):@FOR(Activities(i)|S(j,i)#EQ#1:@SUM(Groups(k)|A
(k)#EQ#i:Alpha(j,k))<=1));
!23.5;
@FOR(Students(j):@FOR(Activities(i)|S(j,i)#EQ#1:@FOR(Days(t):@SUM
(Schedules(h)|D(h)#EQ#t:@SUM(Groups(k)|A(k)#EQ#i#AND#P(i,h)#EQ#1:
Alpha(j,k)))<=2)));
!23.6;
@FOR(Groups(k):@SUM(Students(j)|S(j,A(k))#EQ#1:Alpha(j,k))<=M
(k));
!23.1;
@FOR(Groups(k):@FOR(Students(j):@BIN(Alpha(j,k))));
END
```

Solution

The solution is shown in Fig. 5.9.

5.7 Problem 24. Assignment of Subjects to Classrooms

Given a set of university degrees made up of 150 subjects that are to be taught in a school of future construction. To simplify the size of the model, we consider a single day of subject teaching. Each subject has a starting and ending time instant and a number of students.

The classrooms to be built in the new school will be of two possible sizes, large and small, (Q_1, Q_2). The size represents the number of students that can fit into the classroom. *The Q1* capacity is assumed to be sufficient for any subject.

The construction of classrooms carries a cost proportional to their capacity: C_1 for large classrooms, C_2 for small classrooms, $C_1 > C_2$.

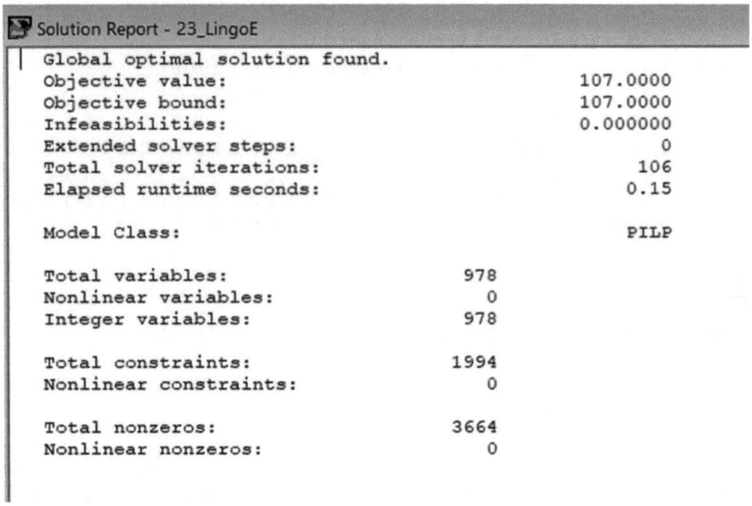

```
Solution Report - 23_LingoE
Global optimal solution found.
Objective value:                              107.0000
Objective bound:                              107.0000
Infeasibilities:                              0.000000
Extended solver steps:                               0
Total solver iterations:                           106
Elapsed runtime seconds:                          0.15

Model Class:                                      PILP

Total variables:                    978
Nonlinear variables:                  0
Integer variables:                  978

Total constraints:                 1994
Nonlinear constraints:                0

Total nonzeros:                    3664
Nonlinear nonzeros:                   0
```

Fig. 5.9 Problem 23: LINGO solution report

The system must decide the classrooms to be built so that we can assign all the subjects. In the assignment, two subjects cannot be given at the same time in the same classroom.

Modeling the problem of calculating the minimum cost of classroom construction.

Modeling

Both the Q_1 and Q_2 capacity classrooms are identical items referred to individually in the system, so we need to calculate an upper bound for each. Since subjects have a starting and ending time, the highest degree of simultaneity of subjects with a number of students less than or equal to Q_2 in the case of small classrooms (HDS_2), and the highest degree of simultaneity of all subjects for the case of large classrooms (HDS_1), we take as the number of classrooms, $m = HDS_1 + HDS_2$.

List of elements in Table 5.7.

Table 5.7 Problem 24: table of elements

			Data				
Elements	Set	QN	Name	Parameter	Type	Belonging	Value
Classrooms	$i = 1\ldots m$	I_U	Capacity	Q_i	I	W	...
			Cost	C_i	C($)	W	...
Subjects	$j, k = 1\ldots 150$	I_U	Starting time	st_j	C(s)	W	...
			Ending time	et_i	C(s)	W	...
			Students[a]	n_i	I	W	...
			Overlap[b]	O_{jk}	B	S	...

[a]Students do not need to make them explicit as collective elements

[b]We define the overlap as calculated data from the starting and ending times of each subject according to the following expression:

$O_{jk} = 1$ *if* $(st_j \leq st_k$ AND $et_j > st_k)$ *OR* $(st_k \leq st_j$ AND $et_k > st_j)$; 0 otherwise

Decision Activities
Assign subjects (*i*) to classrooms (*j*)

$$\forall i,j/Q_i \geq n_j : \alpha_{ij} = 1 \text{ if subject } i \text{ is assigned to classroom } j \tag{24.1}$$

We define only the assignments that can take place. If we define them all, then we would have to impose a capacity specification for the assignment subject-classroom. In this way, it is not necessary.

Specifications

– Quantitative selection specification:

Each subject needs a classroom:

Activity	Elements selecting	Selectable elements	Selection	Constraints
α_{ij}	Subjects ($j = 1 \ldots n$)	Classrooms ($i = 1 \ldots m$)	=1	$\forall j : \sum_{i/Q_i \geq n_j} \alpha_{ij} = 1$ (24.2)

– Subjects overlapping in time cannot go in the same classroom (propositional logic specification):

$$\begin{aligned}&\forall i,j,k/j < k\&O_{jk} = 1\&Q_i \geq n_j\&Q_i \geq n_k : \text{NOT}\big(\alpha_{ij} = 1 \text{ AND } \alpha_{ik} = 1\big)\\ &\Rightarrow f_{37}, f_{11} \Rightarrow \forall i,j,k/j < k\&O_{jk} = 1\&Q_i \geq n_j\&Q_i \geq n_k :\\ &\alpha_{ij} + \alpha_{ik} \leq 1\end{aligned} \tag{24.3}$$

Objective Criterion
Minimize the cost of classrooms built (= classrooms used)
Binary Logical Calculation

$$\forall i : \beta_i = 1 \text{ if classroom } i \text{ has been used; 0 otherwise.} \tag{24.4}$$

$$\begin{aligned}&\forall i : \beta_i = 1 \text{ IF AND ONLY IF } \sum_{j/Q_i \geq n_j} \alpha_{ij} > 0\\ &\Rightarrow f_{S_V} \Rightarrow \forall i : \text{ IF } \sum_{j/Q_i \geq n_j} \alpha_{ij} > 0 \text{ THEN } \beta_i = 1\\ &\Rightarrow f_3, f_7 \Rightarrow \forall i : \text{ IF } 1 - \beta_i = 1 \text{ THEN } \sum_{j/Q_i \geq n_j} \alpha_{ij} \leq 0\end{aligned}$$

$$\Rightarrow f_{14}\left(\text{UB} \sum_{j/Q_i \geq n_j} \alpha_{ij} = 150\right) \Rightarrow \forall i : \sum_{j/Q_i \geq n_j} \alpha_{ij} \leq 150\beta_i \tag{24.5}$$

Objective Function

$$\text{Min} \sum_{i=1}^{m} C_i \beta_i \tag{24.6}$$

Finally, for the solution to be structured, we are going to impose that the classrooms are activated in order of their ordinal, for each of the sizes (bounds imposition):

$$\forall i / i < \text{HDS}_1 : \beta_i \geq \beta_{i+1} \tag{24.7}$$

$$\forall i / i > \text{HDS}_1 \& i < \text{HDS}_2 : \beta_i \geq \beta_{i+1} \tag{24.8}$$

Complete Model

$$\text{Min} \sum_{i=1}^{m} C_i \beta_i \tag{24.6}$$

s.t.

$$\forall j : \sum_{i/Q_i \geq n_j} \alpha_{ij} = 1 \tag{24.2}$$

$$\forall i, j, k / j < k \& O_{jk} = 1 \& Q_i \geq n_j \& Q_i \geq n_k : \alpha_{ij} + \alpha_{ik} \leq 1 \tag{24.3}$$

$$\forall i : \sum_{j/Q_i \geq n_j} \alpha_{ij} \leq 150 \beta_i \tag{24.5}$$

$$\forall i / i < \text{HDS}_1 : \beta_i \geq \beta_{i+1} \tag{24.7}$$

$$\forall i / i > \text{HDS}_1 \& i < \text{HDS}_2 : \beta_i \geq \beta_{i+1} \tag{24.8}$$

$$\forall i, j / Q_i \geq n_j : \alpha_{ij} \text{ binaries} \tag{24.1}$$

$$\forall i : \beta_i \text{ binaries} \tag{24.4}$$

Implementation in LINGO

We store the information in the database "24_DB.accdb."

```
MODEL:
DATA:
 HDS1 =@ODBC('24_ODBC','System','HDS1');
 HDS2 =@ODBC('24_ODBC','System','HDS2');
ENDDATA
CALC:
 m =HDS1+HDS2;
ENDCALC
SETS:
 Classrooms/1..m/:Q,C,Beta ;
 Subjects/1..150/:n;
 CS(Classrooms,Subjects):Alpha;
 Overlaps(Subjects,Subbects) /@ODBC
```

```
('24_ODBC','Overlaps','A1','A2')/;
ENDSETS
DATA:
 Q =@ODBC('24_ODBC','Classrooms','Capacity');
 C =@ODBC('24_ODBC','Classrooms','Cost');
 n=@ODBC('24_ODBC','Subjects','students');
 @ODBC('24_ODBC','Subjects','Assigned Classroom')= @WRITEFOR
(Subjects(j):@SUM(Classrooms(i):i*Alpha(i,j)));
ENDDATA
!24.6;
MIN=@SUM(Classrooms(i):C(i)*Beta(i));
!24.2;
@FOR(Subjects(j):@SUM(Classrooms(i)|Q(i)#GE#n(j):Alpha(i,j))=1);
!24.3;
@FOR(Classrooms(i):@FOR(Overlaps(j,k)|Q(i)#GE#n(j)#AND#Q(i)#GE#n
(k):Alpha(i,j)+Alpha(i,k)<=1));
!24.5;
@FOR(Classrooms(i):@SUM(Subjects(j)|Q(i)#GE#n(j):Alpha(i,j))
<=150*Beta(i));
!24.7;
@FOR(Classrooms(i)|i#LT#HDS1:Beta(i)>=Beta(i+1));
!24.8;
@FOR(Classrooms(i)|i#GT#HDS1#AND#i#LT#HDS2:Beta(i)>=Beta(i+1));
!24.1;
@FOR(Classrooms(i):@FOR(Subjects(j)|Q(i)#GE#n(j):@BIN(Alpha(i,
j))));
!24.4;
@FOR(Classrooms(i):@BIN(Beta(i)));
END
```

Solution

It is shown in Figs. 5.10 and 5.11.

5.8 Problem 25. Disciplines Management

Let there be a group of *n* people with whom we are going to create *m* working groups. Each group will work in two disciplines to be decided. Each of the 10 existing disciplines must be worked on by at least one group. Each group will be made up of a minimum and a maximum of people.

How each person handles each discipline is known by a continuous index. If a person has a handling index lower than 0.5, he cannot go in a group where that discipline is worked on.

A person can go in a maximum of three groups, but they cannot be groups that share the same discipline.

The two disciplines of each group must be compatible (we have information on the compatibility between pairs of disciplines).

```
Solution Report - 24_LingoE
Global optimal solution found.
Objective value:                                    60.00000
Objective bound:                                    60.00000
Infeasibilities:                                    0.000000
Extended solver steps:                                     0
Total solver iterations:                               42288
Elapsed runtime seconds:                                5.77


Export Summary Report
---------------------
Transfer Method:        ODBC BASED
ODBC Data Source:       24_ODBC
Data Table Name:        Subjects
Columns Specified:             1
   Assigned Classroom
LINGO Column Length:         150
Database Column Length:      150

Model Class:                                            MILP

Total variables:                    4530
Nonlinear variables:                   0
Integer variables:                  3880

Total constraints:                  1089
Nonlinear constraints:                 0

Total nonzeros:                     9576
Nonlinear nonzeros:                    0
```

Fig. 5.10 Problem 24: LINGO solution report

Todos los obj...
Buscar...
Tablas
Classrooms
Overlaps
Subjects
System

System | Subjects

Id	students	Assigned classroom
1	20	11
2	10	1
3	15	12
4	15	12
5	25	12
6	50	1
7	60	12
8	60	12
9	80	12
10	90	11

Fig. 5.11 Problem 24: Database

The objective in the formation of groups is to maximize the sum of the management of the disciplines of all the groups (the management of the disciplines of a group is obtained by adding the handling index of the people of the group with the disciplines assigned to that group).

Modeling
List of elements in Table 5.8.

Table 5.8 Problem 25: Table of elements

Elements	Set	QN	Data				
			Name	Parameter	Type	Belonging	Value
People	$i = 1 \ldots n$	I_U	Handling index	IN_{ik}	C	S	...
Work group	$j = 1 \ldots m$	I_U	Minimum	mi_j	I	W	...
			Maximum	Ma_j	I	W	...
Disciplines	$k, k' = 1 \ldots 10$	I_U	Compatibility	$C_{kk'}$	B	S	...
				IN_{ik}			

Decision Activities
Assign people (*i*) to work groups (*j*)

$$\forall i, j : \alpha_{ij} = 1 \text{ if person } i \text{ is assigned to work group } j; 0 \text{ otherwise} \tag{25.1}$$

Assign disciplines (k) to work groups (*j*)

$$\forall k, j : \beta_{kj} = 1 \text{ if discipline } k \text{ is assigned to work group } j; \quad 0 \text{ otherwise} \tag{25.2}$$

Specifications

– Quantitative selection specification:

Activity	Elements selecting	Selectable elements	Selection	Constraints
α_{ij}	Work groups ($j = 1 \ldots m$)	People ($i = 1 \ldots n$)	$\leq mi_j$	$\forall j : \sum_{i=1}^{n} \alpha_{ij} \geq mi_j$ (25.3)
			$\geq Ma_j$	$\forall j : \sum_{i=1}^{n} \alpha_{ij} \leq Ma_j$(25.4)
	People ($i = 1 \ldots n$)	Work groups ($j = 1 \ldots m$)	≤ 3	$\forall i : \sum_{j=1}^{m} \alpha_{ij} \leq 3$ (25.5)
β_{kj}	Disciplines ($k = 1 \ldots 10$)	Work groups ($j = 1 \ldots m$)	≥ 1	$\forall k : \sum_{j=1}^{m} \beta_{kj} \geq 1$ (25.6)
	Work groups ($j = 1 \ldots m$)	Disciplines ($k = 1 \ldots 10$)	$=2$	$\forall j : \sum_{k=1}^{10} \beta_{kj} = 2$ (25.7)

– Compatibility of disciplines (propositional logic specification):

$$\forall j,k,k'/k < k'\&C_{kk'} = 0 : \text{NOT}\left(\beta_{kj} = 1 \text{ AND } \beta_{k'j} = 1\right)$$
$$\Rightarrow f_{37}, f_{11} \Rightarrow \forall j,k,k'/k < k'\&C_{kk'} = 0 : \beta_{kj} + \beta_{k'j} \leq 1 \quad (25.8)$$

– If a person has a handling index lower than 0.5 in a discipline, they cannot go into a group where that discipline is worked (propositional logic specification):

$$\forall j,i,k/\text{IN}_{ik} < 0.5 : \text{NOT}\left(\alpha_{ij} = 1 \text{ AND } \beta_{kj} = 1\right)$$
$$\Rightarrow f_{37}, f_{11} \Rightarrow \forall j,i,k/\text{IN}_{ik} < 0.5 : \alpha_{ij} + \beta_{kj} \leq 1 \quad (25.9)$$

It could also have been expressed as:

$$\forall j,i,k : \text{IF } \text{IN}_{ik} < 1 \text{ AND } \beta_{kj} = 1 \text{ THEN } \alpha_{ij} = 0$$

– A person cannot go in groups that share the same discipline (propositional logic specification):

$$\forall i,j,j'/j < j',k : \text{NOT}\left(\alpha_{ij} = 1 \text{ AND } \alpha_{ij'} = 1 \text{ AND } \beta_{kj} = 1 \text{ AND } \beta_{kj'} = 1\right)$$
$$\Rightarrow f_{37}, f_{11} \Rightarrow \forall i,j,j'/j < j',k : \alpha_{ij} + \alpha_{ij'} + \beta_{kj} + \beta_{kj'} \leq 3 \quad (25.10)$$

Objective Criterion
Maximize the management of the disciplines of each group.
We calculate the handling of disciplines of each Group j:
Binary Logical Calculation
We need a logical calculation to know if a person i and a discipline k are part of a group j.

$$\forall i,j,k : \delta_{ijk} = 1 \text{ if person } i \text{ and discipline } k \text{ are part of the work group } j; 0 \text{ otherwise.} \quad (25.11)$$

$$\forall i,j,k : \delta_{ijk} = 1 \text{ IF AND ONLY IF } \alpha_{ij} = 1 \text{ AND } \beta_{kj} = 1$$
$$\Rightarrow f_{37} \Rightarrow \forall i,j,k : \delta_{ijk} = 1 \text{ IF AND ONLY IF } \alpha_{ij} + \beta_{kj} \geq 2$$
$$\Rightarrow f_{S_V} \Rightarrow \forall i,j,k : \text{IF } \delta_{ijk} = 1 \text{ THEN } \alpha_{ij} + \beta_{kj} \geq 2$$
$$\Rightarrow f_{15}\left(\text{LB}_{\alpha_{ij} + \beta_{kj}} = 0\right) \Rightarrow \forall i,j,k : \alpha_{ij} + \beta_{kj} \geq 2\delta_{ijk} \quad (25.12)$$

Auxiliary Calculation

$\forall j : z_j =$ management of the disciplines of the group j

$$\forall j : z_j = \sum_{i=1}^{n} \sum_{k=1}^{10} IN_{ik}\delta_{ijk} \tag{25.13}$$

Objective Function

$$\text{Max} \sum_{j=1}^{m} z_j \tag{25.14}$$

Complete Model

$$\text{Max} \sum_{j=1}^{m} z_j \tag{25.14}$$

$$\text{s.t.}$$

$$\forall j : \sum_{i=1}^{n} \alpha_{ij} \geq mi_j \tag{25.3}$$

$$\forall j : \sum_{i=1}^{n} \alpha_{ij} \leq Ma_j \tag{25.4}$$

$$\forall i : \sum_{j=1}^{m} \alpha_{ij} \leq 3 \tag{25.5}$$

$$\forall k : \sum_{j=1}^{m} \beta_{kj} \geq 1 \tag{25.6}$$

$$\forall j : \sum_{k=1}^{10} \beta_{kj} = 2 \tag{25.7}$$

$$\forall j, k, k'/k < k' \& C_{kk'} = 0 : \beta_{kj} + \beta_{k'j} \leq 1 \tag{25.8}$$

$$\forall j, i, k/\text{IN}_{ik} < 0.5 : \alpha_{ij} + \beta_{kj} \leq 1 \tag{25.9}$$

$$\forall i, j, j'/j < j', k : \alpha_{ij} + \alpha_{ij'} + \beta_{kj} + \beta_{kj'} \leq 3 \tag{25.10}$$

$$\forall i, j, k : \alpha_{ij} + \beta_{kj} \geq 2\delta_{ijk} \tag{25.12}$$

$$\forall j : z_j = \sum_{i=1}^{n} \sum_{k=1}^{10} \text{IN}_{ik}\delta_{ijk} \tag{25.13}$$

$$\forall i,j : \alpha_{ij} \text{ binaries} \tag{25.1}$$

$$\forall j,k : \beta_{jk} \text{ binaries} \tag{25.2}$$

$$\forall i,j,k : \delta_{ijk} \text{ binaries} \tag{25.11}$$

Implementation in LINGO

Data are collected in File "25_Text.txt" (Fig. 5.12).

```
MODEL:
SETS:
People/1..@FILE('C:\Files_\25_Text.txt')/:;
Groups/1..@FILE('C:\Files_\25_Text.txt')/:mi,Ma,z;
Disciplines/1..10/;
PG(People,Groups):Alpha;
PD(People,Disciplines):IN;
DD(Disciplines,Disciplines):C;
DG(Disciplines,Groups):Beta;
PGD(People,Groups,Disciplines):Delta;
ENDSETS
DATA:
 mi=@FILE('C:\Files_\25_Text.txt');
 Ma=@FILE('C:\Files_\25_Text.txt');
 IN =@FILE('C:\Files_\25_Text.txt');
 C =@FILE('C:\Files_\25_Text.txt');
ENDDATA
!25.14;
MAX = @SUM(Groups(j):z(j));
!25.3;
@FOR(Groups(j):@SUM(People(i):Alpha(i,j))>=mi(j));

!25.4;
@FOR(Groups(j):@SUM(People(i):Alpha(i,j))<=Ma(j));
!25.5;
@FOR(People(i):@SUM(Groups(j):Alpha(i,j))<=3);
!25.6;
@FOR(Disciplines(k):@SUM(Groups(j):Beta(k,j))>=1);
!25.7;
@FOR(Groups(j):@SUM(Disciplines(k):Beta(k,j))=2);
!25.8;
@FOR(Groups(j):@FOR(Disciplines(k):@FOR(Disciplines(kk)|k#LT#kk
#AND# C(k,kk)#EQ#0:Beta(k,j)+Beta(kk,j)<=1)));
!25.9;
@FOR(Groups(j):@FOR(People(i):@FOR(Disciplines(k)|IN(i,k)#LT#0.5:
Alpha(i,j)+Beta(k,j)<=1)));
!25.10;
@FOR(People(i):@FOR(Groups(j):@FOR(Groups(jj)|j#LT#jj:@FOR
(Disciplines(k): Alpha(i,j)+Alpha(i,jj)+Beta(k,j)+Beta(k,jj)
<=3))));
!25.12;
@FOR(People(i):@FOR(Groups(j):@FOR(Disciplines(k):Alpha(i,j)+Beta
(k,j)>=2*Delta(i,j,k))));
!25.13;
@FOR(Groups(j):z(j)=@SUM(People(i):@SUM(Disciplines(k):IN(i,k)
```

```
25_Text.txt - Notepad
File Edit Format View Help
!n;
20 ~
!m;
5 ~
!mi;
2 2 5 3 5 ~
!MA;
7 8 7 6 10 ~
!IN;
1 0.1 0.3 0.5 0.7 0.8 0.4 0.9 1 1 0 1 0 1 0 0 0.4 0.1 0.1 0.8
0 0 1 0 1 0.6 0.4 1 0 0 1 0.1 0.3 0.5 0.7 0.8 0.4 0.9 1 1
1 0 1 0 1 0 1 0 0 0 1 0 0 1 0 1 0 0 0 1
0 0 1 0 0 0 1 0 0 0 1 0.9 1 1 1  1 1 1 0 0
0 0.7 1 0 0.7 0 0 0 1 0 1 0 0 1 0 0 0 0 0 1
1 0 0 1 0 1 0 0 1 0 0 1 0 1 0 1 0 0 0 0
0.7 0 1 0 1 0.3 0 1 0 0 0 1 0 1 0 1 0 0 1 1
0 0.7 0 0 1 0 1 0 0 0  0 0 0 0 0 1 0 1 0 1
 1 1 0 0 0 0 0 0 0 0 0.6 0.6 1 0 0 1 0 1 0 0
0 0 0 0.7 0 0 1 0 1 0  0 0 0.7 0 0 0 1 0 0 1
~
!C;
1 1 0 0 0 1 1 0 1 1
1 1 0 1 1 0 1 0 0 1
0 0 1 0 1 0 1 0 1 1
0 1 0 1 1 1 1 0 1 1
0 0 0 0 1 0 0 0 0 0
1 0 1 0 0 1 0 1 0 1
1 1 0 1 0 1 1 0 1 1
0 0 1 0 0 0 0 1 0 0
1 0 0 1 0 1 1 0 0 1
1 0 1 0 0 0 1 0 0 1  ~
```

Fig. 5.12 "25_text.txt"

```
*Delta(i,j,k))));
!25.1;
@FOR(People(i):@FOR(Groups(j):@BIN(Alpha(i,j))));
!25.2;
@FOR(Disciplines(k):@FOR(Groups(j):@BIN(Beta(k,j))));
!25.11;
@FOR(People(i):@FOR(Groups(j):@FOR(Disciplines(k):@BIN(Delta(i,j,
k)))));
END
```

Solution

It is shown in Fig. 5.13.

5.9 Problem 26. Installation of Agencies

Given a region represented as an undirected graph *G* (*N*, *A*). The nodes represent populations, and the edges represent the connection between each two populations. Each edge has a distance data associated with it. Each town has a number of inhabitants.

```
Solution Report - 25_LingoE
Global optimal solution found.
Objective value:                                    41.10000
Objective bound:                                    41.10000
Infeasibilities:                                    0.000000
Extended solver steps:                                  5143
Total solver iterations:                             3467715
Elapsed runtime seconds:                              438.43

Model Class:                                            MILP

Total variables:                      1155
Nonlinear variables:                     0
Integer variables:                    1150

Total constraints:                    3756
Nonlinear constraints:                   0

Total nonzeros:                      13280
Nonlinear nonzeros:                      0
```

Fig. 5.13 Problem 25: LINGO solution report

The public administration wants to install agencies in these towns that provide coverage to the inhabitants of all towns in the region. Each population must be assigned to a population with an installed agency. The maximum number of inhabitants assigned to an agency is M inhabitants (a value greater than the number of inhabitants of any municipality). Up to two agencies can be installed in a population, but that population must have at least R inhabitants.

The cost of setting up an agency in each town i is C_i. There is a spending cap of T $\$$.

Minimize the total distance that all inhabitants of the region will travel (if a population is assigned to itself, it is assumed that its inhabitants do not travel any distance, $d_{ii} = 0$).

Modeling

It is not necessary to consider the inhabitants as an element because there are no decisions or specifications about them.

List of elements in Table 5.9.

Table 5.9 Problem 26: table of elements

			Data				
Elements	Set	QN	Name	Parameter	Type	Belonging	Value
Populations (nodes)	$i = 1 \ldots n$	I_U	Inhabitants	H_i	I	W	...
			Agency installation cost	C_i	C ($)	S	
Edges	(i, j)	I_U	Distance	d_{ij}	C (Km)	W	...
Agencies	–	C_I		C_i			
System	–	I_U	Max no inhabitants	M	I	W	...
			Min inhabitants two agencies	R	I	W	...
			Spending cap	T	C ($)	W	...

Decision Activities
Install agencies in populations (*i*)

$$\forall i : x_i = \text{Number of agencies installed in population } i \tag{26.1}$$

Assign population to population [with agency]

$$\forall i,j : \alpha_{ij} = 1 \text{ if we assign population } i \text{ to population } j \tag{26.2}$$

[with agency]: It is a condition to establish in the specification section.
Specifications

- Quantitative selection specification:

Activity	Elements selecting	Selectable elements	Selection	Constraints
α_{ij}	Populations ($i = 1\ldots n$)	Populations ($j = 1\ldots n$)	=1	$\forall i : \sum_{j=1}^{n} \alpha_{ij} = 1$ (26.3)

- Assign populations to populations with agency (propositional logic specification).

$$\begin{aligned} &\forall i,j : \text{IF } \alpha_{ij} = 1 \text{ THEN } x_j > 0 \\ &\Rightarrow f_5 \Rightarrow \forall i,j : \text{IF } \alpha_{ij} = 1 \text{ THEN } x_j \geq 1 \\ &\Rightarrow f_{15}(\text{LB}_{x_j} = 0) \Rightarrow \forall i,j : x_j \geq \alpha_{ij} \end{aligned} \tag{26.4}$$

- Maximum of two agencies per population (upper bound imposition on measurable activity)

$$\forall i : \; x_i \leq 2 \tag{26.5}$$

- If two agencies are installed, there are at least R inhabitants (propositional logic specification).

$$\begin{aligned} &\forall i : \text{IF } x_i = 2 \text{ THEN } H_i \geq R \\ &\Rightarrow f_3 \Rightarrow \forall i : \text{IF } \; H_i < R \text{ THEN } x_i \leq 1 \\ &\Rightarrow f_{45} \Rightarrow \forall i / H_i < R : \; x_i \leq 1 \end{aligned} \tag{26.6}$$

- Maximum inhabitants per agency (capacity specification)

We must apply the specification to each population. It is a capacity specification where both consumption and capacity are variable.

$$\forall i : \sum_{k=1}^{n} H_k \alpha_{ki} \leq M x_i \tag{26.7}$$

This specification makes it unnecessary to consider specification (26.4).

– Spending cap (capacity specification)

$$\sum_{i=1}^{n} C_i x_i \leq T \tag{26.8}$$

Objective Function

Minimize the total distance that all inhabitants of the region will travel.

$$\text{Min } \sum_{i=1}^{n} \sum_{j=1}^{n} d_{ij} H_i \alpha_{ij} \tag{26.9}$$

Complete Model

$$\text{Min } \sum_{i=1}^{n} \sum_{j=1}^{n} d_{ij} H_i \alpha_{ij} \tag{26.9}$$

s.t.

$$\forall i : \sum_{j=1}^{n} \alpha_{ij} = 1 \tag{26.3}$$

$$\forall i : \ x_i \leq 2 \tag{26.5}$$

$$\forall i / H_i < R : \ x_i \leq 1 \tag{26.6}$$

$$\forall i : \sum_{k=1}^{n} H_k \alpha_{ki} \leq M x_i \tag{26.7}$$

$$\sum_{i=1}^{n} C_i x_i \leq T \tag{26.8}$$

$$\forall i : x_i \geq 0 \text{ integers} \tag{26.1}$$

$$\forall i, j : \alpha_{ij} \text{ binaries} \tag{26.2}$$

Implementation in LINGO

We collect the information from the Excel sheet “26_Excel.xlsx.”

```
MODEL:
DATA:
 n=@OLE('C:\Files_\26_Excel.xlsx','n');
ENDDATA
```

```
SETS:
 Populations/1..n/:H,C,x;
 Edges(Populations,Populations):D,Alpha;
ENDSETS
DATA:
 H=@OLE('C:\Files_\26_Excel.xlsx','H');
 C=@OLE('C:\Files_\26_Excel.xlsx','Costs');
 D=@OLE('C:\Files_\26_Excel.xlsx','Distances');
 M=@OLE('C:\Files_\26_Excel.xlsx','M');
 R=@OLE('C:\Files_\26_Excel.xlsx','Min');
 T=@OLE('C:\Files_\26_Excel.xlsx','T');
ENDDATA
CALC:
@FOR(Populations(i):@FOR(Populations(j)|i#LT#j: D(j,i)=D(i,j)));
ENDCALC
!26.9;
MIN = @SUM(Populations(i):@SUM(Populations(j):d(i,j)*H(i)*Alpha(i,
j)));
!26.3;
@FOR(Populations(i):@SUM(Populations(j): Alpha(i,j))=1);
!26.5;
@FOR(Populations(i): x(i)<=2);
!26.6;
@FOR(Populations(i)|H(i)#LT#R: x(i)<=1);
!26.7;
@FOR(Populations(i):@SUM(Populations(k):H(k)*Alpha(k,i))<=M*x
(i));

!26.8;
@SUM(Populations(i):C(i)*x(i))<=T;

!26.1;
@FOR(Populations(i):@GIN(x(i)));
!26.2; @FOR(Populations(i):@FOR(Populations(j):@BIN(Alpha(i,j))));
END
```

```
Solution Report - 26_LingoE
Global optimal solution found.
Objective value:                              251351.0
Objective bound:                              251351.0
Infeasibilities:                              0.000000
Extended solver steps:                               0
Total solver iterations:                           209
Elapsed runtime seconds:                          0.25

Model Class:                                      PILP

Total variables:                   650
Nonlinear variables:                 0
Integer variables:                 650

Total constraints:                  97
Nonlinear constraints:               0

Total nonzeros:                   1945
Nonlinear nonzeros:                  0
```

Fig. 5.14 Problem 26: LINGO solution report

Solution
The solution is shown in Fig. 5.14.

References

Castro B, Diez H, Urrutia AM (2011) Ejercicios resueltos de investigación o p e r a t i v a. Servicio Editorial de la Universidad del País Vasco, Bilbao. isbn:978-84-694-5889-1
García JM (2021) Modelling and types of specifications. In: Modelling in mathematical programming, Springer. Cham

Chapter 6
Medium-High Level Problems

6.1 Introduction

The problems in this chapter present more practical and realistic scenarios. One of the problems is a classic in optimization, the Steiner problem. Its inclusion is based on wanting to provide an explanation to different versions of its modelling. A particular problem also appears within the field of scheduling, and a problem for warehouse management focused on the collection of products, such as the IKEA purchasing system.

The characteristics of the problems, taking into account the methodology (Garcia 2021), are shown below.

Input files with a lot of data are not displayed due to their size. They are available for download.

Problems:

Number	Problem
27	Production of pieces
28	Grouping of people
29	Task Scheduling with precedences and setup times
30	Steiner problem
31	Management of pharmacies on guard duty
32	Project task planning
33	Product warehouse management
34	Assignment of engineers to projects
35	Excursion planning
36	Iron bars production
37	Discount management

J. M. García Sánchez, *Building and Solving Mathematical Programming Models*,
International Series in Operations Research & Management Science 329,
https://doi.org/10.1007/978-3-030-97626-2_6

Versions and elements:

Problem	Versions	Elements				
		Collective elements	Measurable individual elements	Grouping of items	Upper bound in the number of elements	Data handling
27	1	✓				Medium
28	1					High
29	2					High
30	2					High
31	1					High
32	1		✓		✓	High
33	1	✓				High
34	1	✓				High
35	2					High
36	1	✓				Medium
37	1	✓				Medium

Calculations:

Problem	Binary logical calculations	Not-binary logical calculations	Bounds calculations	Conditioned bounds calculations
27	✓	✓		
28	✓			
29	✓			
30	✓			
31			✓	
32	✓			
33	✓			
34	✓			
35	✓			
36	✓			
37	✓			

Specifications:

Problem	QSS[a]	Capacity	Demand	Bound imposition	Allocation/assignment
27			✓	✓	
28	✓				
29	✓				
30	✓			✓	
31	✓			✓	
32	✓			✓	
33	✓	✓			
34	✓			✓	
35	✓			✓	

(continued)

Problem	QSS[a]	Capacity	Demand	Bound imposition	Allocation/assignment
36		✓	✓		
37			✓		

[a]*QSS* quantitative selection specifications

Problem	Flow balance	PLS[a]—One operator	PLS—several operators	PLS—Attributes	Variable costs cases
27					
28					
29			✓		
30		✓	✓		
31		✓	✓		
32		✓	✓		
33		✓			
34		✓			
35		✓			
36		✓			
37					

[a]*PLS* propositional logic specifications

Objective criterion, input data and output data:

	Objective criterion		Input data				
Problem	Calculations	Particular cases	Integrated data	Text file	Spreadsheet	Database	Output data
27	✓		✓				
28	✓					✓	
29	✓				✓		
30					✓		✓
31	✓			✓	✓		
32	✓					✓	
33	✓					✓	✓
34					✓		
35	✓				✓		
36	✓	✓				✓	✓
37	✓	✓	✓				

6.2 Problem 27. Production of Pieces

Let us assume a piece production system. For the production of pieces, we have six machines and an undetermined number of operators. Each machine i has a production rate R_i units/hour.

Assignment of an operator is mandatory in the case of machine 4, if the machine is used. For the rest of the machines, the assignment of operators is not mandatory, but in the case of assigning an operator the rate of production of parts increases by 20%. In no case can more than one operator be assigned.

The cost of an operator in each machine i is C_i \$, $i = 1\ldots6$.

Assign the production of 10,000 pieces, proposing as an objective to minimize costs, bearing in mind that if the production time in a machine is greater than 8 h, there is an extra cost of F\$.

Modeling

List of elements in Table 6.1.

Table 6.1 Problem 27: table of elements

Elements	Set	QN	Data				
			Name	Parameter	Type	Belonging	Value
Machines	$i = 1\ldots6$	I_U	Normal production rate	R_i	C(uds./h)	S	...
			Rapid production rate	RR_i	C(uds./h)	S	
			Operator cost	C_i	C(\$)	S	...
Pieces	–	C_D	Quantity	Q	I	W	10,000
				R_i, RR_i			
Operators	–	C_I		C_i			
System	–	I_U	Extra cost	F	C(\$)	W	

The rapid rate (RR_i) has been calculated as:

$RR_i = R_i*1,20\ i = 1, 2, 3, 5, 6$
$RR_4 = R_4$

Decision Activities

Produce pieces in machines (*i*)

$$\forall i : x_i = \text{Number of pieces produced in machine } i. \tag{27.1}$$

Assign operators to machines (*i*)

$$\forall i/i \neq 4 : y_i = \text{Number of operators assigned to machines } i. \tag{27.2}$$

Specifications

– An operator at most per machine (upper bound imposition):

$$\forall i/i \neq 4 : y_i \leq 1 \tag{27.3}$$

We can define y_i as binary and eliminate this constraint in the final model. We can assume that y_i is a binary variable in the propositional logic specifications in which it participates.

– Pieces demand (demand specification):

$$\sum_{i=1}^{6} x_i = 10{,}000 \tag{27.4}$$

Objective Criterion
Minimize Costs

$$\text{Costs} = \text{Operators Assignment Costs} + \text{Machine 4 Cost} + \text{Extraordinary Cost}$$

$$\text{Operators Assignment Costs} = \sum_{i/i \neq 4} C_i y_i$$

Machine 4 Cost:
It is necessary to calculate whether machine 4 has been used.
Binary Logical Calculation

$$\begin{aligned}
y_4 &= 1 \text{ if machine 4 has been used.} \\
y_4 &= 1 \text{ IF AND ONLY IF } x_4 > 0 \\
\Rightarrow f_{S_V} &\Rightarrow \text{IF } x_4 > 0 \text{ THEN } y_4 = 1 \\
\Rightarrow f_5 &\Rightarrow \text{IF } x_4 \geq 1 \text{ THEN } y_4 = 1 \\
\Rightarrow f_3, f_7 &\Rightarrow \text{IF } 1 - y_4 = 1 \text{ THEN } x_4 \leq 0 \\
&\Rightarrow f_{14}(\text{UB}_{x_4} = Q) \Rightarrow x_4 \leq Q y_4 \qquad (27.5) \\
&\text{Cost} = C_4 y_4
\end{aligned}$$

Extraordinary Cost:
To calculate whether the extra cost has occurred, we must first calculate the time that each machine has used in the production process, both in the normal and in rapid options. If any of these times exceed 8 h, then we will have an extraordinary cost.
No Binary Logical Calculation

$$\forall i : tn_i = \text{Calculation of the production time of each machine } i \text{ at normal speed}$$

$$\Rightarrow \text{IF } y_i = 0 \text{ THEN } tn_i = \frac{x_i}{R_i};\ \text{ELSE } tn_i = 0$$

$$\Rightarrow f_{S_V} \Rightarrow \text{IF } y_i = 0 \text{ THEN } tn_i = \frac{x_i}{\text{RN}_i}$$

We can assume that y_i for all i as a binary variable.

$$\Rightarrow f_7 \Rightarrow \text{IF } 1 - y_i = 1 \text{ THEN } tn_i = \frac{x_i}{R_i}$$

$$\Rightarrow f_{\text{UB}} \Rightarrow \text{IF } 1 - y_i = 1 \text{ THEN } tn_i \geq \frac{x_i}{R_i}$$

$$\Rightarrow \text{IF } 1 - y_i = 1 \text{ THEN } tn_i - \frac{x_i}{R_i} \geq 0$$

$$\Rightarrow f_{15}\left(\text{LB}_{tn_i - \frac{x_i}{R_i}} = -\frac{Q}{R_i}\right) \Rightarrow \quad tn_i - \frac{x_i}{R_i} \geq -\frac{Q}{R_i} y_i$$

$$\Rightarrow \quad tn_i \geq \frac{x_i}{\text{RN}_i} - \frac{Q}{\text{RN}_i} y_i \tag{27.6}$$

$\forall i : tr_i =$ Calculation of the production time of each machine i at a fast pace

$$\Rightarrow \text{IF } y_i = 1 \text{ THEN } tr_i = \frac{x_i}{\text{RR}_i}; \text{ ELSE } tr_i = 0$$

$$\Rightarrow f_{S_V} \Rightarrow \text{IF } y_i = 1 \text{ THEN } tr_i = \frac{x_i}{\text{RR}_i}$$

$$\Rightarrow f_{\text{UB}} \Rightarrow \text{IF } y_i = 1 \text{ THEN } tr_i \geq \frac{x_i}{\text{RR}_i}$$

$$\Rightarrow \text{IF } y_i = 1 \text{ THEN } tr_i - \frac{x_i}{\text{RR}_i} \geq 0$$

$$\Rightarrow f_{15}\left(\text{LB}_{tr_i - \frac{x_i}{RR_i}} = -\frac{Q}{\text{RR}_i}\right) \Rightarrow \quad tr_i - \frac{x_i}{\text{RR}_i} \geq -\frac{Q}{\text{RR}_i}(1 - y_i)$$

$$\Rightarrow \quad tr_i \geq \frac{x_i}{\text{RR}_i} - \frac{Q}{\text{RR}_i}(1 - y_i) \tag{27.7}$$

In case any of these times exceed 8 h, we incur the extraordinary cost. We must calculate if they exceed 8 h. We can group the normal and fast times of each machine, since both cannot occur at the same time:

Binary Logical Calculation

$$\forall i : \alpha_i = 1 \text{ machine } i \text{ time exceeds 8 h; 0 otherwise.} \tag{27.8}$$

$$\forall i : \alpha_i = 1 \text{ IF AND ONLY IF } tn_i + tr_i > 8$$

$$\Rightarrow f_{S_V} \Rightarrow \forall i : \text{ IF } tn_i + tr_i > 8 \text{ THEN } \alpha_i = 1$$

$$\Rightarrow f_3, f_7 \Rightarrow \forall i : \text{ IF } 1 - \alpha_i = 1 \text{ THEN } tn_i + tr_i \leq 8$$

$$\Rightarrow f_{14}\left(\text{UB}_{tn_i + tr_i} = \frac{Q}{R_i}\right) \Rightarrow \forall i : \quad tn_i + tr_i \leq 8 + \left(\frac{Q}{R_i} - 8\right)\alpha_i \tag{27.9}$$

Objective Function

$$\text{Min } \sum_{i=1}^{6} (C_i y_i + F\alpha_i) \tag{27.10}$$

Complete Model

$$\text{Min } \sum_{i=1}^{6} (C_i y_i + F\alpha_i) \tag{27.10}$$

s.t.

$$\sum_{i=1}^{6} x_i = 10{,}000 \tag{27.4}$$

$$x_4 \leq Q y_4 \tag{27.5}$$

$$\forall i: \; tn_i \geq \frac{x_i}{R_i} - \frac{Q}{R_i} y_i \tag{27.6}$$

$$\forall i: \; tr_i \geq \frac{x_i}{\text{RR}_i} - \frac{Q}{\text{RR}_i}(1 - y_i) \tag{27.7}$$

$$\forall i: \; tn_i + tr_i \leq 8 + \left(\frac{Q}{R_i} - 8\right)\alpha_i \tag{27.9}$$

$$\forall i: \; tn_i \geq 0; tr_i \geq 0$$

$$\forall i: \alpha_i \text{ binaries} \tag{27.8}$$

$$\forall i: x_i \text{ integers} \tag{27.1}$$

$$\forall i: y_i \text{ binaries} \tag{27.2, 27.3}$$

Implementation in LINGO

```
MODEL:
SETS:
 Machines/1..6/:R,RR,C,x,y,Alpha,tn,tr;
ENDSETS
DATA:
 R= 500 300 190 160 100 90;
 RR= 600 360 228 160 120 108;
 C= 150 100 130 120 100 100;
 Q=10000;
 F=1500;
ENDDATA

!27.10; MIN= @SUM(Machines(i):C(i)*y(i)+F*Alpha(i));
!27.4; @SUM(Machines(i):x(i))=10000;
!27.5; x(4)<=Q*y(4);
```

```
Solution Report - 27_LingoE
Global optimal solution found.
Objective value:                              120.0000
Objective bound:                              120.0000
Infeasibilities:                              0.000000
Extended solver steps:                               0
Total solver iterations:                           136
Elapsed runtime seconds:                          0.04

Model Class:                                      MILP

Total variables:                    30
Nonlinear variables:                 0
Integer variables:                  18

Total constraints:                  21
Nonlinear constraints:               0

Total nonzeros:                     74
Nonlinear nonzeros:                  0
```

Fig. 6.1 Problem 27: LINGO solution report

```
!27.6; @FOR(Machines(i):tn(i)>=(x(i)/R(i))-(Q/R(i))*y(i));
!27.7; @FOR(Machines(i):tr(i)>=(x(i)/RR(i))-(Q/RR(i))*(1-y(i)));
!27.9; @FOR(Machines(i):tn(i)+tr(i)<=8+((Q/R(i))-8)*Alpha(i));
!27.8; @FOR(Machines(i):@BIn(Alpha(i)));
!27.1; @FOR(Machines(i):@GIN(x(i)));
!27.2;!27.3; @FOR(Machines(i):@BIN(y(i)));
END
```

Solution

The solution is shown in Fig. 6.1.

6.3 Problem 28. Grouping of People

Let there be a group of n people with whom m work groups will be created. Each group will consist of a fixed number of people. Some people know each other. The goal of allocating people to groups is to maximize the number of people who have all the people they know in their group.

Modeling

List of elements in Table 6.2.

Table 6.2 Problem 28: table of elements

			Data				
Elements	Set	QN	Name	Parameter	Type	Belonging	Value
People	$j, k = 1 \ldots n$	I_U	Known	C_{jk}	B	S	...
Work groups	$i = 1 \ldots m$	I_U	Number of people	P_i	I	W	...

Decision Activities
Assign people(j) to groups(i)

$$\forall i : \alpha_{ij} = 1 \text{ if person } j \text{ is assigned to group } i; 0 \text{ otherwise.} \tag{28.1}$$

Specifications

– Quantitative selection specifications:

Activity	Elements selecting	Selectable elements	Selection	Constraints
α_{ij}	People ($j = 1 \ldots n$)	Groups ($i = 1 \ldots m$)	$=1$	$\forall j : \sum_{i=1}^{m} \alpha_{ij} = 1$ (28.2)
	Groups ($i = 1 \ldots m$)	People ($j = 1 \ldots n$)	$=P_i$	$\forall i : \sum_{j=1}^{n} \alpha_{ij} = P_i$ (28.3)

Objective Criterion
Maximize the number of people having all the people they know in their group.

It is necessary to collect in a logical calculation whether or not two known people are in the same group.

Binary Logical Calculation

$$\begin{aligned} &\forall j,k/j < k\&C_{jk} = 1 : \beta_{jk} = 1 \text{ IF AND ONLY IF} \\ &\left(\alpha_{1j} = 1 \text{ AND } \alpha_{1k} = 1\right) \text{ OR } \left(\alpha_{2j} = 1 \text{ AND } \alpha_{2k} = 1\right) \\ &\text{OR} \ldots \text{OR } \left(\alpha_{mj} = 1 \text{ AND } \alpha_{mk} = 1\right) \end{aligned} \tag{28.4}$$

$$\Rightarrow f_{S_V} \Rightarrow \forall j,k/j < k\&C_{jk} = 1 :$$
$$IF\left(\alpha_{1j} = 1 \text{ AND } \alpha_{1k} = 1\right) \text{ OR } \ldots \text{OR } \left(\alpha_{mj} = 1 \text{ AND } \alpha_{mk} = 1\right) \text{ THEN } \beta_{jk} = 1$$

$$\begin{aligned} &\Rightarrow f_{37} \Rightarrow \forall j,k/j < k\&C_{jk} = 1 : \text{IF}\left(\alpha_{1j} + \alpha_{1k} \geq 2\right) \\ &\text{OR } \ldots \text{OR } \left(\alpha_{mj} + \alpha_{mk} \geq 2\right) \text{ THEN } \beta_{jk} = 1 \end{aligned}$$

$$\begin{aligned} &\Rightarrow \forall j,k/j < k\&C_{jk} = 1 : \omega_{1jk} = 1 \text{ IF AND ONLY IF } \left(\alpha_{1j} + \alpha_{1k} \geq 2\right) \\ &\Rightarrow \ldots \\ &\Rightarrow \forall j,k/j < k\&C_{jk} = 1 : \omega_{mjk} = 1 \text{ IF AND ONLY IF } \left(\alpha_{mj} + \alpha_{mk} \geq 2\right) \end{aligned} \tag{28.5}$$

$$\begin{aligned} &\Rightarrow f_{S_v} \Rightarrow \forall j,k/j < k\&C_{jk} = 1 : \text{IF } \omega_{1jk} = 1 \text{ THEN}\left(\alpha_{1j} + \alpha_{1k} \geq 2\right) \\ &\Rightarrow \ldots \\ &\Rightarrow f_{S_v} \Rightarrow \forall j,k/j < k\&C_{jk} = 1 : \text{IF } \omega_{mjk} = 1 \text{ THEN } \left(\alpha_{mj} + \alpha_{mk} \geq 2\right) \end{aligned}$$

$$
\begin{aligned}
&\Rightarrow f_{15}\left(\mathrm{LB}_{\alpha_{1j}+\ \alpha_{1k}}=0\right) \Rightarrow \forall j,k/j<k\&C_{jk}=1:\alpha_{1j}+\ \ \alpha_{1k}\geq 2\omega_{1jk}\\
&\Rightarrow \ldots \\
&\Rightarrow f_{15}\left(\mathrm{LB}_{\alpha_{mj}+\ \alpha_{mk}}=0\right) \Rightarrow \forall j,k/j<k\&C_{jk}=1:\alpha_{mj}+\ \ \alpha_{mk}\geq 2\omega_{mjk}
\end{aligned} \tag{28.6}
$$

$$
\begin{aligned}
&\Rightarrow \forall j,k/j<k\&C_{jk}=1:\text{IF }\ \omega_{1jk}=1\ \text{ OR }\ \ldots\text{OR }\ \omega_{mjk}=1\ \text{ THEN }\ \beta_{jk}=1\\
&\Rightarrow f_{34}\Rightarrow \forall j,k/j<k\&C_{jk}=1:\text{IF }\ \omega_{1jk}+\ldots+\ \omega_{mjk}\geq 1\ \text{ THEN }\ \beta_{jk}=1\\
&\Rightarrow f_3,f_7\Rightarrow \forall j,k/j<k\&C_{jk}=1:\text{IF }\ 1-\beta_{jk}=1\ \text{ THEN }\ \omega_{1jk}+\ldots+\ \omega_{mjk}\leq 0\\
&\Rightarrow f_{14}\left(\mathrm{UB}_{\omega_{1jk}+\ldots+\ \omega_{mjk}}=1\right)\Rightarrow \forall j,k/j<k\&C_{jk}=1:\omega_{1jk}+\ldots+\ \omega_{mjk}\\
&\qquad\qquad\qquad\qquad\qquad \leq \beta_{jk}
\end{aligned} \tag{28.7}
$$

Now we impose for each person a logical calculation to know if all his acquaintances are in the same group as him.

Binary Logical Calculation

$$
\forall j:\delta_j=1\ \ \text{if all the acquaintances of } j \text{ are in his group; 0 otherwise} \tag{28.8}
$$

$$
\begin{aligned}
&\forall j:\delta_j=1\ \text{ IF AND ONLY IF }\ \sum_{k/j<k\wedge C_{jk}=1}\beta_{jk}+\sum_{k/j>k\wedge C_{kj}=1}\beta_{kj}\\
&=\sum_{k/j<k\wedge C_{jk}=1}C_{jk}+\sum_{k/j>k\wedge C_{kj}=1}C_{kj}\\
&\Rightarrow f_{\mathrm{UB}}\Rightarrow \forall j:\delta_j\\
&=1\ \text{ IF AND ONLY IF }\ \sum_{k/j<k\wedge C_{jk}=1}\beta_{jk}\\
&+\sum_{k/j>k\wedge C_{kj}=1}\beta_{kj}\geq \sum_{k/j<k\wedge C_{jk}=1}C_{jk}+\sum_{k/j>k\wedge C_{kj}=1}C_{kj}\\
&\Rightarrow f_{S_V}\Rightarrow \forall j:\text{IF }\ \delta_j=1\ \text{ THEN }\ \sum_{k/j<k\wedge C_{jk}=1}\beta_{jk}+\sum_{k/j>k\wedge C_{kj}=1}\beta_{kj}\\
&\geq \sum_{k/j<k\wedge C_{jk}=1}C_{jk}+\sum_{k/j>k\wedge C_{kj}=1}C_{kj}\\
&\Rightarrow f_{15}\left(\mathrm{LB}_{\sum_{k/j<k\wedge C_{jk}=1}\beta_{jk}+\sum_{k/j>k\wedge C_{kj}=1}\beta_{kj}}=0\right)\Rightarrow\\
&\forall j:\ \sum_{k/j<k\wedge C_{jk}=1}\beta_{jk}+\sum_{k/j>k\wedge C_{kj}=1}\beta_{kj}\geq\left(\sum_{k/j<k\wedge C_{jk}=1}C_{jk}+\sum_{k/j>k\wedge C_{kj}=1}C_{kj}\right)\delta_j
\end{aligned} \tag{28.9}
$$

Objective Function

$$\text{Max} \sum_{j=1}^{n} \delta_j \tag{28.10}$$

Complete Model

$$\text{Max} \sum_{j=1}^{n} \delta_j \tag{28.10}$$

s.t.

$$\forall j : \sum_{i=1}^{m} \alpha_{ij} = 1 \tag{28.2}$$

$$\forall i : \sum_{j=1}^{n} \alpha_{ij} = P_i \tag{28.3}$$

$$\forall i, j, k/j < k \& C_{jk} = 1 : \alpha_{ij} + \ \alpha_{ik} \geq 2\omega_{ijk} \tag{28.6}$$

$$\forall j, k/j < k \& C_{jk} = 1 : \omega_{1jk} + \ldots + \ \omega_{mjk} \leq \beta_{jk} \tag{28.7}$$

$$\forall j : \quad \sum_{k/j<k \wedge C_{jk}=1} \beta_{jk} + \sum_{k/k>j \wedge C_{kj}=1} \beta_{kj} \geq \left(\sum_{k/j<k \wedge C_{jk}=1} C_{jk} + \sum_{k/j>k \wedge C_{kj}=1} C_{kj} \right) \delta_j \tag{28.9}$$

$$\forall i, j : \alpha_{ij} \ \text{binaries} \tag{28.1}$$

$$\forall j, k/j < k \& C_{jk} = 1 : \beta_{jk} \ \text{binaries} \tag{28.4}$$

$$\forall i, j, k/j < k \& C_{jk} = 1 : \omega_{ijk} \ \text{binaries} \tag{28.5}$$

$$\forall j : \delta_j \ \text{binaries} \tag{28.8}$$

Implementation in LINGO

```
MODEL:
DATA:
 n=@ODBC('28_ODBC','System','N');
 m=@ODBC('28_ODBC','System','M');
ENDDATA
SETS:
 People/1..n/: C,Delta;
 Groups/1..m/: P;
 PP(People,People)/ @ODBC('28_ODBC','Known','Person1','Person2')/:
Beta;
 PG(Groups,People): Alpha;
 GPP(Groups,People,People):Omega;
ENDSETS
```

```
DATA:
 P =@ODBC('28_ODBC','Groups','People');
ENDDATA
!28.10;
MAX = @SUM(Groups(i):Delta(i));
!28.2;
@FOR(People(j): @SUM(Groups(i): Alpha(i,j))=1);
!28.3;
@FOR(Groups(i): @SUM(People(j): Alpha(i,j))=P(i));
!28.6;
@FOR(Groups(i):@FOR(PP(j,k):Alpha(i,j)+Alpha(i,k)>=2*Omega(i,j,
k)));

!28.7;
@FOR(PP(j,k):@SUM(Groups(i):Omega(i,j,k))<=Beta(j,k));
!28.9;
@FOR(People(j):@SUM(PP(j,k):Beta(j,k))+ @SUM(PP(k,j):Beta(k,j))>=
(@SUM(PP(j,k):1)+@SUM(PP(k,j):1))*Delta(j));
!28.1;
@FOR(Groups(i):@FOR(People(j): @BIN(Alpha(i,j))));
!28.4;
@FOR(PP(j,k):@BIN(Beta(j,k)));
!28.5;
@FOR(Groups(i):@FOR(PP(j,k):@BIN(Omega(i,j,k))));
!28.8;
@FOR(People(j):@BIN(Delta(j)));
END
```

Solution

It is shown in Fig. 6.2.

6.4 Problem 29. Task Scheduling with Precedence and Setup Times

Let a set of *n* production tasks be. Each task has a process time. We are going to schedule the tasks on a production machine. For this, *n* positions are established in the processing sequence, so that each task must be assigned to a position. In addition:

- There are conditional precedents between tasks: Task *i* must be processed after *j* if task *t* has been processed before *i*. This is collected in the binary attribute A_{ijt}.
- If a task *i* is in position *k* and a task *j is* in position *k* + 1, an additional machine setup time is added, s_{ij}.

The problem aims to minimize the total production time.

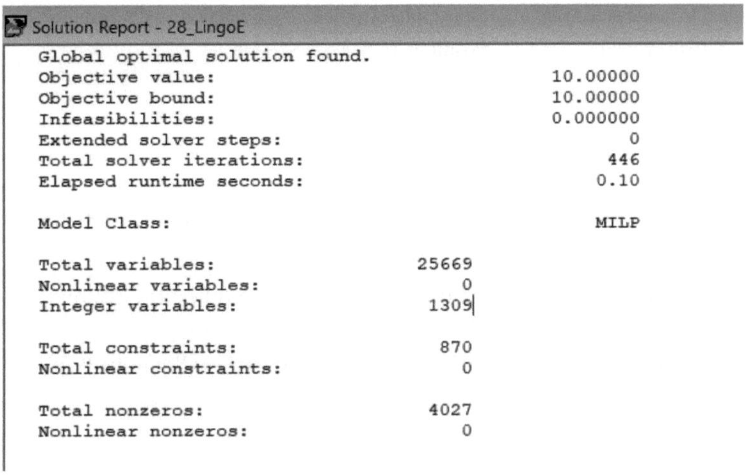

```
Solution Report - 28_LingoE
Global optimal solution found.
Objective value:                              10.00000
Objective bound:                              10.00000
Infeasibilities:                              0.000000
Extended solver steps:                               0
Total solver iterations:                           446
Elapsed runtime seconds:                          0.10

Model Class:                                      MILP

Total variables:                  25669
Nonlinear variables:                  0
Integer variables:                 1309

Total constraints:                  870
Nonlinear constraints:                0

Total nonzeros:                    4027
Nonlinear nonzeros:                   0
```

Fig. 6.2 Problem 28: LINGO solution report

List of elements in Table 6.3.

Table 6.3 Problem 29: Table of elements

			Data				
Elements	Set	QN	Name	Parameter	Type	Belonging	Value
Tasks	$i, j, t = 1 \ldots n$	I_U	Processing time	tp_i	C(s)	W	...
			Precedence	A_{ijt}	B	S	...
			Setup time	s_{ij}	C(s)	S	...
Positions	$k = 1 \ldots n$	I_U					
Machine	–	I_U					

Decision Activities

Assign tasks (*i*) to positions (k).

$$\forall i, k : \alpha_{ik} = 1 \text{ if task } i \text{ is assigned to position } k; 0 \text{ otherwise.} \qquad (29.1)$$

Specifications

– Quantitative selection specifications:

A position for each task.
A task for each position.

Activity	Elements selecting	Selectable elements	Selection	Constraints
α_{ik}	Tasks ($i = 1 \ldots n$)	Positions ($k = 1 \ldots n$)	=1	$\forall i : \sum_{k=1}^{n} \alpha_{ik} = 1$ (29.2)
	Positions ($k = 1 \ldots n$)	Tasks ($i = 1 \ldots n$)	=1	$\forall k : \sum_{i=1}^{n} \alpha_{ik} = 1$ (29.3)

– Specification associated with conditional precedence (propositional logic specification):

The propositional statement of this specification could be the following:
If $A_{ijt} = 1$ then if task t is processed before i, j is processed after i:
Mathematical formulation: We are going to propose two versions to formulate the specification.

Version 1

$$\forall i,j/j \neq i, t/t \neq i \& t \neq j, k_1, k_2, k_3 : \text{IF } Aijt = 1 \text{ THEN}$$
$$\left(\text{IF } \alpha_{tk_1} = 1 \text{ AND } \alpha_{ik_2} = 1 \text{ AND } \alpha_{jk3} = 1 \text{ AND } k_1 < k_2 \text{ THEN } k_3 > k_2\right)$$
$$\Rightarrow f_{45} \Rightarrow \forall i,j/j \neq i, t/t \neq i \& t \neq j / A_{ijt} = 1, k_1, k_2, k_3 :$$
$$\text{IF } \alpha_{tk_1} = 1 \text{ AND } \alpha_{ik_2} = 1 \text{ AND } \alpha_{jk3} = 1 \text{ AND } k_1 < k_2 \text{ THEN } k_3 > k_2$$
$$\Rightarrow f_{45} \Rightarrow \forall i,j/j \neq i, t/t \neq i \& t \neq j / A_{ijt} = 1, k_1, k_2/k_1 < k_2, k_3 :$$
$$\text{IF } \alpha_{tk_1} = 1 \text{ AND } \alpha_{ik_2} = 1 \text{ AND } \alpha_{jk3} = 1 \text{ THEN } k_3 > k_2$$
$$\Rightarrow f_{37} \Rightarrow \forall i,j/j \neq i, t/t \neq i \& t \neq j / A_{ijt} = 1, k_1, k_2/k_1 < k_2, k_3 :$$
$$\text{IF } \alpha_{tk_1} + \alpha_{ik_2} + \alpha_{jk3} \geq 3 \text{ THEN } k_3 > k_2$$
$$\Rightarrow f_3, f_{11} \Rightarrow \forall i,j/j \neq i, t/t \neq i \& t \neq j / A_{ijt} = 1, k_1, k_2/k_1 < k_2, k_3 :$$
$$\text{IF } k_3 \leq k_2 \text{ THEN } \alpha_{tk_1} + \alpha_{ik_2} + \alpha_{jk3} \leq 2$$

$$\Rightarrow f_{45} \Rightarrow \forall i,j/j \neq i, t/t \neq i \& t \neq j / A_{ijt} = 1, k_1, k_2/k_1 < k_2, k_3/k_3 \leq k_2 : \\ \alpha_{tk_1} + \alpha_{ik_2} + \alpha_{jk3} \leq 2 \tag{29.4}$$

Version 2

We are going to perform auxiliary calculations to collect the ordinal position of the trios of tasks with conditional precedence:

$$\forall i : x_i = \text{Position of task } i. \tag{29.5}$$

$$\forall i : x_i = \sum_{k=1}^{n} k\alpha_{ik} \tag{29.6}$$

$$\forall i,j/j \neq i, t/t \neq i \& t \neq j : \text{IF } A_{ijt} = 1 \text{ THEN}\left(\text{IF } x_t < x_i \text{ THEN } x_j > x_i\right)$$
$$\Rightarrow f_{45} \Rightarrow \forall i,j/j \neq i, t/t \neq i \& t \neq j / A_{ijt} = 1 : \text{IF } x_t < x_i \text{ THEN } x_j > x_i$$

$$\Rightarrow f_4, f_5 \Rightarrow \forall i,j/j \neq i, t/t \neq i\&t \neq j/A_{ijt} = 1 : IF \;\; x_t \leq x_i - 1 \;\; \text{THEN} \;\; x_j \geq x_i + 1$$
$$\Rightarrow \forall i,j/j \neq i, t/t \neq i\&t \neq j/A_{ijt} = 1 : IF \;\; x_t - x_i \leq -1 \;\; \text{THEN} \;\; x_j - x_i \geq 1$$
$$\Rightarrow f_{17}(\text{LB}_{x_t - x_i} = -(n-1))$$
$$\Rightarrow \forall i,j/j \neq i, t/t \neq i\&t \neq j/A_{ijt} = 1 : x_t - x_i \geq \;\; -(n-1)\omega_{ijt}$$
$$\Rightarrow f_{21}\left(\text{LB}_{x_j - x_i} = -(n-1)\right)$$
$$\Rightarrow \forall i,j/j \neq i, t/t \neq i\&t \neq j/A_{ijt} = 1 : \;\; x_j - x_i \geq \omega_{ijt} - (n-1)\left(1 - \omega_{ijt}\right)$$
$$\Rightarrow \forall i,j/j \neq i, t/t \neq i\&t \neq j/A_{ijt} = 1 : x_t - x_i \geq \;\; -(n-1)\omega_{ijt} \quad (29.7)$$
$$\Rightarrow \forall i,j/j \neq i, t/t \neq i\&t \neq j/A_{ijt} = 1 : \;\; x_j - x_i \geq -(n-1) + n\omega_{ijt} \quad (29.8)$$

Objective Criterion
Minimize the total production time.

$$\text{Total production time} = \text{process times} + \text{setup times}.$$

$$\text{Process times are constant} = \sum_{i=1}^{n} tp_i$$

To collect the setup times to be allocated, it is necessary to perform a logical calculation to know the next task for each task in each position, except the last one that has no next task.
Binary Logical Calculation

$$\forall i,j/i \neq j, k < n : \beta_{ijk}$$
$$= 1 \;\; \text{if task } i \text{ is scheduled at position } k \text{ and task } j \text{ at position } k+1. \quad (29.9)$$

$$\forall i,j/i \neq j, k < n : \beta_{ijk} = 1 \;\; \text{IF AND ONLY IF} \;\; \alpha_{ik} = 1 \;\; \text{AND} \;\; \alpha_{jk+1} = 1$$
$$\Rightarrow f_{S_V} \Rightarrow \forall i,j/i \neq j, k < n : \text{IF} \;\; \alpha_{ik} = 1 \;\; Y \;\; \alpha_{jk+1} = 1 \;\; \text{THEN} \;\; \beta_{ijk} = 1$$
$$\Rightarrow f_{37} \Rightarrow \forall i,j/i \neq j, k < n : \text{IF} \;\; \alpha_{ik} + \;\; \alpha_{jk+1} \geq 2 \;\; \text{THEN} \;\; \beta_{ijk} = 1$$
$$\Rightarrow f_3, f_7, f_{11} \Rightarrow \forall i,j/i \neq j, k < n : \text{IF} \;\; 1 - \beta_{ijk} = 1 \;\; \text{THEN} \;\; \alpha_{ik} + \;\; \alpha_{jk+1} \leq 1$$
$$\Rightarrow f_{14} \Rightarrow \forall i,j/i \neq j, k < n : \alpha_{ik} + \;\; \alpha_{jk+1} \leq 1 + \beta_{ijk} \quad (29.10)$$

Objective Function

$$\text{Min} \quad \sum_{i=1}^{n} \sum_{j \neq i} \sum_{k=1}^{n-1} s_{ij}\beta_{ijk} \quad (29.11)$$

Complete Model
Version 1

$$\text{Min} \quad \sum_{i=1}^{n} \sum_{j \neq i} \sum_{k=1}^{n-1} s_{ij}\beta_{ijk} \tag{29.11}$$

s.t.

$$\forall i : \sum_{k=1}^{n} \alpha_{ik} = 1 \tag{29.2}$$

$$\forall k : \sum_{i=1}^{n} \alpha_{ik} = 1 \tag{29.3}$$

$$\begin{aligned} &\forall i,j/j \neq i, t/t \neq i \& t \neq j/A_{ijt} = 1, k_1, k_2/k_1 < k_2, k_3/k_3 \leq k_2 : \\ &\alpha_{tk_1} + \ \alpha_{ik_2} + \ \alpha_{jk3} \leq 2 \end{aligned} \tag{29.4}$$

$$\forall i,j/i \neq j, k < n : \alpha_{ik} + \ \alpha_{jk+1} \leq 1 + \beta_{ijk} \tag{29.10}$$

$$\forall i,k : \alpha_{ik} \ \text{binaries} \tag{29.1}$$

$$\forall i,j/i \neq j, k < n : \beta_{ijk} \ \text{binaries} \tag{29.9}$$

Version 2

$$\text{Min} \quad \sum_{i=1}^{n} \sum_{j \neq i} \sum_{k=1}^{n-1} s_{ij}\beta_{ijk} \tag{29.11}$$

s.t.

$$\forall i : \sum_{k=1}^{n} \alpha_{ik} = 1 \tag{29.2}$$

$$\forall k : \sum_{i=1}^{n} \alpha_{ik} = 1 \tag{29.3}$$

$$\forall i : x_i = \sum_{k=1}^{n} k\alpha_{ik} \tag{29.6}$$

$$\forall i,j/j \neq i, t/t \neq i \& t \neq j/A_{ijt} = 1 : x_t - x_i \geq \ -(n-1)\omega_{ijt} \tag{29.7}$$

$$\forall i,j/j \neq i, t/t \neq i \& t \neq j/A_{ijt} = 1 : \ x_j - x_i \geq -(n-1) + n\omega_{ijt} \tag{29.8}$$

$$\forall i,j/i \neq j, k < n : \alpha_{ik} + \ \alpha_{jk+1} \leq 1 + \beta_{ijk} \tag{29.10}$$

$$\forall i,k : \alpha_{ik} \ \text{binaries} \tag{29.1}$$

$$\forall i : x_i \ \text{integers} \tag{29.5}$$

	A	B	C	D	E	F	G	H	I	J	K	L
1												
2	N	10										
3												
4		1	2	3	4	5	6	7	8	9	10	
5	Times	12	14	25	10	13	18	7	9	11	18	
6												
7	Precedences	1	2	3								
8		4	6	3								
9		3	10	8								
10		8	7	1								
11		10	5	8								
12												
13			1	2	3	4	5	6	7	8	9	10
14	Setup times	1	0	1	2	3	4	5	6	7	8	9
15		2	1	0	2	2	2	2	1	1	0	9
16		3	1	1	0	2	3	4	5	6	7	8
17		4	2	2	3	0	1	2	2	13	12	11
18		5	4	1	10	20	0	2	1	1	2	2
19		6	4	15	6	3	0	0	1	10	1	20
20		7	1	2	2	2	0	0	0	1	1	0
21		8	1	3	3	3	0	0	0	0	1	2
22		9	9	1	1	1	4	3	3	4	0	3
23		10	8	2	4	4	4	4	4	4	7	0
24												
25												
26												

Fig. 6.3 "29_Excel.xlsx"

$$\forall i,j/i \neq j, k < n : \beta_{ijk} \text{ binaries} \tag{29.9}$$

$$\forall i,j/j \neq i, t/t \neq i\&t \neq j/A_{ijt} = 1 : \quad \omega_{ijt} \text{ binaries} \tag{29.12}$$

Implementation in LINGO

We implement the first version. The input data (29_Excel.xlsx) are shown in Fig. 6.3.

```
MODEL:
DATA:
 N=@OLE('C:\Files_\29_Excel.xlsx','N');
ENDDATA
SETS:
 Tasks/1..N/:tp,x;
 Positions/1..N/;
 TaPo(Tasks,Positions):Alpha;
 TT(Tasks,Tasks):s;
A(Tasks,Tasks,Tasks)/@OLE('C:\Files_\29_Excel.xlsx','Pre')/;
 TTP(Tasks,Tasks,Positions):Beta;
ENDSETS
DATA:
 tp=@OLE('C:\Files_\29_Excel.xlsx','Times');
 s=@OLE('C:\Files_\29_Excel.xlsx','Setup');
ENDDATA
```

```
Solution Report - 29_LingoE
Global optimal solution found.
Objective value:                                    9.000000
Objective bound:                                    9.000000
Infeasibilities:                                    0.000000
Extended solver steps:                                 37416
Total solver iterations:                             1137485
Elapsed runtime seconds:                               67.61

Model Class:                                            MILP

Total variables:                   1110
Nonlinear variables:                  0
Integer variables:                  910

Total constraints:                 2481
Nonlinear constraints:                0

Total nonzeros:                    8318
Nonlinear nonzeros:                   0
```

Fig. 6.4 Problem 29 (version 1): LINGO solution report

```
!29.11; MIN = @SUM(Tasks(i):@SUM(Tasks(j)|j#NE#i:@SUM(Positions(k)|
k#LT#n:s(i,j)*Beta(i,j,k))));
!29.2; @FOR(Tasks(i):@SUM(Positions(k):Alpha(i,k))=1);
!29.3; @FOR(Positions(k):@SUM(Tasks(i):Alpha(i,k))=1);
!29.4; @FOR(A(i,j,t):@FOR(Positions(k1): @FOR(Positions(k2)|
k1#LT#k2: @FOR(Positions(k3)|k3#LE#k2:
Alpha(t,k1)+Alpha(i,k2)+Alpha(j,k3)<=2))));
!29.10; @FOR(Tasks(i):@FOR(Tasks(j)|j#NE#i:@FOR(Positions(k)|
k#LT#n:Alpha(i,k)+Alpha(j,k+1)<=1+Beta(i,j,k))));
!29.1; @FOR(Tasks(i):@FOR(Positions(k):@BIN(Alpha(i,k))));
!29.9; @FOR(Tasks(i):@FOR(Tasks(j)|j#NE#i:@FOR(Positions(k)|
k#LT#n:@BIN(Beta(i,j,k)))));
END
```

Solution

The solution is shown in Fig. 6.4.

6.5 Problem 30. Steiner Problem

Given an undirected graph $G(N, E)$, we need to obtain a minimum cost network that connects all the terminal nodes of the graph (Hwang et al. 1992).

The nodes of the graph are of two types:

- Terminal: They must be part of the network.
- Steiner: It is not mandatory that they are part of the final network.

Modeling

The versions that we are going to implement are of arborescent type; we change edges by arcs to select a tree from a directed graph. A terminal node must be identified as the root node of the tree.

We create a directed graph $G(N, A)$, where two arcs are generated from each edge of the original undirected graph:

$$\text{Edge } (i,j) \in A \Rightarrow \text{Arc } (i,j) \in A; \text{Arc } (j,i) \in A$$

We also perform a data calculation, the number of incoming arcs (NE_i) of each node. It will serve as the upper bound in the modeling process.

In the tree structure, the main decision is the selection of arcs; the specifications will control that we have a tree structure of those arcs. However, the model also needs to make a decision about how deep each node is in the tree. The depth variables will help us to control a specification necessary to create a tree, the nonexistence of cycles.

Although depth could become a non-binary logical calculation in other tree creation problems, in Steiner's problem, we implement it as a decision activity since there is an interval for the decision of the depth of each node, since we do not know how many nodes will be part of the tree.

Specifications:

Every terminal node, except the root node, must have a parent node.

Each selected Steiner node (which is the parent of some node) must have a parent.

The minimum depth of a node other than the root node is 1.

The maximum depth of a node other than the root node is $n - 1$.

Anti-looping specification:

- MTZ (Miller–Tucker–Zemlin) model (Miller et al. 1960): If you select the arc (i, j), then the depth of i is greater than that of j (at least one more unit).

  Thanks to this specification and the last two previous specifications, the root node will end up taking depth 0 and it is not necessary to consider this anti-looping specification in the arcs in which that node appears. If it is considered, nothing happens, but we increase the number of constraints.
- DL (Desroches–Laporte) model (Desroches and Laporte 1991): If you select arc (i, j) or arc (j, i), then the depth of i plus the selection of the arc (j, i) is equal to the depth of j plus the selection of the arc (i, j). The same happens with the root node as in the case of the MTZ model.

Table of Elements

For this problem, to ease our modeling, we are going to identify the arcs with two ordered indices (i, j), where i and j are nodes (Table 6.4).

Table 6.4 Problem 30: table of elements

Elements	Set	QN	Data				
			Name	Parameter	Type	Belonging	Value
Nodes	$i, j = 1 \ldots n$	I_M	Terminal	T_i	B	W	–
			No. of incoming arcs	NE_i	I	W	–
Arcs	(i, j)	I_U	Cost	C_{ij}	C($)	W	–
Tree	–	I_U	Root	R	O	W	–

Decision Activities

Select arcs (*i*, *j*)

$$\forall (i,j) \in A : \alpha_{ij} = 1 \text{ if we select the arc } (i,j); 0 \text{ otherwise.} \tag{30.1}$$

Assign depth to nodes (*i*)

$$\forall i / i \neq R : x_i = \text{depth of node } i$$

Specifications

– Every terminal node has a parent, except the root node (quantitative selection specification).

Activity	Elements selecting	Selectable elements	Selection	Constraints
α_{ij}	Nodes $\forall i/$ $T_i = 1$ & $R \neq i$	Nodes $j/(i, j) \in A$	=1	$\forall i/T_i = 1 \& R \neq i : \sum_{j/(i,j)\in A} \alpha_{ij} = 1$ (30.2)

– Each selected Steiner node (which is the parent of some node) must have a parent (binary logical calculation + propositional logic specification):

First, we declare a binary logic calculation to know if each Steiner is selected:
Binary Logical Calculation

$$\forall j / T_j = 0 : \beta_j = 1 \text{ IF AND ONLY IF } \sum_{i/(i,j)\in A} \alpha_{ij} > 0 \tag{30.3}$$

$$\Rightarrow f_{S_V} \Rightarrow \forall j / T_j = 0 : \text{IF } \sum_{i/(i,j)\in A} \alpha_{ij} > 0 \text{ THEN } \beta_j = 1$$

$$\Rightarrow f_3, f_7 \Rightarrow \forall j / T_j = 0 : \text{IF } 1 - \beta_j = 1 \text{ THEN } \sum_{i/(i,j)\in A} \alpha_{ij} \leq 0$$

$$\Rightarrow f_{14}\left(\mathrm{UB}_{\sum\limits_{i/(i,j)\in A}\alpha_{ij}}=\mathrm{NE}_j\right)\Rightarrow \forall j/T_j=0:\ \sum_{i/(i,j)\in A}\alpha_{ij}\le \mathrm{NE}_j\beta_j \tag{30.4}$$

Propositional logic specification: Each selected Steiner node must have a parent:

$$\forall i/T_i=0:\text{IF }\ \beta_i=1\ \text{ THEN }\ \sum_{j/(i,j)\in A}\alpha_{ij}=1$$

$$\Rightarrow f_{\mathrm{UB}}\Rightarrow \forall i/T_i=0:\text{IF }\ \beta_i=1\ \text{ THEN }\ \sum_{j/(i,j)\in A}\alpha_{ij}\ge 1$$

$$\Rightarrow f_{15}\Rightarrow \forall i/T_i=0:\ \sum_{j/(i,j)\in A}\alpha_{ij}\ge \beta_i \tag{30.5}$$

– Maximum depth (upper bound imposition):

$$\forall i/i\ne R:\ x_i\le n-1 \tag{30.6}$$

– Minimum depth (lower bound imposition):

$$\forall i/i\ne R:\ x_i\ge 1 \tag{30.7}$$

– Anti-loops specification (propositional logic specification).

MTZ model:

$$\forall(i,j)/i\ne R\&j\ne R:\text{IF }\ \alpha_{ij}=1\ \text{ THEN }\ x_i\ge x_j+1$$
$$\forall(i,j)/i\ne R\&j\ne R:\text{IF }\ \alpha_{ij}=1\ \text{ THEN }\ x_i-x_j\ge 1$$
$$\Rightarrow f_{15}\left(\mathrm{LB}_{x_i-x_j}=-(n-2)\right)\Rightarrow$$
$$\Rightarrow \forall(i,j)/i\ne R\&j\ne R:x_i-x_j\ge \alpha_{ij}-(n-2)\left(1-\alpha_{ij}\right)$$
$$\Rightarrow \forall(i,j)/i\ne R\&j\ne R:x_j-x_i+(n-1)\alpha_{ij}\le (n-2) \tag{30.8}$$

DL model:

A condition is applied to each edge. Since the problem is configured with arcs, we indicate it as:

$$\forall(i,j)/i<j\&i\ne R\&j\ne R:\text{IF }\ \alpha_{ij}+\alpha_{ji}=1\ \text{ THEN }\ x_i+\alpha_{ji}=x_j+\alpha_{ij}$$
$$\forall(i,j)/i<j\&i\ne R\&j\ne R:\text{IF }\ \alpha_{ij}+\alpha_{ji}=1\ \text{ THEN }\ x_i+\alpha_{ji}-x_j-\alpha_{ij}=0$$

We assume α_{ij} + α_{ji}as a binary expression, so we apply f_{16}:

$$\Rightarrow f_{16}\left(\mathrm{UB}_{x_i+\alpha_{ji}-x_j-\alpha_{ij}}=n-2\right)\Rightarrow\forall(i,j)/i<j\&i\neq R\&j\neq R:$$
$$x_i+\alpha_{ji}-x_j-\alpha_{ij}\leq(n-2)\left(1-\alpha_{ij}-\alpha_{ji}\right)$$
$$\Rightarrow\forall(i,j)/i<j\&i\neq R\&j\neq R:$$
$$x_i+\alpha_{ji}-x_j-\alpha_{ij}\leq(n-2)-(n-2)\alpha_{ij}-(n-2)\alpha_{ji}$$
$$\Rightarrow\forall(i,j)/i<j\&i\neq R\&j\neq R:$$
$$x_i-x_j\leq(n-2)-(n-3)\alpha_{ij}-(n-1)\alpha_{ji}$$

$$\begin{gathered}\Rightarrow\forall(i,j)/i<j\&i\neq R\&j\neq R:\\ x_i-x_j+(n-3)\alpha_{ij}+(n-1)\alpha_{ji}\leq(n-2)\end{gathered}\tag{30.9}$$

$$\Rightarrow f_{16}\left(\mathrm{LB}_{x_i+\alpha_{ji}-x_j-\alpha_{ij}}=-(n-2)\right)\Rightarrow\forall(i,j)/i<j\&i\neq R\&j\neq R:$$
$$x_i+\alpha_{ji}-x_j-\alpha_{ij}\geq-(n-2)\left(1-\alpha_{ij}-\alpha_{ji}\right)$$
$$\Rightarrow\forall(i,j)/i<j\&i\neq R\&j\neq R:$$
$$x_i+\alpha_{ji}-x_j-\alpha_{ij}\geq-(n-2)+(n-2)\alpha_{ij}+(n-2)\alpha_{ji}$$
$$\Rightarrow\forall(i,j)/i<j\&R_i=0\&R_j=0:$$
$$x_i-x_j\geq-(n-2)+(n-1)\alpha_{ij}+(n-3)\alpha_{ji}$$

$$\begin{gathered}\Rightarrow\forall(i,j)/i<j\&R_i=0\&R_j=0:\\ x_j-x_i+(n-1)\alpha_{ij}+(n-3)\alpha_{ji}\leq(n-2)\end{gathered}\tag{30.10}$$

Objective Function
Minimize the total cost of selected arcs.

$$\text{Min}\sum_{(i,j)\in A}C_{ij}\alpha_{ij}\tag{30.11}$$

Complete Model

$$\text{Min}\sum_{(i,j)\in A}C_{ij}\alpha_{ij}\tag{30.11}$$

s.t.

$$\forall i/T_i=1\&i\neq R:\quad\sum_{j/(i,j)\in A}\alpha_{ij}=1\tag{30.2}$$

$$\forall j/T_j=0:\ \sum_{i/(i,j)\in A}\alpha_{ij}\leq\mathrm{NE}_j\beta_j\tag{30.4}$$

$$\forall i/T_i=0:\quad\sum_{j/(i,j)\in A}\alpha_{ij}\geq\beta_i\tag{30.5}$$

$$\forall i/i \neq R: \quad x_i \leq n-1 \tag{30.6}$$

$$\forall i/i \neq R: \quad x_i \geq 1 \tag{30.7}$$

Version MTZ:

$$\forall (i,j)/i \neq R\&j \neq R: x_j - x_i + (n-1)\alpha_{ij} \leq (n-2) \tag{30.8}$$

Version DL:

$$\forall (i,j)/i < j\&i \neq R\&j \neq R: \quad x_i - x_j + (n-3)\alpha_{ij} + (n-1)\alpha_{ji} \leq (n-2) \tag{30.9}$$

$$\forall (i,j)/i < j\&i \neq R\&j \neq R: \quad x_j - x_i + (n-1)\alpha_{ij} + (n-3)\alpha_{ji} \leq (n-2) \tag{30.10}$$

$$\forall (i,j) \in A: \alpha_{ij} \text{ binaries} \tag{30.1}$$

$$\forall j/T_j = 0: \beta_j \text{ binaries} \tag{30.3}$$

$$\forall i/i \neq R: x_i \geq 0$$

Implementation in LINGO

```
MODEL:
DATA:
 n=@OLE('c:\Files_\30_Excel.xlsx','N');
ENDDATA
SETS:
Nodes/1..n/:T,IA,x,Beta;
Arcs(Nodes,Nodes)/@OLE('c:\Files_\30_Excel.xlsx','Arcs')/:c,
Alpha;
ENDSETS
DATA:
R=@OLE('c:\Files_\30_Excel.xlsx','Root');
IA=@OLE('c:\Files_\30_Excel.xlsx','IA');
T=@OLE('c:\Files_\30_Excel.xlsx','Terminals');
c = @OLE('c:\Files_\30_Excel.xlsx','Costs');
@OLE('c:\Files_\30_Excel.xlsx','Solution') =@WRITEFOR(Arcs(i,j):
@IF(Alpha(i,j)#EQ#1,'X',' '));
ENDDATA

!30.11; MIN= @SUM(Arcs(i,j):c(i,j)*Alpha(i,j));
!30.2; @FOR(Nodes(i)|i#NE#R#AND#T(i)#EQ#1:@SUM(Arcs(i,j):Alpha(i,
j))=1);
!30.4; @FOR(Nodes(j)|T(j)#EQ#0:@SUM(Arcs(i,j):Alpha(i,j))<=IA(j)
*Beta(j));
!30.5; @FOR(Nodes(i)|T(i)#EQ#0:@SUM(Arcs(i,j):Alpha(i,j))>=Beta
(i));
!30.6; @FOR(Nodes(i)|i#NE#R:x(i)<=n-1);
!30.7; @FOR(Nodes(i)|i#NE#R:x(i)>=1);
!30.8; @FOR(Arcs(i,j)|i#NE#R#AND#j#NE#R:x(j)-x(i)+(n-1)*Alpha(i,j)
<=n-2);
```

```
Solution Report - 30_LingoE
Global optimal solution found.
Objective value:                                   101.0000
Objective bound:                                   101.0000
Infeasibilities:                                   0.000000
Extended solver steps:                                33598
Total solver iterations:                             631074
Elapsed runtime seconds:                              60.75



Export Summary Report
---------------------
Transfer Method:          OLE BASED
Workbook:                 c:\Files_\30_Excel.xlsx
Ranges Specified:                  1
    Solution
Ranges Found:                      1
Range Size Mismatches:             0
Values Transferred:              306

Model Class:                                           MILP

Total variables:                         454
Nonlinear variables:                       0
Integer variables:                       357

Total constraints:                       569
Nonlinear constraints:                     0

Total nonzeros:                         1976
Nonlinear nonzeros:                        0
```

Fig. 6.5 Problem 30: LINGO solution report

```
!30.1; @FOR(Arcs(i,j):@BIN(Alpha(i,j)));
!30.3; @FOR(Nodes(i)|T(i)#EQ#0:@BIN(Beta(i)));
END
```

Solution

The solution is shown in Fig. 6.5.

6.6 Problem 31. Management of Pharmacies on Guard Duty

A given city has 52 pharmacies distributed in 5 districts ($j = 1\ldots5$). We want to do the annual guard assignment. There are two types of guard:

- Daytime: 9: 00-21: 00 (for holidays days)
- Night: 21: 00-9: 00 (for every day)

Pharmacies are of two types:

– Normal: Opening and closing at normal hours
– Pharmacies 12 h: They are open every day of the year until 10 p.m.

Obviously, the full calendar of holidays for the year is known.

The days that are considered for the assignment of daytime guards are all holidays of the year. For night guards, every day of the year is considered.

The assignment specifications are as follows:

Legal regulations require that there be in each district a number of pharmacies on guard duty during the daytime (for holidays) and a number of pharmacies on guard duty at night (every day). In the number of daytime guards, already the 12-h pharmacies in each district are taken into account.

A normal pharmacy cannot be on guard duty for more than 24 consecutive hours.

A pharmacy (normal or 12 h) cannot be on guard duty two consecutive holidays.

The difference in the number of guard services in each normal pharmacy cannot be greater than three.

Balance the total number of guards in each pharmacy each month, imposing that the largest difference does not exceed 2 units.

The objective is to balance the ratio of the number of night guards of a pharmacy divided by the number of pharmacies in its district.

Modeling

List of elements in Table 6.5.

Table 6.5 Problem 31: table of elements

			Data				
Elements	Set	QN	Name	Parameter	Type	Belonging	Value
Pharmacies	$i = 1\ldots52$	I_U	Pharmacy type	TF_i	B {1 = Normal; 0 = 12 h-Pharm.}	W	...
			District	P_i	O	W	...
Districts	$j = 1\ldots5$	I_U	Number of guards	N_{jk}	I	S	...
Guard types	$k = 1, 2$ {daytime, night}	I_U		N_{jk}			
Days	$t = 1\ldots365$	I_U	Holiday	F_t	B	W	...
			Month	M_t	O	W	...
Months	$h = 1\ldots12$	I_U					

Decision Activities

Assign guard type (*k*) to pharmacies (*i*) on days (*t*)

Only the following subset of events should be considered:

$$k = 1 \Rightarrow \forall i/\mathrm{TF}_i = 1 \Rightarrow \forall t/F_t = 1;$$
$$k = 2 \Rightarrow \forall i \Rightarrow \forall t$$

$$(k = 1, \forall i/\text{TF}_i = 1, \forall t/F_t = 1) \& k = 2, \forall i,$$
$$\forall t : \alpha_{ikt} = 1 \text{ } if \text{ guard type } k \text{ is assigned to pharmacy } i \text{ on day } t. \tag{31.1}$$

Specifications

- Quantitative selection specifications:

Activity	Elements selecting			Selectable elements	Selection	Constraints
α_{ikt}	Guard types ($k = 1...2$)	Days ($t = 1...365$)	Districts ($j = 1...5$)	Pharmacies ($i/P_i = j$)	$=N_{jk}$	*

$$^* : \forall k, t/(F_t = 1 \& k = 1) \vee (k = 2), j : \sum_{i/P_i = j \& ((k=1 \& TF_i = 1) \vee k=2)} \alpha_{ikt} = N_{jk} \tag{31.2}$$

- A normal pharmacy cannot be on guard duty for more than 24 consecutive hours (propositional logic specification).

$$\begin{aligned} &\forall i/\text{TF}_i = 1, t/t < 365 \wedge F_t = 1 \wedge F_{t+1} = 1 : \\ &\text{NOT}(\alpha_{i1t} = 1 \text{ AND } \alpha_{i2t} = 1 \text{ AND } \alpha_{i1t+1} = 1) \\ &\Rightarrow f_{37}, f_{11} \Rightarrow \forall i/T_i = 1, t/t < 365 \wedge F_t = 1 \wedge F_{t+1} = 1 : \\ &\alpha_{i1t} + \alpha_{i2t} + \alpha_{i1t+1} \leq 2 \end{aligned} \tag{31.3}$$

$$\begin{aligned} &\forall i/\text{TF}_i = 1, t/t < 365 \wedge F_{t+1} = 1 : \\ &\text{NOT}(\alpha_{i2t} = 1 \text{ AND } \alpha_{i1t+1} = 1 \text{ AND } \alpha_{i2t+1} = 1) \\ &\Rightarrow f_{37}, f_{11} \Rightarrow \forall i/\text{TF}_i = 1, t/t < 365 \wedge F_{t+1} = 1 : \\ &\alpha_{i2t} + \alpha_{i1t+1} + \alpha_{i2t+1} \leq 2 \end{aligned} \tag{31.4}$$

- A pharmacy (normal or 12 h) cannot be on guard duty for two consecutive holidays (propositional logic specification).

$$\forall i/\text{TF}_i = 1, t/t < 365 \wedge F_t = 1, t' > t \wedge F_{t'} = 1 \wedge \sum_{t''=t+1}^{t'-1} F_{t''} = 0 :$$
$$\text{NOT}((\alpha_{i1t} = 1 \text{ OR } \alpha_{i2t} = 1) \text{ AND } (\alpha_{i1t'} = 1 \text{ OR } \alpha_{i2t'} = 1))$$

$$\Rightarrow f_{34} \Rightarrow \forall i/\text{TF}_i = 1, t/t < 365 \wedge F_t = 1, t' > t \wedge F_{t'} = 1 \wedge \sum_{t''=t+1}^{t'-1} F_{t''} = 0:$$

NOT$((\alpha_{i1t} + \alpha_{i2t} \geq 1)$ AND $(\alpha_{i1t'} + \alpha_{i2t'} \geq 1))$

$\Rightarrow f_{35} \Rightarrow \lambda_{it} = 1$ IF AND ONLY IF $\alpha_{i1t} + \alpha_{i2t} \geq 1$

$\Rightarrow f_{S_V} \Rightarrow$ IF $\alpha_{i1t} + \alpha_{i2t} \geq 1$ THEN $\lambda_{it} = 1$

$$\Rightarrow f_3, f_7, f_{14} \Rightarrow \alpha_{i1t} + \alpha_{i2t} \leq 2\lambda_{it} \tag{31.5}$$

$$\Rightarrow \forall i/\text{TF}_i = 1, t/t < 365 \wedge F_t = 1, t' > t \wedge F_{t'} = 1 \wedge \sum_{t''=t+1}^{t'-1} F_{t''} = 0:$$

NOT$(\lambda_{it} = 1$ AND $\lambda_{it'} = 1)$

$$\Rightarrow f_{37}, f_{11} \Rightarrow \forall i/TF_i = 1, t/t < 365 \wedge F_t = 1, t' > t \wedge F_{t'} = 1 \wedge \sum_{t''=t+1}^{t'-1} F_{t''} = 0:$$

$$\lambda_{it} + \lambda_{it'} \leq 1 \tag{31.6}$$

If we want to avoid the use of variables λ_{it}, we can raise the specification in this way:

$$\forall i/\text{TF}_i = 1, t/t < 365 \wedge F_t = 1, t' > t \wedge F_{t'} = 1 \wedge \sum_{t''=t+1}^{t'-1} F_{t''} = 0:$$

NOT$(\alpha_{i1t} = 1$ AND $\alpha_{i1t'} = 1)$ AND NOT$(\alpha_{i1t} = 1$ AND $\alpha_{i2t'} = 1)$ AND NOT$(\alpha_{i2t} = 1$ AND $\alpha_{i1t'} = 1)$AND NOT$(\alpha_{i2t} = 1$ AND $\alpha_{i2t'} = 1)$

$$\alpha_{i1t} + \alpha_{i1t'} \leq 1 \tag{31.7}$$

$$\Rightarrow f_{37}, f_{11} \Rightarrow \forall i/TF_i = 1, t/t < 365 \wedge F_t = 1, t' > t \wedge F_{t'} = 1 \wedge \sum_{t''=t+1}^{t'-1} F_{t''} = 0 : \alpha_{i1t} + \alpha_{i2t'} \leq 1 \tag{31.8}$$

$$\Rightarrow f_{37}, f_{11} \Rightarrow \forall i/\text{TF}_i = 1, t/t < 365 \wedge F_t = 1, t' > t \wedge F_{t'} = 1 \wedge \sum_{t''=t+1}^{t'-1} F_{t''} = 0 : \alpha_{i2t} + \alpha_{i1t'} \leq 1 \tag{31.9}$$

$$\Rightarrow f_{37}, f_{11} \Rightarrow \forall i/\mathrm{TF}_i = 1, t/t < 365 \wedge F_t = 1, t' > t \wedge F_{t'} = 1 \wedge \sum_{t''=t+1}^{t'-1} F_{t''} = 0 : \alpha_{i2t} + \alpha_{i2t'} \leq 1 \tag{31.10}$$

On the other hand, for 12-h pharmacies:

$$\forall i/\mathrm{TF}_i = 0, t/t < 365 \wedge F_t = 1, t' > t \wedge F_{t'} = 1 \wedge \sum_{t''=t+1}^{t'-1} F_{t''} = 0 :$$
$$\mathrm{NOT}(\alpha_{i2t} = 1 \ \mathrm{AND}\ \alpha_{i2t'} = 1)$$
$$\Rightarrow f_{37}, f_{11} \Rightarrow \forall i/\mathrm{TF}_i = 0, t/t < 365 \wedge F_t = 1, t' > t \wedge F_{t'} = 1 \wedge \sum_{t''=t+1}^{t'-1} F_{t''} = 0 : \alpha_{i2t} + \alpha_{i2t'} \leq 1 \tag{31.11}$$

– The difference in the number of total guards in normal pharmacies cannot be greater than three (bound calculations + upper bound imposition).

  Bound Calculations
  x_{MAX}: Upper bound of the number of total guards of normal pharmacies
  x_{MIN}: Lower bound of the number of total guards of normal pharmacies
  Definition:

$$\forall i/\mathrm{TF}_i = 1 : x_{\mathrm{MAX}} \geq \sum_{t=1}^{365} \sum_{k=1}^{2} \alpha_{ikt} \tag{31.12}$$

$$\forall i/\mathrm{TF}_i = 1 : x_{\mathrm{MIN}} \leq \sum_{t=1}^{365} \sum_{k=1}^{2} \alpha_{ikt} \tag{31.13}$$

  Upper bound imposition:

$$x_{\mathrm{MAX}} - x_{\mathrm{MIN}} \leq 3 \tag{31.14}$$

– Balance the total number of guards in each pharmacy each month, imposing that the greatest difference does not exceed 2 units (bound calculations + upper bound imposition).

  Bound Calculations
  z_i^{MAX} : Upper bound of the monthly number of total guards of normal pharmacies
  z_i^{MIN} : Lower bound of the monthly number of total guards of normal pharmacies
  Definition:

$$\forall i/\mathrm{TF}_i = 1, \forall h = 1..12 : z_i^{\mathrm{MAX}} \geq \sum_{t/M_t=h} \sum_{k=1}^{2} \alpha_{ikt} \tag{31.15}$$

$$\forall i/\mathrm{TF}_i = 0, \forall h = 1..12 : z_i^{\mathrm{MAX}} \geq \sum_{t/M_t=h} \alpha_{i2t} \tag{31.16}$$

$$\forall i/\mathrm{TF}_i = 1, \forall h = 1..12 : z_i^{\mathrm{MIN}} \leq \sum_{t/M_t=h} \sum_{k=1}^{2} \alpha_{ikt} \tag{31.17}$$

$$\forall i/\mathrm{TF}_i = 0, \forall h = 1..12 : z_i^{\mathrm{MIN}} \leq \sum_{t/M_t=h} \alpha_{i2t} \tag{31.18}$$

Upper bound imposition:

$$\forall i : z_i^{\mathrm{MAX}} - z_i^{\mathrm{MIN}} \leq 2 \tag{31.19}$$

Objective Criterion

Balance the ratio of the number of night guards of a pharmacy divided by the number of pharmacies in its district.

Bound Calculation

y_{MAX}: Upper bound for the number of night guards in pharmacies

y_{MIN}: Lower bound for the number of night guards in pharmacies

Definition:

$$\forall i : y_{\mathrm{MAX}} \geq \frac{\sum_{t=1}^{365} \alpha_{i2t}}{\sum_{i'/P_{i'}=P_i} 1} \tag{31.20}$$

$$\forall i : y_{\mathrm{MIN}} \leq \frac{\sum_{t=1}^{365} \alpha_{i2t}}{\sum_{i'/P_{i'}=P_i} 1} \tag{31.21}$$

Objective Function

$$\mathrm{Min}\ y_{\mathrm{MAX}} - y_{\mathrm{MIN}} \tag{31.22}$$

Complete Model

$$\mathrm{Min}\ y_{\mathrm{MAX}} - y_{\mathrm{MIN}} \tag{31.22}$$

s.t.

$$\forall k, t/(F_t = 1\&k = 1) \vee (k = 2), j : \sum_{i/P_i=j\&((k=1\&\text{TF}_i=1)\vee k=2)} \alpha_{ikt} = N_{jk} \tag{31.2}$$

$$\forall i/\text{TF}_i = 1, t/t < 365 \wedge F_t = 1 \wedge F_{t+1} = 1 : \alpha_{i1t} + \alpha_{i2t} + \alpha_{i1t+1} \leq 2 \tag{31.3}$$

$$\forall i/\text{TF}_i = 1, t/t < 365 \wedge F_{t+1} = 1 : \alpha_{i2t} + \alpha_{i1t+1} + \alpha_{i2t+1} \leq 2 \tag{31.4}$$

$$\forall i/\text{TF}_i = 1, t/t < 365 \wedge F_t = 1, t' > t \wedge F_{t'} = 1 \wedge \sum_{t''=t+1}^{t'-1} F_{t''} = 0$$

$$: \alpha_{i1t} + \alpha_{i1t'} \leq 1 \tag{31.7}$$

$$\forall i/\text{TF}_i = 1, t/t < 365 \wedge F_t = 1, t' > t \wedge F_{t'} = 1 \wedge \sum_{t''=t+1}^{t'-1} F_{t''} = 0$$

$$: \alpha_{i1t} + \alpha_{i2t'} \leq 1 \tag{31.8}$$

$$\forall i/\text{TF}_i = 1, t/t < 365 \wedge F_t = 1, t' > t \wedge F_{t'} = 1 \wedge \sum_{t''=t+1}^{t'-1} F_{t''} = 0$$

$$: \alpha_{i2t} + \alpha_{i1t'} \leq 1 \tag{31.9}$$

$$\forall i/\text{TF}_i = 1, t/t < 365 \wedge F_t = 1, t' > t \wedge F_{t'} = 1 \wedge \sum_{t''=t+1}^{t'-1} F_{t''} = 0$$

$$: \alpha_{i2t} + \alpha_{i2t'} \leq 1 \tag{31.10}$$

$$\forall i/\text{TF}_i = 0, t/t < 365 \wedge F_t = 1, t' > t \wedge F_{t'} = 1 \wedge \sum_{t''=t+1}^{t'-1} F_{t''} = 0$$

$$: \alpha_{i2t} + \alpha_{i2t'} \leq 1 \tag{31.11}$$

$$\forall i/\text{TF}_i = 1 : x_{\text{MAX}} \geq \sum_{t=1}^{365} \sum_{k=1}^{2} \alpha_{ikt} \tag{31.12}$$

$$\forall i/\text{TF}_i = 1 : x_{\text{MIN}} \leq \sum_{t=1}^{365} \sum_{k=1}^{2} \alpha_{ikt} \tag{31.13}$$

$$x_{\text{MAX}} - x_{\text{MIN}} \leq 3 \tag{31.14}$$

$$\forall i/\text{TF}_i = 1, \forall h = 1 \ldots 12 : z_i^{\text{MAX}} \geq \sum_{t/M_t=h} \sum_{k=1}^{2} \alpha_{ikt} \tag{31.15}$$

$$\forall i/\text{TF}_i = 0, \forall h = 1 \ldots 12 : z_i^{\text{MAX}} \geq \sum_{t/M_t=h} \alpha_{i2t} \tag{31.16}$$

$$\forall i/\mathrm{TF}_i = 1, \forall h = 1\ldots12 : z_i^{\mathrm{MIN}} \le \sum_{t/M_t=h} \sum_{k=1}^{2} \alpha_{ikt} \quad (31.17)$$

$$\forall i/\mathrm{TF}_i = 0, \forall h = 1\ldots12 : z_i^{\mathrm{MIN}} \le \sum_{t/M_t=h} \alpha_{i2t} \quad (31.18)$$

$$\forall i : z_i^{\mathrm{MAX}} - z_i^{\mathrm{MIN}} \le 2 \quad (31.19)$$

$$\forall i : y_{\mathrm{MAX}} \ge \frac{\sum_{t=1}^{365} \alpha_{i2t}}{\sum_{i'/P_{i'}=P_i} 1} \quad (31.20)$$

$$\forall i : y_{\mathrm{MIN}} \le \frac{\sum_{t=1}^{365} \alpha_{i2t}}{\sum_{i'/P_{i'}=P_i} 1} \quad (31.21)$$

$$(k = 1, \forall i/\mathrm{TF}_i = 1, \forall t/F_t = 1) \& k = 2, \forall i, \forall t : \alpha_{ikt} \text{ binaries} \quad (31.1)$$

$$y_{\mathrm{MAX}} \ge 0, y_{\mathrm{MIN}} \ge 0, x_{\mathrm{MAX}} \ge 0, x_{\mathrm{MIN}} \ge 0$$

$$\forall i : z_i^{\mathrm{MAX}} \ge 0, z_i^{\mathrm{MIN}} \ge 0$$

Implementation in Lingo

```
MODEL:
SETS:
 Pharmacies/1..52/:TF,P,zMAX,zMIN;
 Districts/1..5/;
 Guards/1..2/;
 DG(Districts,Guards):N;
 Days/1..365/:M,F;
 Months/1..12/;
 FGD(Pharmacies,Guards,Days): Alpha;
ENDSETS

DATA:
 TF=@OLE('C:\Files_\31_Excel.xlsx','TF');
 P=@OLE('C:\Files_\31_Excel.xlsx','P');
 N=@OLE('C:\Files_\31_Excel.xlsx','Guards');
 M= @ODBC('31_ODBC','Days','Month');
 F= @ODBC('31_ODBC','Days','Holidays');
ENDDATA
!31.22;
MIN = yMAX - yMIN;
!31.2;
@FOR(Guards(k):@FOR(Days(t)|(F(t)#EQ#1 #AND# k#EQ#1) #OR# (k#EQ#2):
@FOR(Districts(j):@SUM(Pharmacies(i)|P(i)#EQ#j #AND#
((k#EQ#1#AND#TF(i)#EQ#1) #OR# k#EQ#2): Alpha(i,k,t))=N(j,k))));
!31.3;
@FOR(Pharmacies(i)|TF(i)#EQ#1:@FOR(Days(t)|t#LT#365#AND#F(t)
```

```
#EQ#1:@FOR(Days(tt)|tt#EQ#t+1#AND#F(tt)#EQ#1: Alpha(i,1,t) + Alpha
(i,2,t) + Alpha(i,1,tt)<=2)));

!31.4;
@FOR(Pharmacies(i)|TF(i)#EQ#1:@FOR(Days(t)|t#LT#365:@FOR(Days
(tt)|tt#EQ#t+1#AND#F(tt)#EQ#1: Alpha(i,2,t) + Alpha(i,1,tt) + Alpha
(i,2,tt)<=2)));
!31.7;
@FOR(Pharmacies(i)|TF(i)#EQ#1:@FOR(Days(t)|t#LT#365#AND#F(t)
#EQ#1:@FOR(Days(tt)|tt#GT#t#AND#F(tt)#EQ#1#AND#@SUM(Days(ttt)|
ttt#GT#t#AND#ttt#LT#tt:F(ttt))#EQ#0: Alpha(i,1,t) + Alpha(i,1,tt)
<=1)));
!31.8;
@FOR(Pharmacies(i)|TF(i)#EQ#1:@FOR(Days(t)|t#LT#365#AND#F(t)
#EQ#1:@FOR(Days(tt)|tt#GT#t#AND#F(tt)#EQ#1#AND#@SUM(Days(ttt)|
ttt#GT#t#AND#ttt#LT#tt:F(ttt))#EQ#0: Alpha(i,1,t) + Alpha(i,2,tt)
<=1)));
!31.9;
@FOR(Pharmacies(i)|TF(i)#EQ#1:@FOR(Days(t)|t#LT#365#AND#F(t)
#EQ#1:@FOR(Days(tt)|tt#GT#t#AND#F(tt)#EQ#1#AND#@SUM(Days(ttt)|
ttt#GT#t#AND#ttt#LT#tt:F(ttt))#EQ#0: Alpha(i,2,t) + Alpha(i,1,tt)
<=1)));
!31.10;
@FOR(Pharmacies(i)|TF(i)#EQ#1:@FOR(Days(t)|t#LT#365#AND#F(t)
#EQ#1:@FOR(Days(tt)|tt#GT#t#AND#F(tt)#EQ#1#AND#@SUM(Days(ttt)|
ttt#GT#t#AND#ttt#LT#tt:F(ttt))#EQ#0: Alpha(i,2,t) + Alpha(i,2,tt)
<=1)));
!31.11;
@FOR(Pharmacies(i)|TF(i)#EQ#0:@FOR(Days(t)|t#LT#365#AND#F(t)
#EQ#1:@FOR(Days(tt)|tt#GT#t#AND#F(tt)#EQ#1#AND#@SUM(Days(ttt)|
ttt#GT#t#AND#ttt#LT#tt:F(ttt))#EQ#0: Alpha(i,2,t) + Alpha(i,2,tt)
<=1)));
!31.12;
@FOR(Pharmacies(i)|TF(i)#EQ#1: xMAX>=@SUM(Days(t):@SUM(Guards(k):
Alpha(i,k,t))));
!31.13;
@FOR(Pharmacies(i)|TF(i)#EQ#1: xMIN<=@SUM(Days(t):@SUM(Guards(k):
Alpha(i,k,t))));
!31.14;
xMAX-xMIN<=2;
!31.15;
@FOR(Pharmacies(i)|TF(i)#EQ#1:@FOR(Months(h):zMAX(i)>=@SUM(Days
(t)|M(t)#EQ#h:@SUM(Guards(k):Alpha(i,k,t)))));
!31.16;
@FOR(Pharmacies(i)|TF(i)#EQ#0:@FOR(Months(h):zMAX(i)>=@SUM(Days
(t)|M(t)#EQ#h:Alpha(i,2,t))));
!31.17;
@FOR(Pharmacies(i)|TF(i)#EQ#1:@FOR(Months(h):zMIN(i)<=@SUM(Days
(t)|M(t)#EQ#h:@SUM(Guards(k):Alpha(i,k,t)))));
!31.18;
@FOR(Pharmacies(i)|TF(i)#EQ#0:@FOR(Months(h):zMIN(i)<=@SUM(Days
(t)|M(t)#EQ#h:Alpha(i,2,t))));
!31.19;
```

```
Solution Report - 31_LingoE
Feasible solution found.
Objective value:                                  4.571429
Objective bound:                                  4.432451
Infeasibilities:                                  0.000000
Extended solver steps:                               22030
Total solver iterations:                            302851
Elapsed runtime seconds:                            241.66

Model Class:                                          MILP

Total variables:                      38068
Nonlinear variables:                      0
Integer variables:                    35770

Total constraints:                    22104
Nonlinear constraints:                    0

Total nonzeros:                      241482
Nonlinear nonzeros:                       0
```

Fig. 6.6 Problem 30: LINGO solution report

```
@FOR(Pharmacies(i):zMAX(i)-zMIN(i)<=2);
!31.20;
@FOR(Pharmacies(i):@SUM(Pharmacies(ii)|P(ii)#EQ#P(i):1)
*yMAX>=@SUM(Days(t):Alpha(i,2,t)));
!31.21;
@FOR(Pharmacies(i):@SUM(Pharmacies(ii)|P(ii)#EQ#P(i):1)
*yMIN<=@SUM(Days(t):Alpha(i,2,t)));
!31.1;
@FOR(Pharmacies(i):@FOR(Guards(k)|((k#EQ#1#AND#TF(i)#EQ#1) #OR#
k#EQ#2):@FOR(Days(t): @BIN(Alpha(i,k,t)))));

END
```

Solution

The solution is shown in Fig. 6.6. For this problem, due to its size, we present an admissible solution.

6.7 Problem 32. Project Task Planning

A construction project is to be planned. The responsible engineers have designed the project task scheme and carried out the PERT planning of the project. From this study, the interval in which each of the tasks must be performed is known.

However, the resources to do the tasks must be contracted and they present relations of technical incompatibility with the tasks. It also happens that the number of resources required by each task does not have to be one unit. In other words, each task requires a certain number of resources.

Consider that we have n tasks in the project. Each task has a time and an interval within which it must be processed, as well as the number of resources it needs.

Resources can be up to m types, so that each type has a binary compatibility relationship with each task (1 is compatible; 0 is not). Resources of the same type have the same hiring cost. Note, as a specification, that a resource cannot perform two tasks at the same time.

The objective is to minimize the cost of hiring.

Modeling
List of elements in Table 6.6.

Table 6.6 Problem 32: table of elements

			Data				
Elements	Set	QN	Name	Parameter	Type	Belonging	Value
Project		I_U					
Tasks	$i = 1 \ldots n$	I_M	Necessary resources	R_i	I	W	–
			Time	T_i	C (m)	W	–
			Interval start	II_i	C (m)	W	–
			Interval end	FI_i	C (m)	W	–
			Compatibility	C_{ij}	B	S	–
Resources types	$j = 1 \ldots m$	I_U	Cost	CC_j	C($)	W	–
				C_{ij}			
Resources	$k = 1 \ldots \text{UB}$	I_U	Type	S_k	O	W	–

We need to set an upper bound on the number of resources of each type. Although they are identical, they are individually referenced in the system.

As an upper bound, we can consider, for each type j, the total resources necessary for compatible tasks:

$$\text{UB} = \sum_{j=1}^{m} \sum_{i=1}^{n} R_i C_{ij}$$

Decision Activities
Schedule tasks (*i*)

$$\forall i : x_i = \text{Time instant when the task is scheduled } i.$$

Assign resources (*k*) to tasks (*i*)

$$\forall i,k : \alpha_{ik} = 1 \text{ if we assign resource } k \text{ to task } i; 0 \text{ otherwise} \tag{32.1}$$

Specifications

– Quantitative selection specifications:

Activity	Elements selecting	Selectable elements	Selection	Constraints
α_{ik}	Tasks ($i = 1\ldots n$)	Compatible resources $k/C_{iS_k} = 1$	$= R_i$	$\forall i: \sum_{k/C_{iS_k}=1} \alpha_{ik} = R_i$ (32.2)

– Tasks carried out within its interval (bound imposition):

$$\forall i : \text{II}_i \le x_i \tag{32.3}$$

$$\forall i : x_i + T_i \le \text{FI}_i \tag{32.4}$$

– A resource cannot perform two overlapping tasks (propositional logic specification):

$$\begin{aligned}
&\forall i,i'/i \neq i', k/C_{iS_k} = 1 \& C_{i'S_k} = 1 : \\
&\text{IF } x_i \le x_{i'} \text{ AND } x_i + T_i > x_{i'} \text{ THEN NOT}(\alpha_{ik} = 1 \text{ AND } \alpha_{i'k} = 1)
\end{aligned}$$

$$\begin{aligned}
&\Rightarrow f_{37}, f_{11} \Rightarrow \forall i,i'/i \neq i', k/C_{iS_k} = 1 \& C_{i'S_k} = 1 : \\
&\text{IF } x_i \le x_{i'} \text{ AND } x_i + T_i > x_{i'} \text{ THEN } \alpha_{ik} + \alpha_{i'k} \le 1
\end{aligned}$$

$$\begin{aligned}
&\forall i,i'/i \neq i', k/C_{iS_k} = 1 \& C_{i'S_k} = 1 : \\
&\omega_{1ii'} = 1 \text{ IF AND ONLY IF } x_i \le x_{i'}
\end{aligned} \tag{32.5}$$

$$\begin{aligned}
&\forall i,i'/i \neq i', k/C_{iS_k} = 1 \& C_{i'S_k} = 1 : \\
&\omega_{2ii'} = 1 \text{ IF AND ONLY IF } x_i + T_i > x_{i'}
\end{aligned} \tag{32.6}$$

$$\begin{aligned}
&\Rightarrow f_{S_V} \Rightarrow \forall i,i'/i \neq i', k/C_{iS_k} = 1 \& C_{i'S_k} = 1 : \\
&\text{IF } x_i \le x_{i'} \text{ THEN } \omega_{1ii'} = 1 \\
&\Rightarrow f_{3,} f_7, f_9 \Rightarrow \mathit{IF} \quad 1 - \omega_{1ii'} = 1 \text{ THEN } x_i \ge x_{i'} + \xi \\
&\Rightarrow \forall i,i'/i \neq i', k/C_{iS_k} = 1 \& C_{i'S_k} = 1 : \mathit{IF} \quad 1 - \omega_{1ii'} = 1 \text{ THEN } x_i - x_{i'} \ge \xi
\end{aligned}$$

$$\begin{aligned}
&\Rightarrow f_{15}\left(\text{LB}_{x_i - x_{i'}} = -\text{II}_{i'}\right) \Rightarrow \forall i,i'/i \neq i', k/C_{iS_k} = 1 \& C_{i'S_k} = 1 : \\
&x_i - x_{i'} \ge \xi(1 - \omega_{1ii'}) - \text{II}_{i'}\omega_{1ii'}
\end{aligned} \tag{32.7}$$

$$\Rightarrow f_{S_V} \Rightarrow \forall i,i'/i \neq i',k/C_{iS_k} = 1\&C_{i'S_k} = 1:$$
$$\text{IF } x_i + T_i > x_{i'} \text{ THEN } \omega_{2ii'} = 1$$
$$\Rightarrow f_3, f_7, f_{10} \Rightarrow \forall i,i'/i \neq i',k/C_{iS_k} = 1\&C_{i'S_k} = 1:$$
$$\text{IF } 1 - \omega_{2ii'} = 1 \text{ THEN } x_i + T_i \leq x_{i'}$$
$$\Rightarrow \forall i,i'/i \neq i',k/C_{iS_k} = 1\&C_{i'S_k} = 1:$$
$$\text{IF } 1 - \omega_{2ii'} = 1 \text{ THEN } x_i - x_{i'} \leq -T_i$$

$$\Rightarrow f_{14}\left(\text{UB}_{x_i - x_{i'}} = \text{FI}_i - T_i\right) \Rightarrow \forall i,i'/i \neq i',k/C_{iS_k} = 1\&C_{i'S_k} = 1: \quad x_i - x_{i'} \leq -T_i + \text{FI}_i\omega_{2ii'} \tag{32.8}$$

$$\Rightarrow \forall i,i'/i \neq i',k/C_{iS_k} = 1\&C_{i'S_k} = 1:$$
$$\text{IF } \omega_{1ii'} = 1 \text{ AND } \omega_{2ii'} = 1 \text{ THEN } \alpha_{ik} + \alpha_{i'k} \leq 1$$
$$\Rightarrow f_{37} \Rightarrow \forall i,i'/i \neq i',k/C_{iS_k} = 1\&C_{i'S_k} = 1:$$
$$\mathit{IF} \ \omega_{1ii'} + \omega_{2ii'} \geq 2 \text{ THEN } \alpha_{ik} + \alpha_{i'k} \leq 1$$

$$\Rightarrow f_{18}\left(\text{UB}_{\omega_{1ii'} + \omega_{2ii'}} = 2\right) \Rightarrow \forall i,i'/i \neq i',k/C_{iS_k} = 1\&C_{i'S_k} = 1: \quad \omega_{1ii'} + \omega_{2ii'} \leq 1 + \lambda_{ii'} \tag{32.9}$$

$$\Rightarrow f_{20}\left(\text{UB}_{\alpha_{ik} + \alpha_{i'k}}\right) \Rightarrow \forall i,i'/i \neq i',k/C_{iS_k} = 1\&C_{i'S_k} = 1: \quad \alpha_{ik} + \alpha_{i'k} \leq 2 - \lambda_{ii'} \tag{32.10}$$

Objective Criterion
Minimize hiring cost
Binary Logical Calculation: Resource hired

$$\forall k : \beta_k = 1 \text{ if resource } k \text{ is hired; 0 otherwise.} \tag{32.11}$$

$$\forall k : \beta_k = 1 \text{ IF AND ONLY IF } \sum_{i=1}^{n} \alpha_{ik} > 0 \Rightarrow$$

$$\Rightarrow f_{S_V} \Rightarrow \forall k : \text{IF } \sum_{i=1}^{n} \alpha_{ik} > 0 \text{ THEN } \beta_k = 1$$

$$\Rightarrow f_3, f_7, f_{14}\left(\text{UB}_{\sum_{i=1}^{n} \alpha_{ik}} = n\right) \Rightarrow \forall k : \sum_{i=1}^{n} \alpha_{ik} \leq n\beta_k \tag{32.12}$$

Objective Function

$$\text{Min } \sum_{k=1}^{cs} \sum_{j/S_k = j} \text{CC}_j\beta_k \tag{32.13}$$

Complete Model

$$\text{Min} \sum_{k=1}^{cs} \sum_{j/S_k=j} \text{CC}_j \beta_k \tag{32.13}$$

s.t.

$$\forall i: \sum_{k/C_{iS_k}=1} \alpha_{ik} = R_i \tag{32.2}$$

$$\forall i: \text{II}_i \le x_i \tag{32.3}$$

$$\forall i: x_i + T_i \le \text{FI}_i \tag{32.4}$$

$$\forall i, i'/i \ne i', k/C_{iS_k} = 1 \& C_{i'S_k} = 1 : x_i - x_{i'} \ge \xi(1 - \omega_{1ii'}) - II_{i'}\omega_{1ii'} \tag{32.7}$$

$$\forall i, i'/i \ne i', k/C_{iS_k} = 1 \& C_{i'S_k} = 1 : x_i - x_{i'} \le -T_i + \text{FI}_i\omega_{2ii'} \tag{32.8}$$

$$\forall i, i'/i \ne i', k/C_{iS_k} = 1 \& C_{i'S_k} = 1 : \omega_{1ii'} + \omega_{2ii'} \le 1 + \lambda_{ii'} \tag{32.9}$$

$$\forall i, i'/i \ne i', k/C_{iS_k} = 1 \& C_{i'S_k} = 1 : \alpha_{ik} + \alpha_{i'k} \le 2 - \lambda_{ii'} \tag{32.10}$$

$$\forall k: \sum_{i=1}^{n} \alpha_{ik} \le n\beta_k \tag{32.12}$$

$$\forall i: x_i \ge 0$$

$$\forall i, k: \alpha_{ik} \text{ binaries} \tag{32.1}$$

$$\forall i, i'/i \ne i': \omega_{1ii'}, \ \omega_{2ii'} \text{ binaries} \tag{32.5, 32.6}$$

$$\forall k: \beta_k \text{ binaries} \tag{32.11}$$

$$\forall i, i'/i \ne i': \lambda_{ii'} \text{ binaries} \tag{32.14}$$

Implementation in LINGO

```
MODEL:
DATA:
 n=@ODBC('32_ODBC','System','n');
 m=@ODBC('32_ODBC','System','m');
 UB=@ODBC('32_ODBC','System','UB');
ENDDATA
SETS:
 Tasks/1..n/:R,T,II,FI,x;
 ResourcesTypes/1..m/:CC;
 Resources/1..UB/:S,Beta;
TT(Tasks,ResourcesTypes)/@ODBC
('32_ODBC','Comp','Task','Resource')/;
 TR(Tasks,Resources):Alpha;
 Tata(Tasks,Tasks):Omega1,Omega2,Lamda;
ENDSETS
DATA:
 R=@ODBC('32_ODBC','Tasks','Resources');
```

```
 T=@ODBC('32_ODBC','Tasks','Time');
 II=@ODBC('32_ODBC','Tasks','Start');
 FI=@ODBC('32_ODBC','Tasks','End');
 CC=@ODBC('32_ODBC','ResourcesTypes','Cost');
 S=@ODBC('32_ODBC','Resources','Type');
 Epsilon=0.001;
ENDDATA
!32.13;
MIN = @SUM(Resources(k):@SUM(ResourcesTypes(j)|j#EQ#S(k):CC(j)*Beta
(k)));
!32.2;
@FOR(Tasks(i):@SUM(TT(i,j):@SUM(Resources(k)|S(k)#EQ#j:Alpha(i,
k)))=R(i));
!32.3;
@FOR(Tasks(i):II(i)<=x(i));
!32.4;
@FOR(Tasks(i):x(i)+T(i)<=FI(i));

!32.7;
@FOR(Tasks(i):@FOR(Tasks(ip)|ip#NE#i:@FOR(TT(i,j):@FOR(TT(ip,jj)|
jj#EQ#j:@FOR(Resources(k)|S(k)#EQ#j:x(i)-x(ip)>=Epsilon*(1-Omega1
(i,ip))-II(ip)*Omega1(i,ip))))));
!32.8;
@FOR(Tasks(i):@FOR(Tasks(ip)|ip#NE#i:@FOR(TT(i,j):@FOR(TT(ip,jj)|
jj#EQ#j:@FOR(Resources(k)|S(k)#EQ#j:x(i)-x(ip)<=-T(i)+FI(i)
*Omega2(i,ip))))));
!32.9;
@FOR(Tasks(i):@FOR(Tasks(ip)|ip#NE#i:@FOR(TT(i,j):@FOR(TT(ip,jj)|
jj#EQ#j:@FOR(Resources(k)|S(k)#EQ#j:Omega1(i,ip)+Omega2(i,ip)
<=1+Lamda(i,ip))))));
!32.10;
@FOR(Tasks(i):@FOR(Tasks(ip)|ip#NE#i:@FOR(TT(i,j):@FOR(TT(ip,jj)|
jj#EQ#j:@FOR(Resources(k)|S(k)#EQ#j:Alpha(i,k)+Alpha(ip,k)<=2-
Lamda(i,ip))))));
!32.12;
@FOR(Resources(k):@SUM(Tasks(i):ALpha(i,k))<=n*Beta(k));
!32.1;
@FOR(Tasks(i):@FOR(Resources(k):@BIN(Alpha(i,k))));
!32.5;
@FOR(Tasks(i):@FOR(Tasks(ip)|i#NE#ip:@BIN(Omega1(i,ip))));
!32.6;
@FOR(Tasks(i):@FOR(Tasks(ip)|i#NE#ip:@BIN(Omega2(i,ip))));
!32.11;
@FOR(Resources(k):@BIN(Beta(k)));
!32.14;
@FOR(Tasks(i):@FOR(Tasks(ip)|i#NE#ip:@BIN(Lamda(i,ip))));
END
```

Solution

The solution is shown in Fig. 6.7.

```
Solution Report - 32_LingoE
Global optimal solution found.
Objective value:                                    1490.000
Objective bound:                                    1490.000
Infeasibilities:                                    0.000000
Extended solver steps:                                  8908
Total solver iterations:                              531653
Elapsed runtime seconds:                               51.99

Model Class:                                            MILP

Total variables:                    10151
Nonlinear variables:                    0
Integer variables:                   9951

Total constraints:                  85706
Nonlinear constraints:                  0

Total nonzeros:                    260307
Nonlinear nonzeros:                     0
```

Fig. 6.7 Problem 30: LINGO solution report

6.8 Problem 33. Product Warehouse Management

We want to structure a warehouse to collect units of products, similar to IKEA. The warehouse is made up of five streets. On each street, there is a defined set of sections. A section has its own width and height.

The products that have to be stored have a width and a height (depth does not need to be considered). A product is housed in a single section.

Unit columns can be placed in a section as long as they fit in the width and height of the section. We can also include multiple products in a section, but they are not mixed in the columns, and they must be of the same category. The company has 30 categories of products.

The list of products to be placed and the number of units of each product are known.

The goal of placing products is to improve order collection. For this, the company has the data of the last 30 orders placed. An order is made up of a set of products. The intention is that the collection of an order be as fast and dynamic as possible. For this, the company seeks to minimize the number of streets visited to pick up the products of an order. Therefore, set the objective of minimizing the total number of streets visited.

Modeling

List of elements in Table 6.7.

Table 6.7 Problem 33: table of elements

			Data				
Elements	Set	QN	Name	Parameter	Type	Belonging	Value
Streets	$i = 1\ldots5$	I_U					
Sections	$j = 1\ldots n$	I_U	Width	WS_i	C (m)	W	...
			Height	HS_i	C (m)	W	...
			Street	S_j	O	W	...
				NC_{jk}			
Products	$k = 1\ldots m$	C_D	Units	U_k	I	W	...
			Width	W_k	C (m)	W	...
			Height	H_k	C (m)	W	...
			Category	C_k	O	W	...
			No columns in section j	NC_{jk}	I	S	...
			Product_Order	A_{kp}	B	S	...
Orders	$p = 1\ldots30$	I_U		A_{kp}			
Categories	$r = 1\ldots30$	I_U					

Despite the fact that the products are a collective element, all of their units are managed in an integral way. We do not make partial use of them because a product is stored in a single section. Therefore, the participation of a product in the decisions will always be made over the continent object, not over the content.

To control the placement of products in sections, we have calculated the number of columns that each product would occupy in each section. To do this, we first calculate the integer number of units per column and then divide the total number of units of the product by the number of units in a column:

$$\forall k, j : \mathrm{NU}_{kj} = \left\lfloor \frac{\mathrm{HS}_j}{H_k} \right\rfloor$$

$$\forall k, j : \mathrm{NC}_{kj} = \left\lceil \frac{U_k}{\mathrm{NU}_{kj}} \right\rceil$$

Decision Activities

Assign products(k) to sections(j)

It is equivalent to consider "Assign Sections to Articles," so the type of decision is logical, and in no case measurable.

We are going to include in the definition of events that only valid combinations are allowed with respect to the width and height of the product with respect to the dimensions of the session. Thus, it will not be necessary to establish those specifications later.

$$\forall k, j / W_k \le \mathrm{WS}_j \& H_k \le \mathrm{HS}_j : \alpha_{kj} = 1 \text{ if } I \text{ assign article } k \text{ to section } j; 0 \text{ otherwise.} \tag{33.1}$$

Specifications

– Quantitative selection specifications:

Activity	Elements selecting	Selectable elements	Selection	Constraints
α_{kj}	Products ($k = 1 \ldots m$)	Sections ($j = 1 \ldots n$)	=1	$\forall k : \sum_{j/W_k \le \mathrm{WS}_j \& H_k \le \mathrm{HS}_j} \alpha_{kj} = 1$ (33.2)

– Section capacities (capacity specification):

From the data calculations performed in the elements section, the capacity specifications are greatly simplified. We already have the number of columns that each product occupies in each section. It is enough only to impose a specification of capacity with respect to the width, using those data as consumption. Regarding height, it is already included in the definition of variables.

$$\forall j : \sum_{k/W_k \le \mathrm{WS}_j \& H_k \le \mathrm{HS}_j} W_k \mathrm{NC}_{kj} \alpha_{kj} \le \mathrm{WS}_j \tag{33.3}$$

– Products in the same section must have the same category (propositional logic specification).

We put it in the negative: Two products of different categories cannot be in the same section:

$$\forall k, k' / C_k \ne C_{k'}, j / W_k \le \mathrm{WS}_j \& W_{k'} \le \mathrm{WS}_j \& H_k \le \mathrm{HS}_j \& H_{k'} \le \mathrm{HS}_j :$$
$$\mathrm{NOT}\left(\alpha_{jk} = 1 \text{ AND } \alpha_{jk'} = 1\right)$$

$$\Rightarrow f_{37}, f_{11} \Rightarrow \forall k, k' / C_k \ne C_{k'}, j / W_k \le \mathrm{WS}_j \& W_{k'} \le \mathrm{WS}_j \& H_k \le \mathrm{HS}_j \& H_{k'} \le \mathrm{HS}_j :$$

$$\alpha_{kj} + \alpha_{k'j} \le 1 \tag{33.4}$$

Objective Criterion

Minimize the total number of streets visited.

You have to calculate which streets each order visits using a binary logical calculation:

Binary Logical Calculation

$$\forall p, i : \delta_{pi} = 1 \text{ if the order } p \text{ visits street } i. \tag{33.5}$$

$$\forall p,i:\delta_{pi}=1 \text{ IF AND ONLY IF } \sum_{k/A_{kp}=1}\ \sum_{j/S_j=i\&W_k\le \mathrm{WS}_j\&H_k\le \mathrm{HS}_j}\alpha_{kj}>0$$

$$\Rightarrow f_{S_V}\Rightarrow \forall p,i:\text{IF } \sum_{k/A_{kp}=1}\ \sum_{j/S_j=i\&W_k\le \mathrm{WS}_j\&H_k\le \mathrm{HS}_j}\alpha_{kj}>0 \text{ THEN } \delta_{pi}=1$$

$$\Rightarrow f_3,f_7\Rightarrow \forall p,i:\text{IF } 1-\delta_{pi}=1 \text{ THEN } \sum_{k/A_{kp}=1}\ \sum_{j/S_j=i\&W_k\le \mathrm{WS}_j\&H_k\le \mathrm{HS}_j}\alpha_{kj}\le 0$$

$$\Rightarrow f_{14}\left(\mathrm{UB}\sum_{k/A_{kp}=1}\ \sum_{j/S_j=i\&W_k\le \mathrm{WS}_j\&H_k\le \mathrm{HS}_j}\alpha_{kj}=n\right)\Rightarrow$$
$$\forall p,i:\sum_{k/A_{kp}=1}\ \sum_{j/S_j=i\&W_k\le \mathrm{WS}_j\&H_k\le \mathrm{HS}_j}\alpha_{kj}\le n\delta_{pi} \qquad (33.6)$$

Auxiliary Calculation
Simple calculation that adds the streets visited by each order.

$$\forall p:\ y_p=\sum_{i=1}^{5}\delta_{pi} \qquad (33.7)$$

Objective Function

$$\text{Min } \sum_{p=1}^{30} y_p \qquad (33.8)$$

Complete Model

$$\text{Min } \sum_{p=1}^{30} y_p \qquad (33.8)$$

s.t.

$$\forall k:\ \sum_{j/W_k\le \mathrm{WS}_j\&H_k\le \mathrm{HS}_j}\alpha_{kj}=1 \qquad (33.2)$$

$$\forall j:\ \sum_{k/W_k\le \mathrm{WS}_j\&H_k\le \mathrm{HS}_j} W_k\mathrm{NC}_{kj}\alpha_{kj}\le \mathrm{WS}_j \qquad (33.3)$$

$$\forall k,k'/C_k\ne C_{k'},j/W_k\le \mathrm{WS}_j\&W_{k'}\le \mathrm{WS}_j\&H_k\le \mathrm{HS}_j\&H_{k'}\le \mathrm{HS}_j:$$
$$\alpha_{kj}+\ \alpha_{k'j}\le 1 \qquad (33.4)$$

$$\forall p, i : \sum_{k/A_{kp}=1} \sum_{j/S_j=i\&W_k \leq \mathrm{WS}_j\&H_k \leq \mathrm{HS}_j} \alpha_{kj} \leq n\delta_{pi} \tag{33.6}$$

$$\forall p : \quad y_p = \sum_{i=1}^{10} \delta_{pi} \tag{33.7}$$

$$\forall k, j / W_k \leq \mathrm{WS}_j \& H_k \leq \mathrm{HS}_j : \alpha_{kj} \text{ binaries} \tag{33.1}$$

$$\forall p, i : \quad \delta_{pi} \text{ binaries} \tag{33.5}$$

$$\forall p : y_p \geq 0$$

Implementation in LINGO

```
MODEL:
DATA:
 n=@ODBC('33_ODBC','System','n');
 m=@ODBC('33_ODBC','System','m');
ENDDATA
SETS:
 Streets/1..5/;
 Sections/1..m/:WS, HS, S;
 Products/1..n/:U, W, H, C;
 Orders/1..30/:y;
 Categories/1..30/;
PO(Products,Orders)/@ODBC('33_ODBC','AP','Product','Order')/;
 PS(Products,Sections):NNC,NC,Alpha;
 OS(Orders,Streets):Delta;
ENDSETS
DATA:
 WS,HS,S=@ODBC('33_ODBC','Sections','WS','HS','Calle');
 U,W,H,C=@ODBC('33_ODBC','Products','U','W','H','C');
@ODBC('33_ODBC','Products','Section')=@WRITEFOR(Products(k):@SUM
(Sections(j):j*Alpha(k,j)));
ENDDATA
CALC:
 @FOR(Sections(j): @FOR(Products(k):NNC(k,j)=HS(j)/H(k)));
 @FOR(Sections(j): @FOR(Products(k):NC(k,j)=@IF(@INT(U(k)/NNC(k,j))
#GT#0,@INT(U(k)/NNC(k,j)),1+@INT(U(k)/NNC(k,j)))));
ENDCALC
!33.8;
MIN = @SUM(Orders(p):y(p));
!33.2;
@FOR(Products(k):@SUM(Sections(j)|W(k)#LE#WS(j)#AND#H(k)#LE#HS
(j):Alpha(k,j))=1);
!33.3;
@FOR(Sections(j):@SUM(Products(k)|W(k)#LE#WS(j)#AND#H(k)#LE#HS
(j):W(k)*NC(k,j)*Alpha(k,j))<=WS(j));

!33.4;
@FOR(Products(k):@FOR(Products(kk)|C(k)#NE#C(kk):@FOR(Sections
(j)|W(k)#LE#WS(j) #AND# W(kk)#LE#WS(j) #AND# H(k)#LE#HS(j) #AND# H
(kk)#LE#HS(j): Alpha(k,j)+Alpha(kk,j)<=1)));
```

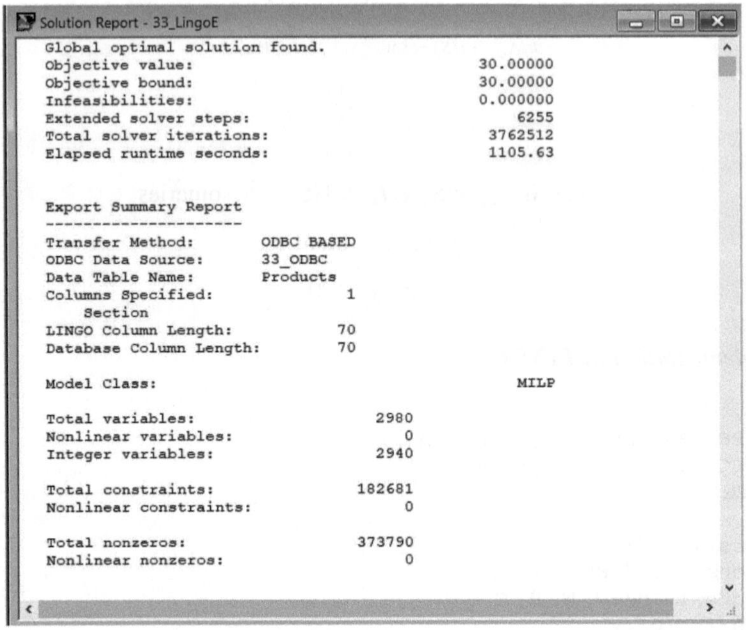

```
Solution Report - 33_LingoE
Global optimal solution found.
Objective value:                              30.00000
Objective bound:                              30.00000
Infeasibilities:                              0.000000
Extended solver steps:                            6255
Total solver iterations:                       3762512
Elapsed runtime seconds:                       1105.63


Export Summary Report
---------------------
Transfer Method:        ODBC BASED
ODBC Data Source:       33_ODBC
Data Table Name:        Products
Columns Specified:              1
    Section
LINGO Column Length:           70
Database Column Length:        70

Model Class:                                      MILP

Total variables:                    2980
Nonlinear variables:                   0
Integer variables:                  2940

Total constraints:                182681
Nonlinear constraints:                 0

Total nonzeros:                   373790
Nonlinear nonzeros:                    0
```

Fig. 6.8 Problem 33: LINGO solution report

```
!33.6;
@FOR(Orders(p):@FOR(Streets(i):@SUM(PO(k,p):@SUM(Sections(j)|W(k)
#LE#WS(j)#AND#H(k)#LE#HS(j)#AND#S(j)#EQ#i:Alpha(k,j)))<=n*Delta
(p,i)));

!33.7;
@FOR(Orders(p):y(p)=@SUM(Streets(i):Delta(p,i)));
!33.1;
@FOR(Products(k):@FOR(Sections(j)|W(k)#LE#WS(j)#AND#H(k)#LE#HS
(j):@BIN(Alpha(k,j))));
!33.5;
@FOR(Orders(p):@FOR(Streets(i):@BIN(Delta(p,i))));

END
```

Solution

The solution is shown in Fig. 6.8.

6.9 Problem 34. Assignment of Engineers to Projects

We plan to assign workers (engineers) to the projects of a company. For the next year, the company has the opportunity to participate in ten engineering projects. Each project has personnel requirements, provides income, and has general costs.

The company's engineers are divided into two categories:

- Senior engineer (with experience)
- Junior engineer (without enough experience)

Currently, the company has a staff available consisting of seven senior engineers and four junior engineers.

Each engineer, whether senior or junior, has a suitability for each of the projects, depending on their training. This suitability is measured as a continuous index between 0 (none) and 1 (optimal).

Each engineer has an annual salary. The company has the possibility of hiring new engineers. The costs that the company estimates for hiring new staff are as follows:

- Senior engineer: $50,000/year
- Junior engineers: $35,000/year

As already mentioned, each project requires a number of engineers. For Project j, A_j engineers (of any type) would be needed, in case the company undertakes the project.

There are two types of assignment for engineers:

- Part-time
- Full-time

The company allows an existing engineer to be assigned part-time or full-time (exclusive) to projects. If he is full-time, he can only be on one project. If you are part-time, you can be in a maximum of two. New engineers will have exclusive dedication. On a project, the salary cost of an engineer assigned to a part-time would be half her salary.

The number of engineers that a project needs is always a value assuming exclusivity. If part-time engineers are assigned, two part-time engineers count as one engineer.

The company wants each project in which it participates to have at least one senior engineer (new or existing) on a full-time basis.

The company wants the total adequacy of the engineers assigned to a project to always be above 50%. For new engineers, the company assigns a suitability of 75% to seniors and 50% to juniors, since personnel will be sought according to the projects. The total suitability calculation is performed as the sum of the suitability of the engineers assigned to the project divided by the number of engineers assigned, whether they are part-time or full-time.

In addition to all this, there are other specifications:

- A project must not have more than three part-time engineers.
- Projects 5 and 6 cannot have any engineer in common in the two projects.
- Participating part-time in Project 2 requires participation in Project 3.

The objective of the company is to maximize profits: Income from projects—Costs (Salary Costs + General Costs).

Modeling
List of elements in Table 6.8.

Table 6.8 Problem 34: table of elements

Elements	Set	QN	Data				
			Name	Parameter	Type	Belonging	Value
Projects	$j = 1\ldots10$	I_U	Required Engineers	A_j	I	W	...
			Costs	C_j	C($)	W	...
			Income	G_j	C($)	W	...
			Suitability	II_{ij}	B	S	...
Engineers	$i = 1\ldots11$	I_U	Senior	S_i	B	W	...
			Junior	JJ_i	B	W	...
			Salary	M_i	C($)	W	...
				II_{ij}			
New senior engineers	–	C_I	Cost	CS	C($)	S	50,000
			Suitability	INS	C%)	W	75%
New junior engineers	–	C_I	Cost	CJ	C($)	S	35,000
			Suitability	INJ	C	W	50%
Assignment mode	$k = 1, 2$ {part-time, full-time]	I_U					

Decision Activities
Assign engineers (*i*) to projects (*j*) in assignment modes (*k*)

$$\forall i, j, k : \alpha_{ijk} = 1 \text{ if engineer } i \text{ is assigned to Project } j \text{ in mode } k \tag{34.1}$$

Hire new senior and junior engineers for projects (*j*)

$$x_j : \text{Number of new senior engineers hired for Project } j \tag{34.2}$$

$$y_j : \text{Number of new junior engineers hired for Project } j \tag{34.3}$$

We are going to consider the participation in or selection of projects, which could be considered as a decision activity, as a logical calculation. If personnel are assigned to the project, then we select the project; otherwise, we do not.

Binary Logical Calculation

$$\forall j : \delta_j = 1 \text{ if we participate in Project } j; 0 \text{ otherwise.} \tag{34.4}$$

$$\forall j : \delta_j = 1 \text{ IF AND ONLY IF } \left(\sum_{i=1}^{11} \sum_{k=1}^{2} \alpha_{ijk} > 0 \text{ OR } x_j + y_j > 0 \right)$$

$$\Rightarrow f_{S_V} \Rightarrow \forall j : \text{IF } \delta_j = 1 \text{ THEN } \sum_{i=1}^{11} \sum_{k=1}^{2} \alpha_{ijk} > 0 \text{ OR } x_j + y_j > 0$$

Since there is a later specification that indicates the total number of engineers in a chosen project and another that indicates that there must be a senior engineer in the project, we can also use one of them as a definition of the calculation and thus simplify the number of constraints.

Specifications

– Quantitative selection specifications:

Activity	Elements selecting		Selectable elements	Selection	Constraints
α_{ijk}	Engineers ($i = 1\ldots11$)	Assignment mode ($k = 1$)	Projects ($j = 1\ldots10$)	≤ 2	$\forall i, k = 1 : \sum_{j=1}^{10} \alpha_{ijk} \leq 2$ (34.5)
		Assignment mode ($k = 2$)	Projects ($j = 1\ldots10$)	≤ 1	$\forall i, k = 2 : \sum_{j=1}^{10} \alpha_{ijk} \leq 1$ (34.6)
	Project ($j = 1\ldots10$)	Assignment mode ($k = 1$)	Engineers ($i = 1\ldots11$)	≤ 3	$\forall j, k = 1 : \sum_{i=1}^{11} \alpha_{ij1} \leq 3$ (34.7)

– If we assign an engineer as full-time, we cannot assign him as part-time (propositional logic specification):

$$\forall i : \text{IF } \sum_{j=1}^{10} \alpha_{ij2} \geq 1 \text{ THEN } \sum_{j=1}^{10} \alpha_{ij1} = 0$$

$$\Rightarrow f_{\text{LB}} \Rightarrow \forall i : \text{IF } \sum_{j=1}^{10} \alpha_{ij2} \geq 1 \text{ THEN } \sum_{j=1}^{10} \alpha_{ij1} \leq 0$$

$\sum_{j=1}^{10} \alpha_{ij2}$ can be considered a binary expression.

$$\Rightarrow f_{14}\left(\mathrm{UB}_{\sum_{j=1}^{10}\alpha_{ij1}}=2\right)\Rightarrow \forall i:\sum_{j=1}^{10}\alpha_{ij1}\leq 2\left(1-\sum_{j=1}^{10}\alpha_{ij2}\right) \tag{34.8}$$

- Number of engineers in each selected project (propositional logic specification):

$$\forall j:\text{IF }\ \delta_j=1\ \text{ THEN }\ \sum_{i=1}^{11}\left(\frac{\alpha_{ij1}}{2}+\alpha_{ij2}\right)+x_j+y_j=A_j$$

And

$$\forall j:\text{IF }\ \delta_j=0\ \text{ THEN }\ \sum_{i=1}^{11}\left(\frac{\alpha_{ij1}}{2}+\alpha_{ij2}\right)+x_j+y_j=0$$

Although the latter would not be necessary for a cost concept, we are going to consider it to collect the final simplification (34.11).

$$\Rightarrow f_{\mathrm{UB}}\Rightarrow \forall j:\text{IF }\ \delta_j=1\ \text{ THEN }\ \sum_{i=1}^{11}\left(\frac{\alpha_{ij1}}{2}+\alpha_{ij2}\right)+x_j+y_j\geq A_j$$

$$\Rightarrow f_{15}\Rightarrow \sum_{i=1}^{11}\left(\frac{\alpha_{ij1}}{2}+\alpha_{ij2}\right)+x_j+y_j\geq A_j\delta_j \tag{34.9}$$

$$\Rightarrow f_7, f_{\mathrm{LB}}\Rightarrow \forall j:\text{IF }\ 1-\delta_j=1\ \text{ THEN }\ \sum_{i=1}^{11}\left(\frac{\alpha_{ij1}}{2}+\alpha_{ij2}\right)+x_j+y_j\leq 0$$

$$\Rightarrow f_{14}\left(\mathrm{UB}_{\sum_{i=1}^{11}\left(\frac{\alpha_{ij1}}{2}+\alpha_{ij2}\right)+x_j+y_j}=A_j\right)\Rightarrow$$
$$\Rightarrow \forall j:\ \sum_{i=1}^{11}\left(\frac{\alpha_{ij1}}{2}+\alpha_{ij2}\right)+x_j+y_j\leq A_j\delta_j \tag{34.10}$$

Joining (34.9) y (34.10):

$$\forall j:\ \sum_{i=1}^{11}\left(\frac{\alpha_{ij1}}{2}+\alpha_{ij2}\right)+x_j+y_j=A_j\delta_j \tag{34.11}$$

- Suitability (bound imposition):

$$\forall j : \sum_{i=1}^{11} \sum_{k=1}^{2} \mathrm{II}_{ij} \alpha_{ijk} + \mathrm{INS} x_j + \mathrm{INJ} y_j \geq 0.5 \left(\sum_{i=1}^{11} \sum_{k=1}^{2} \alpha_{ijk} + x_j + y_j \right) \tag{34.12}$$

– A senior engineer in each selected project (propositional logic specification):

$$\forall j : \text{IF } \delta_j = 1 \text{ THEN } x_j + \sum_{i/S_i=1} \alpha_{ij2} \geq 1$$

$$\Rightarrow f_{15} \left(\mathrm{LB}_{x_j + \sum_{i/S_i=1} \alpha_{ij1}} = 0 \right) \Rightarrow \forall j : \; x_j + \sum_{i/S_i=1} \alpha_{ij2} \geq \delta_j \tag{34.13}$$

– Projects 5 and 6 cannot have any engineer in common (propositional logic specification).

$$\forall i : \text{IF } \sum_{k=1}^{2} \alpha_{i5k} > 0 \text{ THEN } \sum_{k=1}^{2} \alpha_{i6k} = 0$$

$$\Rightarrow f_{\mathrm{LB}}, f_5 \Rightarrow \forall i : \text{IF } \sum_{k=1}^{2} \alpha_{i5k} \geq 1 \text{ THEN } \sum_{k=1}^{2} \alpha_{i6k} \leq 0$$

$$\Rightarrow f_{18} \left(\mathrm{UB}_{\sum_{k=1}^{2} \alpha_{i5k}} = 1 \right) \Rightarrow \forall i : \sum_{i=1}^{11} \sum_{k=1}^{2} \alpha_{i5k} \leq \omega_i \tag{34.14}$$

$$\Rightarrow f_{20} \left(\mathrm{UB}_{\sum_{k=1}^{2} \alpha_{i6k}} = 1 \right) \Rightarrow \forall i : \sum_{i=1}^{11} \sum_{k=1}^{2} \alpha_{i6k} \leq 1 - \omega_i \tag{34.15}$$

– Participation in Project 2 requires participation in Project 3 (propositional logic specification).

$$\text{IF } \delta_2 = 1 \text{ THEN } \delta_3 = 1 \Rightarrow f_{13} \Rightarrow \; \delta_2 \leq \delta_3 \tag{34.16}$$

Objective Function

Maximize income – costs.

$$\text{Income} = \sum_{j=1}^{10} G_j \delta_j$$

$$\text{Costs} = \text{Salary costs} + \text{General cost}$$

$$\text{Salary costs} = \text{Existing engineer cost} + \text{New engineer cost}$$

$$\text{Cost of existing engineers cost} = \sum_{i=1}^{11} \left(\sum_{j=1}^{10} M_i \alpha_{ij1} + \sum_{j=1}^{10} \frac{M_i}{2} \alpha_{ij2} \right)$$

$$\text{New engineers cost} = \sum_{j=1}^{10} \left(\text{CS}x_j + \text{CJ}y_j \right)$$

$$\text{General cost} = \sum_{j=1}^{10} C_j \delta_j$$

Expression:

$$\text{Max} \ \sum_{j=1}^{10} G_j \delta_j - \left(\sum_{i=1}^{11} \left(\sum_{j=1}^{10} M_i \alpha_{ij1} + \sum_{j=1}^{10} \frac{M_i}{2} \alpha_{ij2} \right) + \sum_{j=1}^{10} \left(\text{CS}x_j + \text{CJ}y_j \right) + \sum_{j=1}^{10} C_j \delta_j \right) \tag{34.17}$$

Complete Model

$$\text{Max} \ \sum_{j=1}^{10} G_j \delta_j - \left(\sum_{i=1}^{11} \left(\sum_{j=1}^{10} M_i \alpha_{ij1} + \sum_{j=1}^{10} \frac{M_i}{2} \alpha_{ij2} \right) + \sum_{j=1}^{10} \left(\text{CS}x_j + \text{CJ}y_j \right) + \sum_{j=1}^{10} C_j \delta_j \right) \tag{34.17}$$

s.t.

$$\forall i, k = 1 : \sum_{j=1}^{10} \alpha_{ijk} \leq 2 \tag{34.5}$$

$$\forall i, k = 2 : \sum_{j=1}^{10} \alpha_{ijk} \leq 1 \tag{34.6}$$

$$\forall j, k = 1 : \quad \sum_{i=1}^{11} \alpha_{ij1} \leq 3 \tag{34.7}$$

$$\forall i : \sum_{j=1}^{10} \alpha_{ij1} \leq 2\left(1 - \sum_{j=1}^{10} \alpha_{ij2}\right) \tag{34.8}$$

$$\forall j : \quad \sum_{i=1}^{11} \left(\frac{\alpha_{ij1}}{2} + \alpha_{ij2}\right) + x_j + y_j = A_j \delta_j \tag{34.11}$$

$$\forall j : \sum_{i=1}^{11} \sum_{k=1}^{2} \mathrm{II}_{ij} \alpha_{ijk} + \mathrm{INS} x_j + \mathrm{INJ} y_j \geq 0.5\left(\sum_{i=1}^{11} \sum_{k=1}^{2} \alpha_{ijk} + x_j + y_j\right) \tag{34.12}$$

$$\forall j : \quad x_j + \sum_{i/S_i=1} \alpha_{ij2} \geq \delta_j \tag{34.13}$$

$$\forall i : \sum_{i=1}^{11} \sum_{k=1}^{2} \alpha_{i5k} \leq \omega_i \tag{34.14}$$

$$\forall i : \sum_{i=1}^{11} \sum_{k=1}^{2} \alpha_{i6k} \leq 1 - \omega_i \tag{34.15}$$

$$\delta_2 \leq \delta_3 \tag{34.16}$$

$$\forall i, j, k : \alpha_{ijk} \ \text{binaries} \tag{34.1}$$

$$\forall j : x_j \geq 0, y_j \geq 0 \ \text{integers} \qquad (34.2, 34.3)$$

$$\forall j : \delta_j \ \text{binaries} \tag{34.4}$$

$$\forall i : \omega_i \ \text{binaries} \tag{34.18}$$

Implementation in LINGO

Input file "34_Excel.xlsx" (Fig. 6.9).

```
MODEL:
SETS:
 Projects/1..10/:A,C,G,x,y,Delta;
 Engineers/1..11/:S,JJ,M,Omega;
 Modes/1..2/;
 EP(Engineers,Projects):II,Beta;
 EPM(Engineers,Projects,Modes):Alpha;
ENDSETS
DATA:
 A=@OLE('C:\Files_\34_Excel.xlsx','A');
 C=@OLE('C:\Files_\34_Excel.xlsx','Cost');
 G=@OLE('C:\Files_\34_Excel.xlsx','G');
```

	A	B	C	D	E	F	G	H	I	J	K	L	M	N
1														
2														
3	Projects	1	2	3	4	5	6	7	8	9	10			
4	A	3	3	3	2	2	2	4	4	4	5			
5	C	10000	20000	25000	40000	10000	40000	20000	25000	10000	30000			
6	G	55000	60000	60000	90000	30000	1E+05	65000	50000	50000	60000			
7														
8	Engineers	1	2	3	4	5	6	7	8	9	10	11		
9	S	1	1	0	0	0	1	0	1	0	1	0		
10	JJ	0	0	1	1	1	0	1	0	1	0	1		
11	M	50000	45000	30000	35000	28000	58000	36000	55000	35000	50000	33000		
12														
13			1	2	3	4	5	6	7	8	9	10		
14	Suitability	1	0	1	0,5	0,5	0,5	1	0,3	0,2	0,4	0,8		
15		2	0,5	0,5	1	0,3	0,2	0,4	1	1	0	0		
16		3	1	1	0	0	1	0	0,4	0,5	0,6	0,5		
17		4	2	0,5	0,5	1	0,3	0,2	0,4	2	0,3	0,5		
18		5	0,5	1	0,3	0,2	0,4	1	1	1	0,3	0		
19		6	0,2	0,5	0,5	3	0	0	1	1	1	0,6		
20		7	1	0,6	0,7	0,5	0,5	1	0,3	0,2	0,4	0,6		
21		8	1	3	3	3	0,5	1	0,3	0,2	0,4	1		
22		9	0	1	1	0,5	0,5	1	0,3	0,2	0,4	0,1		
23		10	0,5	1	0,3	0,2	0,5	1	0,3	0,2	0,4	1		
24		11	0,5	1	0,3	0,2	0,5	1	0,3	0,2	0,4	1		
25														
26														

Fig. 6.9 "34_Excel.xlsx"

```
 S=@OLE('C:\Files_\34_Excel.xlsx','S');
 JJ=@OLE('C:\Files_\34_Excel.xlsx','JJ');
 M=@OLE('C:\Files_\34_Excel.xlsx','M');
 II=@OLE('C:\Files_\34_Excel.xlsx','II');
 CS=50000;
 INS=75;
 CJ=35000;
 INJ=50;
ENDDATA
!34.17;
MAX= @SUM(Projects(j):G(j)*Delta(j))-(@SUM(Engineers(i):@SUM
(Projects(j):M(i)*Alpha(i,j,1))+
@SUM(Projects(j):(M(i)/2)*Alpha(i,j,2)))+@SUM(Projects(j):CS*x(j)
+CJ*y(j))
+@SUM (Projects(j):C(j)*Delta(j)));
!34.5;

@FOR(Engineers(i):@SUM(Projects(j):Alpha(i,j,1))<=2);
!34.6;
@FOR(Engineers(i):@SUM(Projects(j):Alpha(i,j,2))<=1);
!34.7;
@FOR(Projects(j):@SUM(Engineers(i):Alpha(i,j,1))<=3);
!34.8;
@FOR(Engineers(i):@SUM(Projects(j):Alpha(i,j,1))<=2*(1-@SUM
(Projects(j):Alpha(i,j,2))));
!34.11;
@FOR(Projects(j):@SUM(Engineers(i):(Alpha(i,j,1)/2)+Alpha(i,j,2))
+x(j)+y(j)=A(j)*Delta(j));
!34.12;
@FOR(Projects(j):@SUM(Engineers(i):@SUM(Modes(k):II(i,j)*Alpha(i,
j,k)))+INS*x(j)+INJ*y(j)>=0.5*(@SUM(Engineers(i):@SUM(Modes(k):
Alpha(i,j,k)))+x(j)+y(j)));
```

```
!34.13;
@FOR(Projects(j):x(j)+@SUM(Engineers(i)|S(i)#EQ#1:Alpha(i,j,2))
>=Delta(j));
!34.14;
@FOR(Engineers(i):@SUM(Modes(k):Alpha(i,5,k))<=Omega(i));
!34.15;
@FOR(Engineers(i):@SUM(Modes(k):Alpha(i,6,k))<=1-Omega(i));

!34.16;
Delta(2)<=Delta(3);
!34.1;
@FOR(Engineers(i):@FOR(Projects(j):@FOR(Modes(k):@BIN(Alpha(i,j,
k)))));
!34.2;
@FOR(Projects(j):@GIN(x(j)));
!34.3;
@FOR(Projects(j):@GIN(y(j)));
!34.4;
@FOR(Projects(j):@BIN(Delta(j)));
!34.18;
@FOR(Engineers(i):@BIN(Omega(i)));
END
```

Solution
It is shown in Fig. 6.10.

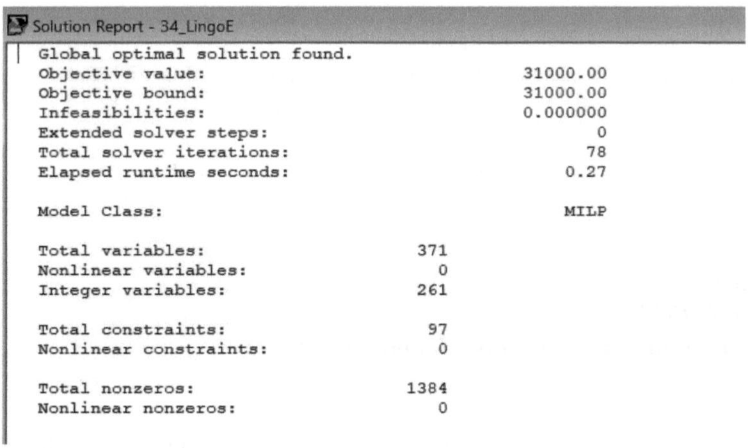

```
Solution Report - 34_LingoE
Global optimal solution found.
Objective value:                              31000.00
Objective bound:                              31000.00
Infeasibilities:                              0.000000
Extended solver steps:                               0
Total solver iterations:                            78
Elapsed runtime seconds:                          0.27

Model Class:                                      MILP

Total variables:                   371
Nonlinear variables:                 0
Integer variables:                 261

Total constraints:                  97
Nonlinear constraints:               0

Total nonzeros:                   1384
Nonlinear nonzeros:                  0
```

Fig. 6.10 Problem 34: LINGO solution report

6.10 Problem 35. Excursion Planning

We have a group of 60 excursionists who have hired the services of the BUSTOUR company for the next 3 days. There are six different excursions that can be done. Each excursionist has chosen a maximum of three excursions. An excursionist can only take one excursion in a day.

The company buses have a capacity or number of seats. It has five buses. A bus in a day can make several excursions depending on the proximity between them. This information will be collected in a binary attribute between excursions (1: they can be done by the same bus, 0: no). However, a bus should not cover more than two excursions in 1 day.

The company wants to plan excursions on 3 days to use the least number of buses (the use of each bus each day is added in the objective function). We should establish an excursionist-to-bus and excursion-to-bus allocation system that ensures that all excursions are conducted for all excursionists.

Modeling

We will propose two versions. The first one with two decision activities that are joined into one in the second version.

List of elements in Table 6.9.

Table 6.9 Problem 35: table of elements

			Data				
Elements	Set	QN	Name	Parameter	Type	Belonging	Value
Excursionists	$i = 1\ldots60$	I_U	Excursion Election	EE_{ij}	B	S	...
Excursions	j, $j' = 1\ldots6$	I_U	Proximity	$Near_{jj'}$	B	S	
				EE_{ij}			
Days	$t = 1\ldots3$	I_U					
Buses	$r = 1\ldots5$	I_U	Capacity	K_r	I	S	60
Seats	–	C_D		K_r			

Version 1

Decision Activities

Assign excursions (*j*) to buses (*r*) on days (*t*)

$$\forall j, t, r : \alpha_{jtr} = 1 \text{ if we assign the excursion } j \text{ to the bus } r \text{ on day } t; 0 \text{ otherwise} \tag{35.1}$$

Assign excursionists (*i*) to buses (*r*) on days (*t*)

$$\forall i,t,r : \beta_{itr} = 1 \text{ if we assign the excursionist } i \text{ to the bus } r \text{ on day } t; 0 \text{ otherwise.} \tag{35.2}$$

Specifications

– Quantitative selection specifications:

Activity	Elements selecting		Selectable elements	Selection	Constraints
α_{jtr}	Days ($t = 1\ldots3$)	Buses ($r = 1\ldots5$)	Excursions ($j = 1\ldots6$)	≤ 2	$\forall t,r : \sum_{j=1}^{6} \alpha_{jtr} \leq 2$ (35.3)
β_{itr}	Excursionists ($i = 1\ldots60$)	Days ($t = 1\ldots3$)	Buses ($r = 1\ldots5$)	≤ 1	$\forall i,t : \sum_{r=1}^{5} \beta_{itr} \leq 1$ (35.4)
	Days ($t = 1\ldots3$)	Buses ($r = 1\ldots5$)	Excursionists ($i = 1\ldots60$)	$\leq K_r$	$\forall t,r : \sum_{i=1}^{60} \beta_{itr} \leq K_r$ (35.5)

– Compatible excursions on the same bus (propositional logic specifications)

$$\forall r, \forall j, j'/j < j' \& \text{Near}_{jj'} = 0, \forall t : \text{NOT}\left(\alpha_{jtr} = 1 \text{ AND } \alpha_{j'tr} = 1\right)$$
$$\Rightarrow f_{37}, f_{11} \Rightarrow \forall r, \forall j, j'/j < j' \& \text{Near}_{jj'} = 0, \forall t : \alpha_{jtr} + \alpha_{j'tr} \leq 1 \tag{35.6}$$

– Each excursionist must make the chosen excursions (binary logical calculation + quantitative selection specification):

It is necessary to define a logical calculation to know if an excursionist i takes bus r on day t with excursion j:

Binary Logical Calculation

$$\forall i, \forall j/\text{EE}_{ij} = 1, \forall t,$$
$$\forall r : \delta_{ijtr} = 1 \text{ if we assign bus } r \text{ on day } t \text{ to excursionist } i \text{ with excursion } j \text{(chosen by } i\text{);}$$
$$0 \text{ otherwise.} \tag{35.7}$$

$$\forall i, \forall j/\text{EE}_{ij} = 1, \forall t, \forall r : \delta_{ijtr} = 1 \text{ IF AND ONLY IF } \beta_{itr} = 1 \text{ AND } \alpha_{jtr} = 1$$
$$\Rightarrow f_{37} \Rightarrow \forall i, \forall j/\text{EE}_{ij} = 1, \forall t, \forall r : \delta_{ijtr} = 1 \text{ IF AND ONLY IF } \beta_{itr} + \alpha_{jtr} \geq 2$$
$$\Rightarrow f_{S_V} \Rightarrow \forall i, \forall j/\text{EE}_{ij} = 1, \forall t, \forall r : \text{IF } \delta_{ijtr} = 1 \text{ THEN } \beta_{itr} + \alpha_{jtr} \geq 2$$
$$\Rightarrow f_{15}\left(\text{LB}_{\beta_{itr}+\alpha_{jtr}} = 0\right) \Rightarrow \forall i, \forall j/\text{EE}_{ij} = 1, \forall t, \forall r : \ \beta_{itr} + \alpha_{jtr} \geq 2\delta_{ijtr} \quad (35.8)$$

Now we impose a quantitative rule on those variables:

$$\forall i, \forall j/\text{EE}_{ij} = 1 : \sum_{r=1}^{5} \sum_{t=1}^{3} \delta_{ijtr} = 1 \quad (35.9)$$

– At most one excursion per day for each excursionist (upper bound imposition).

It is another quantitative selection norm with respect to the previous logical calculation.

$$\forall i, \forall t : \sum_{r=1}^{5} \sum_{j/\text{EE}_{ij}=1} \delta_{ijtr} \leq 1 \quad (35.10)$$

Objective Criterion
We calculate the use of each bus each day.
Binary Logical Calculation

$$\forall t, \forall r : \pi_{tr} = 1 \text{ if we use the bus } r \text{ on day } t; 0 \text{ otherwise.} \quad (35.11)$$

$$\forall t, \forall r : \pi_{tr} = 1 \text{ IF AND ONLY IF } \sum_{j=1}^{6} \alpha_{jtr} > 0$$

$$\Rightarrow f_{S_V} \Rightarrow \forall t, \forall r : \text{IF } \sum_{j=1}^{6} \alpha_{jtr} > 0 \text{ THEN } \pi_{tr} = 1$$

$$\Rightarrow f_3, f_7 \Rightarrow \forall t, \forall r : \text{IF } 1 - \pi_{tr} = 1 \text{ THEN } \sum_{j=1}^{6} \alpha_{jtr} \leq 0$$

$$\Rightarrow f_{14}\left(\text{UB}_{\sum_{j=1}^{6} \alpha_{jtr}} = 2\right) \Rightarrow \forall t, \forall r : \sum_{j=1}^{6} \alpha_{jtr} \leq 2\pi_{tr} \quad (35.12)$$

Objective Function

$$\text{Min} \sum_{t=1}^{3} \sum_{r=1}^{5} \pi_{tr} \tag{35.13}$$

Complete Model

$$\text{Min} \sum_{t=1}^{3} \sum_{r=1}^{5} \pi_{tr} \tag{35.13}$$

s.t.

$$\forall t, r : \sum_{j=1}^{6} \alpha_{jtr} \leq 2 \tag{35.3}$$

$$\forall i, t : \sum_{r=1}^{5} \beta_{itr} \leq 1 \tag{35.4}$$

$$\forall t, r : \sum_{i=1}^{60} \beta_{itr} \leq K_r \tag{35.5}$$

$$\forall r, \forall j, j'/j < j' \& \text{Near}_{jj'} = 0, \forall t : \alpha_{jtr} + \alpha_{j'tr} \leq 1 \tag{35.6}$$

$$\forall i, \forall j/\text{EE}_{ij} = 1, \forall t, \forall r : \quad \beta_{itr} + \alpha_{jtr} \geq 2\delta_{ijtr} \tag{35.8}$$

$$\forall i, j/E_{ij} = 1 : \sum_{r=1}^{5} \sum_{t=1}^{3} \delta_{ijtr} = 1 \tag{35.9}$$

$$\forall i, \forall t : \sum_{r=1}^{5} \sum_{j/\text{EE}_{ij}=1} \delta_{ijtr} \leq 1 \tag{35.10}$$

$$\forall t, \forall r : \sum_{j=1}^{6} \alpha_{jtr} \leq 2\pi_{tr} \tag{35.12}$$

$$\forall j, t, r : \alpha_{jtr} \text{ binaries} \tag{35.1}$$

$$\forall i, t, r : \beta_{itr} \text{ binaries} \tag{35.2}$$

$$\forall i, j/E_{ij} = 1, t, r : \delta_{ijtr} \text{ binaries} \tag{35.7}$$

$$\forall t, r : \pi_{tr} \text{ binaries} \tag{35.11}$$

Implementation in LINGO

The input file "35_Excel.xlsx" is available for download.

```
MODEL:
SETS:
 Excursionists/1..60/;
```

```
 Excursions/1..6/;
 Days/1..3/;
 Buses/1..5/:K;
TE(Excursionists,Excursions)/@OLE('C:\Files_\35_Excel.
xlsx','EE')/;
 EE(Excursions,Excursions):Near;
 EDB(Excursions,Days, Buses):Alpha;
 TDB(Excursionists,Days, Buses):Beta;
 DB (Days, Buses):Pi;
 TEDB(Excursionists,Excursions,Days,Buses):Delta;
ENDSETS
DATA:
 K=@OLE('C:\Files_\35_Excel.xlsx','K');
 Near=@OLE('C:\Files_\35_Excel.xlsx','Near');
ENDDATA
!35.13;
MIN = @SUM(Days(t):@SUM(Buses(r):Pi(t,r)));
!35.3;
@FOR(Days(t):@FOR(Buses(r):@SUM(Excursions(j):Alpha(j,t,r))<=2));
!35.4;
@FOR(Excursionists(i):@FOR(Days(t):@SUM(Buses(r):Beta(i,t,r))
<=1));
!35.5;
@FOR(Days(t):@FOR(Buses(r):@SUM(Excursionists(i):Beta(i,t,r))<=K
(r)));
!35.6;
@FOR(Buses(r):@FOR(Excursions(j):@FOR(Excursions(jj)|
j#LT#jj#AND#Near(j,jj)#EQ#0:@FOR(Days(t):Alpha(j,t,r) + Alpha(jj,t,
r)<=1))));
!35.8;
@FOR(TE(i,j):@FOR(Days(t):@FOR(Buses(r):Beta(i,t,r)+Alpha(j,t,r)
>=2*Delta(i,j,t,r))));
!35.9;
@FOR(TE(i,j):@SUM(Buses(r):@SUM(Days(t):Delta(i,j,t,r)))=1);

!35.10;
@FOR(Excursionists(i):@FOR(Days(t):@SUM(Buses(r):@SUM(TE(i,j):
Delta(i,j,t,r)))<=1));
!35.12;
@FOR(Days(t):@FOR(Buses(r):@SUM(Excursions(j):Alpha(j,t,r))<=2*Pi
(t,r)));
!35.1;
@FOR(Excursions(j):@FOR(Days(t):@FOR(Buses(r):@BIN(Alpha(j,t,
r)))));
!35.2;
@FOR(Excursionists(i):@FOR(Days(t):@FOR(Buses(r):@BIN(Beta(i,t,
r)))));
!35.7;
@FOR(TE(i,j):@FOR(Days(t):@FOR(Buses(r):@BIN(Delta(i,j,t,r)))));
!35.11;
@FOR(Days(t):@FOR(Buses(r):@BIN(Pi(t,r))));
 END
```

```
Solution Report - 35_LingoE
Global optimal solution found.
Objective value:                                 5.000000
Objective bound:                                 5.000000
Infeasibilities:                                 0.000000
Extended solver steps:                               1292
Total solver iterations:                          2762543
Elapsed runtime seconds:                           644.96

Model Class:                                         MILP

Total variables:                    6405
Nonlinear variables:                   0
Integer variables:                  2805

Total constraints:                  2461
Nonlinear constraints:                 0

Total nonzeros:                    11280
Nonlinear nonzeros:                    0
```

Fig. 6.11 Problem 35: LINGO solution report

Solution

The solution is shown in Fig. 6.11.

Version 2: With Union of Activities

Decision Activities

Assign excursionists (*i*) to excursions (*j*) in buses (*r*) on days (*t*)

$\forall i, \forall j/\text{EE}_{ij} = 1, \forall t, \forall r : \delta_{ijtr}=1$ if I assign excursionist *i* to excursion *j* on day *t* in bus *r*; 0 otherwise.

It corresponds to the binary logical calculation performed in the previous version.

Specifications

– Quantitative selection specifications:

Activity	Elements selecting		Selectable elements		Selection	Constraints
$\forall i, \forall j/\text{EE}_{ij} = 1, \forall t, \forall r : \delta_{ijtr}$	Excursionists ($i = 1\ldots60$)	Days ($t = 1\ldots3$)	Excursions ($j = 1\ldots6$)	Buses ($r = 1\ldots5$)	≤ 1	$\forall i, t : \sum_{r=1}^{5} \sum_{j/E_{ij}=1} \delta_{ijtr} \leq 1$
	Buses ($r = 1\ldots5$)	Days ($t = 1\ldots3$)	Excursionists ($i = 1\ldots60$)	Excursions ($j/E_{ij} = 1$)	$\leq K_r$	$\forall r, t : \sum_{i=1}^{60} \sum_{j/E_{ij}=1} \delta_{ijtr} \leq K_r$
	Excursionists ($i = 1\ldots60$)	Excursions ($j/E_{ij} = 1$)	Buses ($r = 1\ldots5$)	Days ($t = 1\ldots3$)	$=1$	$\forall i, j/E_{ij} = 1 : \sum_{r=1}^{5} \sum_{t=1}^{3} \delta_{ijtr} = 1$

– Two excursions at most for each day to each bus (binary logical calculation + bounds imposition)

We make a logical calculation to know if each bus, each day, is assigned each excursion (it coincides with the decision activity of the previous version).

Binary Logical Calculation

$$\forall j,t,r : \alpha_{jtr} = \text{ 1 if bus } r \text{ on day } t \text{ makes excursion } j.$$

$$\forall j,t,r : \alpha_{jtr} = 1 \text{ IF AND ONLY IF } \sum_{i/E_{ij}=1} \delta_{ijtr} > 0$$

$$\Rightarrow f_{S_V} \Rightarrow \forall j,t,r : \text{IF } \sum_{i/E_{ij}=1} \delta_{ijtr} > 0 \text{ THEN } \alpha_{jtr} = 1$$

$$\Rightarrow f_3, f_7 \Rightarrow \forall j,t,r : \text{IF } 1 - \alpha_{jtr} = 1 \text{ THEN } \sum_{i/E_{ij}=1} \delta_{ijtr} \leq 0$$

$$\Rightarrow f_{14}\left(\text{UB}_{\sum_{i/E_{ij}=1} \delta_{ijtr}} = K_r\right) \Rightarrow \forall j,t,r : \sum_{i/E_{ij}=1} \delta_{ijtr} \leq K_r \alpha_{jtr}$$

And now we impose a maximum of two excursions:

$$\forall t,r : \sum_{j=1}^{6} \alpha_{jtr} \leq 2$$

- Compatible excursions on the same bus

  Identical to the previous version.
  Objective Criterion
  Identical to the previous version.

6.11 Problem 36. Iron Bars Production

Let be a factory where cuts are made on a set of bars existing in the warehouse ($j = 1\ldots n$). Each bar j has a length LA_j. Cuts must be used to obtain a set of requested bars ($i = 1\ldots m$). Each set i has a length ld_i and a number of bars D_i.

The market does not demand bars of less than 2 m, so we want to minimize the total length of surplus pieces of less than 2 m, i.e., minimize waste pieces (Shahin and Ossama 2004).

I could also have considered as a specification not allowing pieces smaller than 2 m and as an objective to minimize the number of warehouse bars used, but I have used the first objective because it had a more interesting modeling. We will also add a cost C for each warehouse bar used so that we do not use too many warehouse bars.

Modeling
We raise the problem by measuring the demanded units in the decisions. Warehouse bar measurement is done in the specifications. List of elements in Table 6.10.

Table 6.10 Problem 36: Table of elements

			Data				
Elements	Set	QN	Name	Parameter	Type	Belonging	Value
Warehouse bars	$j = 1 \ldots n$	I_U	Length	LA_j	C (cm)	W	...
Demanded bars	$i = 1 \ldots m$	C_D	Quantity	ND_i	I	W	...
			Length	LD_i	C(cm)	W	...
System	–	I_U	Cost	C	C($)	W	...

Decision Activities

Cut demanded bars (*i*) from warehouse bars (*j*)

$$x_{ij} = \text{Number of demanded bars } i \text{ cut from the warehouse bar } j. \tag{36.1}$$

Specifications

– Demand for bars (demand specification)

$$\forall i : \sum_{j=1}^{n} x_{ij} = \mathrm{ND}_i \tag{36.2}$$

– Length of the warehouse bars (capacity specification)

$$\forall j : \sum_{i=1}^{m} \mathrm{LD}_i x_{ij} \leq \mathrm{LA}_j \tag{36.3}$$

Objective Criterion

Minimize the total length of surplus pieces of less than 2 m.

We start by getting the surplus piece from each bar. It is enough to collect in an auxiliary calculation the difference between the length of the bar and the quantity cut:

Auxiliary Calculation

$$\forall j : S_j = \text{Length of the excess piece of bar } j$$

$$\forall j : S_j = \mathrm{LA}_j - \sum_{i=1}^{m} \mathrm{LD}_i x_{ij} \tag{36.4}$$

This constraint can replace the capacity constraint (36.3).

Now we have to penalize in the objective function only each piece j that is less than 2 m. From two meters, they should not penalize. We have an objective function

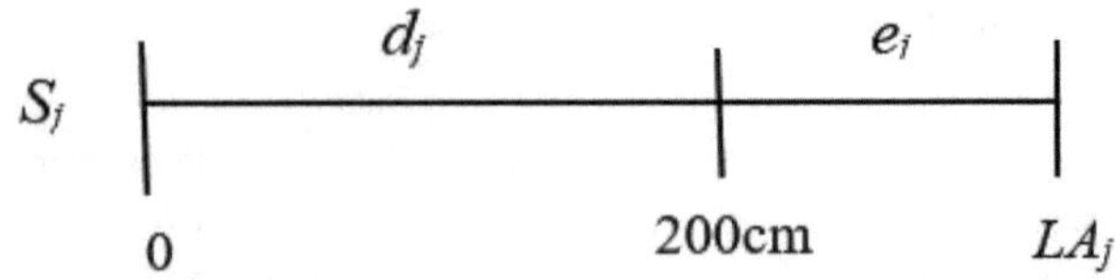

Fig. 6.12 Data and variables for penalizing the defect

that penalizes the defect over a threshold (Sect. 1.6.3.3). Data and variables in Fig. 6.12

$$\forall j : S_j + d_j - e_j = 200 \tag{36.5}$$

However, if the surplus is positive, the cost per use of the bar must be imputed. We get that information from a binary logic calculation:

Binary Logical Calculation

$$\forall j : \alpha_j = 1 \text{ if we use bar } j; 0 \text{ otherwise.} \tag{36.6}$$

$$\forall j : \alpha_j = 1 \text{ IF AND ONLY IF } S_j > 0$$

$$\Rightarrow f_{S_V} \Rightarrow \forall j : \text{ IF } S_j > 0 \text{ THEN } \alpha_j = 1$$

$$\Rightarrow f_3, f_7 \Rightarrow \forall j : \text{ IF } 1 - \alpha_j = 1 \text{ THEN } S_j \leq 0$$

$$\Rightarrow f_{14}(\text{UB}_{S_j} = \text{LA}_j) \Rightarrow \forall j : S_j \leq \text{LA}_j \alpha_j \tag{36.7}$$

Objective Function

$$\text{Min} \sum_{j=1}^{n} d_j + C\alpha_j \tag{36.8}$$

Complete Model

$$\text{Min} \sum_{j=1}^{n} d_j + C\alpha_j \tag{36.8}$$

s.t.

$$\forall i : \sum_{j=1}^{n} x_{ij} = \text{ND}_i \tag{36.2}$$

$$\forall j : S_j = \mathrm{LA}_j - \sum_{i=1}^{m} \mathrm{LD}_i x_{ij} \tag{36.4}$$

$$\forall j : S_j + d_j - e_j = 200 \tag{36.5}$$

$$\forall j : \quad S_j \leq \mathrm{LA}_j \alpha_j \tag{36.7}$$

$$\forall i,j : x_{ij} \geq 0 \text{ integers} \tag{36.1}$$

$$\forall j : S_j \geq 0, d_j \geq 0, e_j \geq 0$$

$$\forall j : \alpha_j \text{ binaries} \tag{36.6}$$

Implementation in LINGO

```
MODEL:
DATA:
 n=@ODBC('36_ODBC','System','n');
 m=@ODBC('36_ODBC','System','m');
ENDDATA
SETS:
 WarehouseBars/1..n/:LA,S,d,e,Alpha;
 DemandedBars/1..m/:LD,ND;
 BB(DemandedBars,WarehouseBars):x;
ENDSETS
DATA:
 LA=@ODBC('36_ODBC','WarehouseBars','Length');
LD,ND=@ODBC('36_ODBC','DemandedBars','Length','Number');
@ODBC('36_ODBC','WarehouseBars','Surplus')=@WRITEFOR
(WarehouseBars(j):S(j));
 C= 200;
ENDDATA
!36.6;
MIN = @SUM(WarehouseBars(j):d(j)+C*Alpha(j));
!36.2;
@FOR(DemandedBars(i):@SUM(WarehouseBars(j):x(i,j))=ND(i));
!36.4;
@FOR(WarehouseBars(j):S(j)=LA(j)-@SUM(DemandedBars(i):LD(i)*x(i,
j)));
!36.5;
@FOR(WarehouseBars(j):S(j)+d(j)-e(j)=200);
!36.7;
@FOR(WarehouseBars(j):S(j)<=LA(j)*Alpha(j));
!36.1;
@FOR(BB(i,j):@GIN(x(i,j)));
!36.6;
@FOR(WarehouseBars(j):@BIN(Alpha(j)));
END
```

Solution

The solution is shown in Fig. 6.13.

```
Solution Report - 36_LingoE
Global optimal solution found.
Objective value:                                    10000.00
Objective bound:                                    10000.00
Infeasibilities:                                    0.000000
Extended solver steps:                                   339
Total solver iterations:                               23501
Elapsed runtime seconds:                                0.44


Export Summary Report
---------------------
Transfer Method:          ODBC BASED
ODBC Data Source:         36_ODBC
Data Table Name:          WarehouseBars
Columns Specified:                 1
    Surplus
LINGO Column Length:              50
Database Column Length:           50

Model Class:                                            MILP

Total variables:                     700
Nonlinear variables:                   0
Integer variables:                   550

Total constraints:                   161
Nonlinear constraints:                 0
```

Fig. 6.13 Problem 36: LINGO solution report

6.12 Problem 37. Discount Management

We have three suppliers (A, B, and C) from whom we can buy units of a product. We have to buy N units.

The first one offers us a proportional discount: Three prices (pA_1 \$/unit, pA_2 \$/unit, pA_3 \$/unit) to apply to all units according to the number of units we request ($pA_1 > pA_2 > pA_3$). Consider, respectively, three intervals: $(0, A_1]$, $(A_1, A_2]$, and $(A_2, N]$.

The second offers us an incremental discount: Three prices (pB_1 \$/unit, pB_2 \$/unit, pB_3 \$/unit) to apply to the units of each interval (($pB_1 > pB_2 > pB_3$). Consider the intervals $(0, B_1]$, (B_1, *B2]*, and (B_2, N], respectively.

The third one offers us a fixed price pC \$/unit and fixed discounts: A discount of D_1\$ for ordering above C_1 units and a second discount that is added to the first of D_2\$ for ordering above C_2 units.

Minimize the cost of the purchase.

Summary

$$\text{Supplier A prices} = \begin{cases} \text{Units purchased} \in 0A_1] : & p_{A1}\$/u \\ \text{Units purchased} \in A_1A_2] : & p_{A2}\$/u \\ \text{Units purchased} \in A_2N] : & p_{A3}\$/u \end{cases}$$

$$\text{Supplier B prices} = \begin{cases} \text{Units purchased} \in 0B_1] : & p_{B1}\$/u \\ \text{Units purchased} \in B_1B_2] : & p_{B1}\$/u \quad \in 0B_1]; \\ \qquad p_{B2}\$/u \in B_1B_2] & \\ \text{Units purchased} \in B_2N] : & \\ p_{B1}\$/u \quad \in 0B_1]; p_{B2}\$/u \in B_1B_2]; p_{B3}\$/u \in B_2N] & \end{cases}$$

$$\text{Supplier C prices} = \begin{cases} \text{Fixed price :} & p_C€/ud \\ \text{Units purchased} \in C_1C_2] : & \text{Discount } D_1\$ \\ \text{Units purchased} \in C_2N] : & \text{Discount } D_2\$ \end{cases}$$

Modeling
List of elements in Table 6.11.

Table 6.11 Problem 37: table of elements

			Data				
Elements	Set	QN	Name	Parameter	Type	Belonging	Value
Product	–	C_D	Demand	N	I	W	–
Supplier A	–	I_U	Price 1	p_{A1}	C($/ud)	W	–
			Threshold 1	A_1	I	W	–
			Price 2	p_{A2}	C($/ud)	W	–
			Threshold 2	A_2	I	W	–
			Price 3	p_{A3}	C($/ud)	W	–
Supplier B	–	I_U	Price 1	p_{B1}	C($/ud)	W	–
			Threshold 1	B_1	I	W	–
			Price 2	p_{B2}	C($/ud)	W	–
			Threshold 2	B_2	I	W	–
			Price 3	p_{B3}	C($/ud)	W	–
Supplier C	–	I_U	Price	p_C	C($/ud)	W	–
			Discount 1	D_1	C($)	W	–
			Threshold 1	C_1	I	W	–
			Discount 2	D_2	C($)	W	–
			Threshold 2	C_2	I	W	–

Decision Activities
Purchase product to supplier A, supplier B, and supplier C

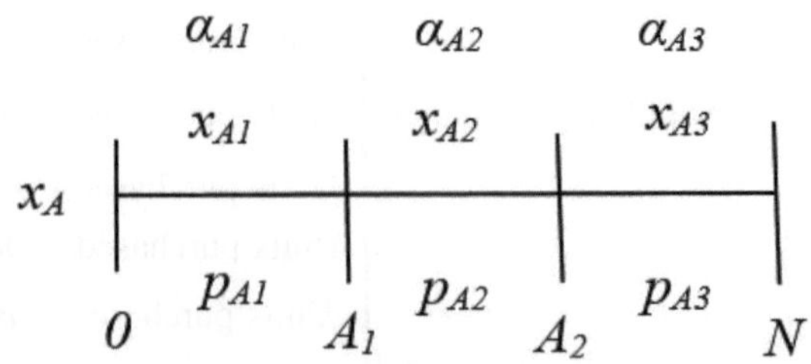

Fig. 6.14 Interval data and variables for supplier A

$$x_A = \text{Units of product purchased from supplier A} \tag{37.1}$$

$$x_B = \text{Units of product purchased from supplier B} \tag{37.2}$$

$$x_C = \text{Units of product purchased from supplier C} \tag{37.3}$$

Specifications

– Purchase N units (demand specification):

$$x_A + x_B + x_C = N \tag{37.4}$$

Objective Criterion
Minimize costs.

$$\text{Costs} = \text{Supplier A Costs} + \text{Supplier B Costs} + \text{Supplier C Costs}$$

- Supplier A costs: Cost according to the interval of values (Sect. 1.6.1). The intervals are shown in Fig. 6.14.

  Constraints:

$$\alpha_{A1} + \alpha_{A2} + \alpha_{A3} = 1 \tag{37.5}$$

$$x_{A1} \leq A_1\alpha_{A1} \tag{37.6}$$

$$x_{A2} \leq A_2\alpha_{A2} \tag{37.7}$$

$$x_{A3} \leq N\alpha_{A3} \tag{37.8}$$

$$x_A \geq (A_1 + 1)\alpha_{A2} \tag{37.9}$$

$$x_A \geq (A_2 + 1)\alpha_{A3} \tag{37.10}$$

$$x_A = x_{A1} + x_{A2} + x_{A3} \tag{37.11}$$

$$\text{Supplier A costs} = p_{A1}x_{A1} + p_{A2}x_{A2} + p_{A3}x_{A3}$$

- Supplier B costs: Case bonus by excess (Sect. 1.6.3.2). Interval data and variables in Fig. 6.15.

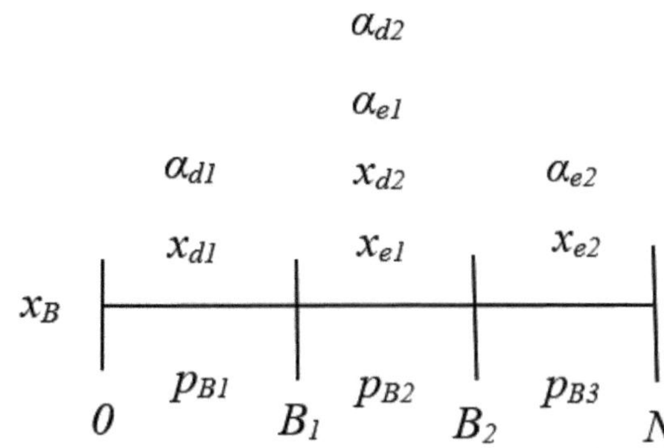

Fig. 6.15 Interval data and variables for supplier B

Constraints:
First incremental discount:

$$x_B + x_{d1} - x_{e1} = B_1 \tag{37.12}$$

$$x_{d1} \leq B_1 \alpha_{d1} \tag{37.13}$$

$$x_{e1} \leq (N - B_1)\alpha_{e1} \tag{37.14}$$

$$\alpha_{d1} + \alpha_{e1} \leq 1 \tag{37.15}$$

Second incremental discount:

$$x_B + x_{d2} - x_{e2} = B_2 \tag{37.16}$$

$$x_{d2} \leq B_2 \alpha_{d2} \tag{37.17}$$

$$x_{e2} \leq (N - B_2)\alpha_{e2} \tag{37.18}$$

$$\alpha_{d2} + \alpha_{e2} \leq 1 \tag{37.19}$$

$$\text{Supplier B cost} = p_{B1}x_B - (p_{B1} - p_{B2})x_{e1} - (p_{B2} - p_{B3})x_{e2}$$

- Supplier C costs: In this case, we just need to collect if the fixed discounts are applied. We use a logical calculation for each discount. The intervals are shown in Fig. 6.16.

  Binary Logical Calculations

$$\alpha_{C2} = 1 \text{ if discount } D_1 \text{ applies; 0 otherwise.} \tag{37.20}$$

$$\alpha_{C2} = 1 \text{ IF AND ONLY IF } x_C > C_1$$

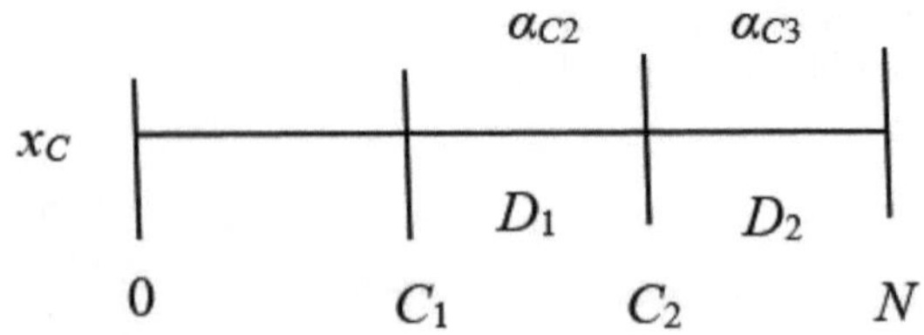

Fig. 6.16 Interval data and variables for supplier C

$$\Rightarrow f_5 \Rightarrow \alpha_{C2} = 1 \text{ IF AND ONLY IF } x_C \geq C_1 + 1$$
$$\Rightarrow f_{S_V} \Rightarrow \text{IF } \alpha_{C2} = 1 \text{ THEN } x_C \geq C_1 + 1$$
$$\Rightarrow f_{15}(\text{LB}_{x_C \geq C_1+1} = 0) \Rightarrow x_C \geq (C_1 + 1)\alpha_{C2} \quad (37.21)$$

$$\alpha_{C3} = 1 \text{ if discount } D_2 \text{ applies; 0 otherwise.} \quad (37.22)$$

$$\alpha_{C3} = 1 \text{ IF AND ONLY IF } x_C > C_2$$
$$\Rightarrow f_5 \Rightarrow \alpha_{C3} = 1 \text{ IF AND ONLY IF } x_C \geq C_2 + 1$$
$$\Rightarrow f_{S_V} \Rightarrow \text{IF } \alpha_{C3} = 1 \text{ THEN } x_C \geq C_2 + 1$$
$$\Rightarrow f_{15}(\text{LB}_{x_C \geq C_2+1} = 0) \Rightarrow x_C \geq (C_2 + 1)\alpha_{C3} \quad (37.23)$$

$$\text{Supplier C cost} = p_C x_C - D_1\alpha_{C2} - D_2\alpha_{C3}$$

Objective Function:

$$\begin{aligned} \text{Min } & (p_{A1}x_{A1} + p_{A2}x_{A2} + p_{A3}x_{A3}) \\ & + (p_{B1}x_B - (p_{B1} - p_{B2})x_{e1} - (p_{B2} - p_{B3})x_{e2}) \\ & + (p_C x_C - D_1\alpha_{C2} - D_2\alpha_{C3}) \end{aligned} \quad (37.24)$$

Complete Model

$$\begin{aligned} \text{Min } & (p_{A1}x_{A1} + p_{A2}x_{A2} + p_{A3}x_{A3}) \\ & + (p_{B1}x_B - (p_{B1} - p_{B2})x_{e1} - (p_{B2} - p_{B3})x_{e2}) \\ & + (p_C x_C - D_1\alpha_{C2} - D_2\alpha_{C3}) \end{aligned} \quad (37.24)$$

s.t.

$$x_A + x_B + x_C = N \quad (37.4)$$

$$\alpha_{A1} + \alpha_{A2} + \alpha_{A3} = 1 \quad (37.5)$$

$$x_{A1} \leq A_1\alpha_{A1} \quad (37.6)$$

$$x_{A2} \leq A_2\alpha_{A2} \quad (37.7)$$

$$x_{A3} \leq N\alpha_{A3} \quad (37.8)$$

$$x_A \geq (A_1 + 1)\alpha_{A2} \quad (37.9)$$

$$x_A \geq (A_2 + 1)\alpha_{A3} \quad (37.10)$$

$$x_A = x_{A1} + x_{A2} + x_{A3} \quad (37.11)$$

$$x_B + x_{d1} - x_{e1} = B_1 \quad (37.12)$$

$$x_{d1} \leq B_1\alpha_{d1} \quad (37.13)$$

$$x_{e1} \leq (N - B_1)\alpha_{e1} \quad (37.14)$$

$$\alpha_{d1} + \alpha_{e1} \leq 1 \quad (37.15)$$

$$x_B + x_{d2} - x_{e2} = B_2 \quad (37.16)$$

$$x_{d2} \leq B_2\alpha_{d2} \quad (37.17)$$

$$x_{e2} \leq (N - B_2)\alpha_{e2} \quad (37.18)$$

$$\alpha_{d2} + \alpha_{e2} \leq 1 \quad (37.19)$$

$$x_C \geq (C_1 + 1)\alpha_{C2} \quad (37.21)$$

$$x_C \geq (C_2 + 1)\alpha_{C3} \quad (37.23)$$

$$\alpha_{C2} \text{ binarie} \quad (37.20)$$

$$\alpha_{C3} \text{ binarie} \quad (37.22)$$

$$x_A, x_B, x_C, x_{A1}, x_{A2}, x_{A3}, x_{d1}, x_{d2}, x_{e1}, x_{e2} \geq 0 \text{ Integers} \quad (37.25)$$

$$\alpha_{A1}, \alpha_{A2}, \alpha_{A3}, \alpha_{d1}, \alpha_{d2}, \alpha_{e1}, \alpha_{e2} \text{ binaries} \quad (37.26)$$

Implementation in LINGO

```
MODEL:
DATA:
 N=1000;
 pA1 =10;
 pA2=9;
 pA3=8;
 A1=200;
 A2=5000;
 pB1 =9.5;
 pB2=9;
 PB3=8.5;
 B1=300;
 B2=700;
 pC=9;
 C1=500;
 C2=800;
 D1=300;
 D2=300;
ENDDATA
!37.24;
MIN = (pA1*xA1+pA2*xA2+pA3*xA3) + (pB1*xB - (pB1-pB2)*xe1 - (pB2-pB3)
```

```
*xe2) + (pc*xc -D1*AlphaC3 -D2*AlphaC3);
!37.4;
xA +xB +xC = N;

!37.5;
AlphaA1 +AlphaA2 + AlphaA3 = 1;
!37.6;
xA1 <= A1* AlphaA1;
!37.7;
xA2 <= A2* AlphaA2;
!37.8;
xA3 <= N* AlphaA3;
!37.9;
xA>=(A1+1)*AlphaA2;
!37.10;
xA>=(A2+1)*AlphaA3;
!37.11;
xA = xA1 + xA2 + xA3;
!37.12;
xB + xd1 - xe1 = B1;
!37.13;
xd1<= B1* Alphad1;
!37.14;
xe1<= (N-B1) * Alphae1;
!37.15;
Alphad1+Alphae1<=1;
!37.16;
xB + xd2 - xe2 = B3;
!37.17;
xd2<= B2* Alphad2;
!37.18;
xe2<= (N-B2) * Alphae2;
```

Solution Report - 37_LingoE

```
Global optimal solution found.
Objective value:                              8400.000
Objective bound:                              8400.000
Infeasibilities:                              0.000000
Extended solver steps:                               0
Total solver iterations:                            19
Elapsed runtime seconds:                          0.03

Model Class:                                      MILP

Total variables:                   20
Nonlinear variables:                0
Integer variables:                 19

Total constraints:                 19
Nonlinear constraints:              0

Total nonzeros:                    51
Nonlinear nonzeros:                 0
```

Fig. 6.17 Problem 37: LINGO solution report

```
!37.19;
Alphad2+Alphae2<=1;
!37.21;
xC>=(C1+1)*AlphaC2;

!37.23;
xC>=(C2+1)*AlphaC3;

!37.20;
@BIN(AlphaC2);
!37.22;
@BIN(AlphaC3);
!37.25;
@GIN(xA);@GIN(xB);@GIN(xC);@GIN(xA1);@GIN(xA2);@GIN(xA3); @GIN
(xd1);@GIN(xe1);@GIN(xd2);@GIN(xe2);
!37.26;
@BIN(AlphaA1);@BIN(AlphaA2);@BIN(AlphaA3);@BIN(Alphad1);@BIN
(Alphae1);@BIN(Alphad2);@BIN(Alphae2);
END
```

Solution

The solution is shown in Fig. 6.17.

References

Desrochers M, Laporte G (1991) Improvements and extensions to the Miller-Tucker-Zemlin subtour elimination constraints. Oper Res Lett 10(1):27–36

García JM (2021) Modelling in mathematical programming: methodology and Technics. Springer, Cham. https://doi.org/10.1007/978-3-030-57250-1

Hwang FK, Richards DS, Winter P (1992) The steiner tree problem. In: Annals of discrete mathematics, vol 53. Elsevier, North-Holland

Miller CE, Tucker AW, Zemlin RA (1960) Integer programming formulations and traveling salesman problems. J ACM 7:326–329

Shahin A, Ossama S (2004) Using genetic algorithms in solving the one-dimensional cutting stock problem in the construction industry. Canadian J Civil Eng 31:321–332

Chapter 7
High-Level Problems

7.1 Introduction

First, we show the characteristics of the problems with respect to methodology (Garcia 2021). The most complex aspects are covered.

Due to the size of the input and output data, in some cases the data are not displayed in each chapter section. They are available for download.

Problems:

Number	Problem
38	Installation of city bike stations
39	Equipment management in hospitals
40	Hiring of employees
41	Element management
42	Creation of sudokus
43	"Unblock me"
44	Distribution of state aid
45	Planning of a formula one race
46	Ready mixed concrete delivery to construction sites
47	Planning of a supermarket car park
48	Assignment of teacher posts
49	Placement of tables
50	Cutting Stock problem

J. M. García Sánchez, *Building and Solving Mathematical Programming Models*,
International Series in Operations Research & Management Science 329,
https://doi.org/10.1007/978-3-030-97626-2_7

Calculations:

Problem	Binary logical calculations	Not-binary logical calculations	Bounds calculations	Conditioned bounds calculations
38	✓		✓	
39	✓			
40	✓			
41	✓	✓		
42	✓			
43	✓		✓	
44	✓	✓	✓	
45	✓			
46	✓			
47	✓	✓		✓
48	✓			
49	✓			
50	✓			

Specifications:

Problem	QSS[a]	Capacity	Demand	Bound imposition	Allocation/assignment
38	✓	✓		✓	
39				✓	✓
40	✓		✓	✓	✓
41	✓				✓
42	✓			✓	✓
43	✓			✓	
44					✓
45	✓			✓	✓
46		✓			
47		✓			✓
48	✓				✓
49	✓			✓	
50	✓				

[a]*QSS* quantitative selection specifications

Problem	Flow balance	PLS[a]—One operator	PLS—Several operators	PLS—Attributes	Variable costs cases
38		✓		✓	✓
39		✓		✓	✓
40		✓			
41		✓		✓	
42					
43		✓	✓		
44		✓			

(continued)

45		✓	✓		
46		✓	✓		
47		✓	✓		
48		✓		✓	
49		✓	✓		
50		✓	✓		

[a]*PLS* propositional logic specifications

Objective criterion, input data and output data:

	Objective criterion		Input data				
Problem	Calculations	Particular cases	Integrated data	Text file	Spreadsheet	Database	Output data
38	✓				✓		✓
39					✓		
40	✓				✓		
41					✓		
42				✓			✓
43	✓			✓			✓
44	✓					✓	
45			✓				
46	✓					✓	
47	✓				✓		
48	✓				✓		✓
49			✓				
50					✓		✓

7.2 Problem 38. Installation of City Bike Stations

In a city, bike stations are going to be installed. There are 40 possible locations in the city for the installation of the stations. Each location allows the installation of a station and a number of bicycle anchorages to be decided up to a maximum number of anchorages known for each location, depending on its size. The locations belong to a sector of the city, a total of 10 sectors. We know the distances between the locations.

The installation of a station involves the cost of the control computer, CO \$, and the cost of each anchorage, which is estimated at CB \$/anchorage.

On the other hand, it will also be necessary to buy bicycles. The cost of each bicycle is CK \$/bike. If more than 500 bicycles are purchased, the bicycle supplier offers a discount of Dt \$ for each bicycle.

The installation requirements are as follows.

All sectors must have some station.

The distance between two locations with a station must be at least 300 m.

Any location with a station must have at least one other location with a station located at a maximum distance of 500 m, to avoid long journeys if there are no bicycles or free anchorages when going to a station.

In the old town sector of the city (sector 1), no more than four stations can be installed.

The number of bicycles purchased must be at least 70% of the total number of installed anchors.

If a station is installed in a location, the number of anchors must be between 8 and its capacity (the anchors capacity of each location is always greater than 8).

The town planning department has a budget of P \$ for the installation project.

Each sector has a potential number of potential clients. The objective is to balance the ratio between potential customers and the anchors installed in the sector.

Modeling

List of elements in Table 7.1.

Table 7.1 Problem 38: table of elements

			Data				
Elements	Set	QN	Name	Parameter	Type	Belonging	Value
Locations	$i, k = 1\ldots40$	I_U	Distance	d_{ik}	C (m)	S	...
			Anchorage capacity	M_i	I	S	...
			Sector	s_i	O	W	...
Stations	–	C_I	Cost	CO	C (\$/u)	W	–
Sectors	$j = 1\ldots10$	I_U	Customers	C_j	I	W	...
Anchors	–	C_I	Cost	CB	C (\$/u)	W	–
				M_i			
Bicycles	–	C_I	Cost	CK	C (\$/u)	W	–
			Discount	Dt	C (\$/u)	W	–
System	–	I_U	Budget	P	C (\$)	W	–

Decision Activities

Install stations at locations (*i*)

$$x_i = \text{Number of stations installed at location } i \quad (38.1)$$

Install anchors at locations (*i*)

$$y_i = \text{Number of anchors installed at location } i \quad (38.2)$$

Purchase bicycles

$$z = \text{Number of bicycles purchased} \tag{38.3}$$

Specifications

– Relationship between activities (propositional logic specification)

$$\forall i : x_i > 0 \text{ IF AND ONLY IF } y_i > 0$$
$$\Rightarrow f_5 \Rightarrow \forall i : x_i \geq 1 \text{ IF AND ONLY IF } y_i \geq 1$$
$$\Rightarrow f_{25}\left(\text{LB}_{y_i} = 0\right) \Rightarrow \forall i : y_i \geq x_i \tag{38.4}$$
$$\Rightarrow f_{25}\left(\text{UB}_{y_i} = M_i\right) \Rightarrow \forall i : y_i \leq M_i x_i \tag{38.5}$$

– Maximum of stations in each location (imposition of upper bound in measurable activities)

$$\forall i : x_i \leq 1 \tag{38.6}$$

This specification makes the variable x_i binary, so with its declaration as a binary variable, we can eliminate this constraint.

– Maximum anchorages at each location (capacity specification)

$$\forall i : y_i \leq M_i \tag{38.7}$$

It is already collected in (38.5).

– All sectors must have some station (lower bound imposition)

$$\forall j : \sum_{i/s_i=j} x_i \geq 1 \tag{38.8}$$

– The distance between two locations with station must be at least 300 m (propositional logic specification)

$$\begin{aligned} &\forall i,k,i<k : \text{IF } x_i > 0 \text{ AND } x_k > 0 \text{ THEN } d_{ik} \geq 300 \\ &\Rightarrow f_5, f_{37} \Rightarrow \forall i,k,i<k : \text{IF } x_i + x_k \geq 2 \text{ THEN } d_{ik} \geq 300 \\ &\Rightarrow f_3 \Rightarrow \forall i,k,i<k : \text{IF } d_{ik} < 300 \text{ THEN } x_i + x_k \leq 1 \\ &\Rightarrow f_{45} \Rightarrow \forall i,k,i<k \& d_{ik} < 300 : \ x_i + x_k \leq 1 \end{aligned} \tag{38.9}$$

– There must be at least one station 500 m from another station. (binary logical calculation + propositional logic specification)

We can use a binary logical calculation to find out if the distance is less than 500 m between pairs of locations that have stations, and then impose a quantitative selection rule with that calculation for each location, conditional on the existence of a station.

Binary Logical Calculation

$$\begin{array}{c}\forall i,k/i<k:\beta_{ik}=1 \text{ IF AND ONLY IF}\\ x_i=1 \text{ AND } x_k=1 \text{ AND } d_{ik}\leq 500\end{array}\tag{38.10}$$

$$\begin{array}{l}\Rightarrow f_{S_V}\Rightarrow \forall i,k/i<k:\text{IF}\beta_{ik}=1 \text{ THEN } x_i=1 \text{ AND } x_k=1 \text{ AND } d_{ik}\leq 500\\ \Rightarrow f_{37}\Rightarrow \forall i,k/i<k:\text{IF}\beta_{ik}=1 \text{ THEN } (x_i+\ x_k\geq 2) \text{ AND } d_{ik}\leq 500\\ \Rightarrow f_3\Rightarrow \forall i,k/i<k:\\ \text{IF NOT}((x_i+\ x_k\geq 2)\text{AND}(d_{ik}\leq 500)) \text{ THEN } \beta_{ik}=0\\ \Rightarrow \forall i,k/i<k:\text{IF NOT}(x_i+\ x_k\geq 2) \text{ OR NOT}(d_{ik}\leq 500) \text{ THEN } \beta_{ik}=0\\ \quad\Rightarrow f_{40}\Rightarrow \forall i,k/i<k:\text{IF NOT}(x_i+\ x_k\geq 2)\ \text{ THEN } \beta_{ik}=0\\ \quad\quad\Rightarrow f_3\Rightarrow \forall i,k/i<k:\text{IF } \beta_{ik}=1 \text{ THEN } x_i+\ x_k\geq 2\\ \quad\quad\Rightarrow f_{15}(\text{LB}_{x_i+\ x_k}=0)\Rightarrow \forall i,k/i<k:\ x_i+\ x_k\geq 2\beta_{ik}\end{array}\tag{38.11}$$

$$\begin{array}{l}\Rightarrow f_{40}\Rightarrow \forall i,k/i<k:\text{IF NOT}(d_{ik}\leq 500) \text{ THEN } \beta_{ik}=0\\ \quad\Rightarrow f_9,f_{45}\Rightarrow \forall i,k/i<k\&d_{ik}>500:\ \beta_{ik}=0\end{array}\tag{38.12}$$

Propositional logic specification:

$$\begin{array}{l}\forall i:\text{IF } x_i=1 \text{ THEN}\sum_{k/k<i}\beta_{ki}+\sum_{k/k>i}\beta_{ik}\geq 1\\ \Rightarrow f_{15}\left(\text{LB}_{\sum_{k/k<i}\beta_{ki}+\sum_{k/k>i}\beta_{ik}}=0\right)\forall i:\sum_{k/k<i}\beta_{ki}+\sum_{k/k>i}\beta_{ik}\geq x_i\end{array}\tag{38.13}$$

- In the old town sector of the city (sector 1), no more than four stations can be installed (upper bound imposition).

$$\sum_{i/S_i=1}x_i\leq 4\tag{38.14}$$

- The number of bicycles purchased must be at least 70% of the total number of anchors installed (lower bound imposition).

$$0.7\sum_{i=1}^{40}y_i\leq z\tag{38.15}$$

- If a station is installed in a location, the number of anchorages must be between 8 and its capacity (propositional logic specification):

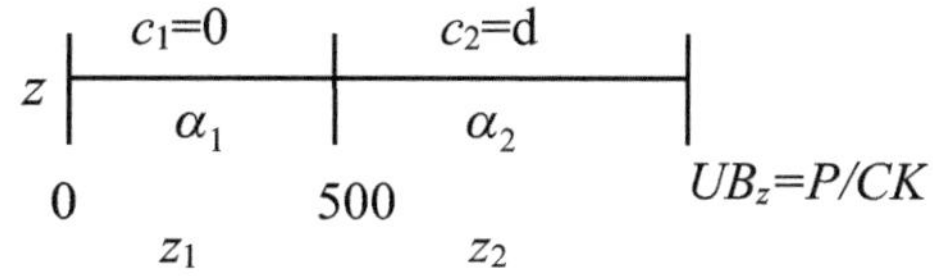

Fig. 7.1 Data and variables for the purchase of bicycles

$$\forall i: \text{IF } x_i = 1 \text{ THEN } y_i \geq 8 \\ \Rightarrow f_{15}(\text{LB}_{y_i} = 0) \Rightarrow \forall i: \ y_i \geq 8x_i \tag{38.16}$$

With (38.16) we can ignore (38.4).

– Available budget (capacity specification)

Budget consumption:

$$\text{Installation of stations}: \sum_{i=1}^{40} \text{CO}x_i$$

$$\text{Installation of anchors}: \sum_{i=1}^{40} \text{CB}y_i$$

Purchase of bicycles: Here, you have to take into account the discount. To do this, we can use the component defined in Sect. 1.6.1 to collect the cost of a variable based on the value it takes. Data and variables are shown in Fig. 7.1.

$$z_1 \leq 500\alpha_1 \tag{38.17}$$

$$z_2 \leq \frac{P}{\text{CK}}\alpha_2 \tag{38.18}$$

$$\alpha_1 + \alpha_2 = 1 \tag{38.19}$$

$$z \geq 501\alpha_2 \tag{38.20}$$

$$z = z_1 + z_2 \tag{38.21}$$

The discount is obtained by using the term $\text{Dt} \cdot z_2$, which we can understand as a capacity contribution. The capacity specification is then:

$$\sum_{i=1}^{40} \text{CO}x_i + \sum_{i=1}^{40} \text{CB}y_i + \text{CK}z \leq P + \text{Dt}z_2 \tag{38.22}$$

Objective Criterion
Balance the relationship between anchors and clients in the sectors.
Auxiliary Calculation

$\forall j : g_j =$ Ratio between anchors and clients in each sector

$$\forall j : g_j = \frac{\sum_{i/s_i=j} y_i}{C_j} \tag{38.23}$$

Upper and Lower Bound Calculations
G_{MAX}: Upper bound of the ratios g_j.
G_{MIN}: Lower bound of the ratios g_j.

$$\forall j : G_{\text{MAX}} \geq g_j \tag{38.24}$$

$$\forall j : G_{\text{MIN}} \leq g_j \tag{38.25}$$

Objective Function

$$\text{Min } G_{\text{MAX}} - G_{\text{MIN}} \tag{38.26}$$

Complete Model

$$\text{Min } G_{\text{MAX}} - G_{\text{MIN}} \tag{38.26}$$

s.t.

$$\forall i : y_i \leq M_i x_i \tag{38.5}$$

$$\forall j : \sum_{i/s_i=j} x_i \geq 1 \tag{38.8}$$

$$\forall i, k, i < k \& d_{ik} < 300 : \ x_i + \ x_k \leq 1 \tag{38.9}$$

$$\forall i, k/i < k : \ x_i + \ x_k \geq 2\beta_{ik} \tag{38.11}$$

$$\forall i, k/i < k \& d_{ik} > 500 : \ \beta_{ik} = 0 \tag{38.12}$$

$$\forall i : \sum_{k/k<i} \beta_{ki} + \sum_{k/k>i} \beta_{ik} \geq x_i \tag{38.13}$$

$$\sum_{i/S_i=1} x_i \leq 4 \tag{38.14}$$

$$0.7 \sum_{i=1}^{40} y_i \leq z \tag{38.15}$$

$$\forall i : \ y_i \geq 8x_i \tag{38.16}$$

$$z_1 \leq 500\alpha_1 \tag{38.17}$$

$$z_2 \leq \frac{P}{\text{CK}} \alpha_2 \tag{38.18}$$

$$\alpha_1 + \alpha_2 = 1 \tag{38.19}$$

$$z \geq 501\alpha_2 \tag{38.20}$$

$$z = z_1 + z_2 \tag{38.21}$$

$$\sum_{i=1}^{40} \text{CO}x_i + \sum_{i=1}^{40} \text{CB}y_i + \text{CK}z \leq P + \text{Dt}z_2 \tag{38.22}$$

$$\forall j : g_j = \frac{\sum_{i/s_i=j} y_i}{C_j} \tag{38.23}$$

$$\forall j : G_{\text{MAX}} \geq g_j \tag{38.24}$$

$$\forall j : G_{\text{MIN}} \leq g_j \tag{38.25}$$

$$\forall i : x_i \text{ binaries} \quad (38.1, 38.6)$$

$$\forall i : y_i \geq 0 \text{ integers} \quad (38.2)$$

$$z \geq 0 \text{ integer}$$

$$\forall i, k / i < k : \beta_{ik} \text{ binaries} \quad (38.10)$$

$$z_1, z_2 \geq 0 \text{ integers} \quad (38.27)$$

$$\alpha_1, \alpha_2 \text{ binaries} \quad (38.28)$$

$$\forall j : g_j \geq 0$$

$$G_{\text{MIN}}, G_{\text{MAX}} \geq 0$$

Implementation in LINGO

```
MODEL:
DATA:
ENDDATA
SETS:
 Locations/1..40/:M,s,x,y;
 Sectors/1..10/:C,g;
 Distances(Locations,Locations):D,Beta;
ENDSETS
DATA:
 M=@OLE('C:\Files_\38_Excel.xlsx','M');
 s=@OLE('C:\Files_\38_Excel.xlsx','s');
 C=@OLE('C:\Files_\38_Excel.xlsx','CC');
 D=@OLE('C:\Files_\38_Excel.xlsx','D');
 CO=@OLE('C:\Files_\38_Excel.xlsx','CO');
 CB=@OLE('C:\Files_\38_Excel.xlsx','CB');
 CK=@OLE('C:\Files_\38_Excel.xlsx','CK');
 Dt=@OLE('C:\Files_\38_Excel.xlsx','Dt');
```

```
P=@OLE('C:\Files_\38_Excel.xlsx','P');
@OLE('C:\Files_\38_Excel.xlsx','Solution')=y;
ENDDATA
!38.26;
MIN = GMAX-GMIN;
!38.5;
@FOR(Locations(i):y(i)<=M(i)*x(i));
!38.8;
@FOR(Sectors(j):@SUM(Locations(i)|s(i)#EQ#j:x(i))>=1);
!38.9;
@FOR(Locations(i):@FOR(Locations(k)|i#LT#k#AND#D(i,k)#LT#300:x(i)
+x(k)<=1));
!38.11;
@FOR(Locations(i):@FOR(Locations(k)|i#LT#k:x(i)+x(k)>=2*Beta(i,
k)));
!38.12;
@FOR(Locations(i):@FOR(Locations(k)|i#LT#k#AND#D(i,k)#GT#500:Beta
(i,k)=0));
!38.13;
@FOR(Locations(i):@SUM(Locations(k)|k#LT#i:Beta(k,i))+@SUM
(Locations(k)|k#GT#i:Beta(i,k))>=x(i));
!38.14;
@SUM(Locations(i)|S(i)#EQ#1:x(i))<=4;
!38.15;
0.7*@SUM(Locations(i):y(i))<=z;
!38.16;
@FOR(Locations(i):y(i)>=8*x(i));
!38.17;
z1<=500*Alpha1;
!38.18;
z2<=(P/CK)*Alpha2;
!38.19;
Alpha1 + Alpha2=1;
!38.20;
z>=501*Alpha2;
!38.21;
z=z1 + z2;
!38.22;
@SUM(Locations(i):CO*x(i))+@SUM(Locations(i):CB*y(i))+CK*z<=P
+Dt*z2;
!38.23;
@FOR(Sectors(j):g(j)=@SUM(Locations(i)|s(i)#EQ#j:y(i))/C(j));
!38.24;
@FOR(Sectors(j): GMAX>=g(j));
!38.25;
@FOR(Sectors(j):GMIN<=g(j));
!38.1; !38.6;
@FOR(Locations(i):@BIN(x(i)));
!38.2;
@FOR(Locations(i):@GIN(y(i)));
!38.3;
@GIN(z);
!38.10;
```

```
@FOR(Locations(i):@FOR(Locations(k)|i#LT#k:@BIN(Beta(i,k))));
!38.27;
@GIN(z1);@GIN(z2);
!38.28;
@BIN(Alpha1);@GIN(Alpha2);
DATA:
 @OLE('C:\Files_\38_Excel.xlsx','bicycles')=z;
ENDDATA
END
```

Solution

The solution is shown in Fig. 7.2.

```
Solution Report - 38_LingoE
Global optimal solution found.
Objective value:                              0.1583333E-01
Objective bound:                              0.1583333E-01
Infeasibilities:                               0.000000
Extended solver steps:                                0
Total solver iterations:                             26
Elapsed runtime seconds:                           0.45


Export Summary Report
---------------------
Transfer Method:         OLE BASED
Workbook:                C:\Files_\38_Excel.xlsx
Ranges Specified:               1
    Solution
Ranges Found:                   1
Range Size Mismatches:          0
Values Transferred:            40


Export Summary Report
---------------------
Transfer Method:         OLE BASED
Workbook:                C:\Files_\38_Excel.xlsx
Ranges Specified:               1
    bicycles
Ranges Found:                   1
Range Size Mismatches:          0
Values Transferred:             1

Model Class:                                       MILP

Total variables:                   1394
Nonlinear variables:                  0
Integer variables:                  562

Total constraints:                 1145
Nonlinear constraints:                0

Total nonzeros:                    3855
Nonlinear nonzeros:                   0
```

Fig. 7.2 Problem 38: LINGO solution report

Table 7.2 Maintenance cost

Equipment	Cost per equipment
1	C_1
>1	C_2

7.3 Problem 39. Equipment Management in Hospitals

A Spanish businessman has decided to donate 20 new technology equipment for MRI (magnetic resonance imaging) scans to public hospitals in Andalusia. The management that must be done focuses on which hospitals to install the equipment and the assignment of citizens to hospitals with equipment. Andalusia has n public hospitals, and the number of citizens belonging to each hospital is known. The location of the hospitals is known; therefore, the distance between them and the province to which they belong is known.

The hiring of technical personnel for the use of the equipment is the responsibility of the Andalusian administration. A technician is necessary in each hospital where there is equipment. The cost of a technician is Ct. The maintenance cost of the equipment varies depending on the number of equipment per hospital according to Table 7.2.

The administration has a budget of P $ for the hiring of technical staff and equipment maintenance.

The Andalusian administration established as a requirement in the assignment that each Andalusian province (eight provinces) has at least one equipment.

In the installation of equipment, the number of citizens belonging to a hospital is a priority. A hospital with more citizens than another cannot have less equipment than the other.

On the other hand, if a hospital does not own equipment, it is necessary to assign the citizens of that hospital to others with equipment. And in that assignment, the ratio between the total number of citizens assigned to a hospital divided by the number of teams must not be greater than M. All citizens who belong to a hospital where some equipment has been installed are assigned to your hospital. It is also necessary to take into account that if citizens of a hospital are assigned to another hospital, the distance between those hospitals should not be more than 50 km.

The objective is to minimize the total distance of citizens' assignments to other hospitals.

Modeling

List of elements in Table 7.3.

Table 7.3 Problem 39: table of elements

Elements	Set	QN	Data				
			Name	Parameter	Type	Belonging	Value
Provinces	$j = 1 \ldots 8$	I_U					
Hospitals	i, $i' = 1 \ldots n$	I_U	Province	P_i	O	W	...
			Number of citizens	H_i	I	S	...
			Distance	$D_{ii'}$	C (Km)	S	...
Citizens	–	C_D		H_i			
Equipment	–	C_D	Quantity	E	I	W	20
			Cost 1	C_1	C	W	...
			Cost 2	C_2	C	W	...
Technicians	–	C_I	Cost	Ct	C	W	...
System	–	I_U	Budget	PP	C	W	...
			Ratio	M	C	W	...

Decision Activities

Install equipment in hospitals (i)

$$\forall i : x_i = \text{number of equipment installed in the hospital } i \tag{39.1}$$

Assign citizens from hospitals (i) to hospitals (i')

$$\forall i, i' : y_{ii'} = \text{number of citizens of hospital } i \text{ assigned to hospital } i' \tag{39.2}$$

Specifications

- Install E equipment (allocation specification)

$$\sum_{i=1}^{n} x_i = E \tag{39.3}$$

- Install equipment in each province (lower bound imposition):

$$\forall j : \sum_{i/P_i=j} x_i \geq 1 \tag{39.4}$$

- Budget cap (logical calculations + capacity specification):

$$\text{Costs} \leq PP$$

$$\text{Costs} = \text{Recruitment Costs} + \text{Maintenance Costs}$$

Recruitment Costs:

We have to make a logical calculation to know if equipment has been installed in each hospital or not, since it is equivalent to knowing if a technician has been hired or not.

Binary Logical Calculation

$$\forall i : \beta_i = 1 \text{ if equipment is installed in the hospital } i; 0 \text{ otherwise.} \tag{39.5}$$

$$\forall i : \beta_i = 1 \text{ IF AND ONLY IF } x_i > 0$$

Having equipment can have positive and negative connotations in the problem, so we do not simplify the definition of values.

$$\Rightarrow f_5 \Rightarrow \forall i : \beta_i = 1 \text{ IF AND ONLY IF } x_i \geq 1$$

$$\Rightarrow f_{25}(\text{LB}_{x_i} = 0) \Rightarrow \forall i : \; x_i \geq \beta_i \tag{39.6}$$

$$\Rightarrow f_{25}(\text{UB}_{x_i} = 13) \Rightarrow \forall i : \; x_i \leq 13\beta_i \tag{39.7}$$

With eight provinces and equipment in all provinces, the maximum number of equipment in a hospital cannot exceed 13.

$$\text{Recruitment Costs} = \sum_{i=1}^{n} \text{Ct}\beta_i$$

Maintenance Costs:

The cost depends on the number of equipment installed in each hospital, that is, the cost according to the interval of values (Sect. 1.6.1). Interval data and variables are shown in Fig. 7.3.

$$\forall i : \alpha_{i1} + \alpha_{i2} = 1 \tag{39.8}$$

$$\forall i : x_{i1} \leq 1\alpha_{i1} \tag{39.9}$$

$$\forall i : x_{i2} \leq 13\alpha_{i2} \tag{39.10}$$

$$\forall i : 2\alpha_{i2} \leq x_i \tag{39.11}$$

$$\forall i : x_i = x_{i1} + x_{i2} \tag{39.12}$$

$$\text{Maintenance Costs} = \sum_{i=1}^{n} C_1 x_{i1} + C_2 x_{i2}$$

Final constraint:

$$\text{Costs} \leq \text{PP} \Rightarrow \sum_{i=1}^{n} \text{Ct}\beta_i + \sum_{i=1}^{n} C_1 x_{i1} + C_2 x_{i2} \leq \text{PP} \tag{39.13}$$

- Priority in the installation of equipment (bounds imposition):

$$\forall i, i' / H_i > H_{i'} : x_i \geq x_{i'} \tag{39.14}$$

- Assign all citizens of each hospital (allocation specification):

$$\forall i : \sum_{i'=1}^{n} y_{ii'} = H_i \tag{39.15}$$

- Maximum ratio (upper bound imposition):

$$\forall i : \sum_{i'=1}^{n} y_{i'i} \leq x_i M \tag{39.16}$$

- Citizens of a hospital with equipment are assigned to their own hospital (propositional logic specification):

$$\begin{aligned} &\forall i : \text{IF } \beta_i = 1 \text{ THEN } y_{ii} = H_i \\ &\Rightarrow f_{\text{UB}} \Rightarrow \forall i : \text{IF } \beta_i = 1 \text{ THEN } y_{ii} \geq H_i \\ &\Rightarrow f_{15}(\text{LB}_{y_{ii}} = 0) \Rightarrow \forall i : y_{ii} \geq H_i \beta_i \end{aligned} \tag{39.17}$$

- Do not exceed 50 km (propositional logic specification):

$$\begin{aligned} &\forall i, i' : \text{IF } y_{ii'} > 0 \text{ THEN } D_{ii'} \leq 50 \\ &\Rightarrow f_3, f_{45} \Rightarrow \forall i, i' / D_{ii'} > 50 : y_{ii'} = 0 \end{aligned} \tag{39.18}$$

Objective Function
Minimize total distances.

$$\text{Min} \sum_{i=1}^{n} \sum_{i'=1}^{n} D_{ii'} y_{ii'} \tag{39.19}$$

Complete Model

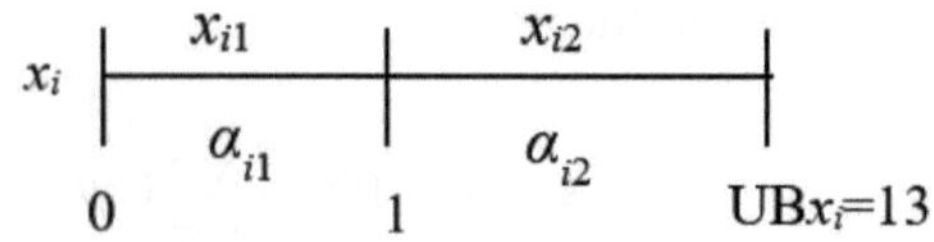

Fig. 7.3 Interval data

$$\text{Min} \sum_{i=1}^{n} \sum_{i'=1}^{n} D_{ii'} y_{ii'} \tag{39.19}$$

s.t.

$$\sum_{i=1}^{n} x_i = E \tag{39.3}$$

$$\forall j : \sum_{i/P_i=j} x_i \geq 1 \tag{39.4}$$

$$\forall i : \ x_i \geq \beta_i \tag{39.6}$$

$$\forall i : \ x_i \leq 13\beta_i \tag{39.7}$$

$$\forall i : \alpha_{i1} + \alpha_{i2} = 1 \tag{39.8}$$

$$\forall i : x_{i1} \leq 1\alpha_{i1} \tag{39.9}$$

$$\forall i : x_{i2} \leq 13\alpha_{i2} \tag{39.10}$$

$$\forall i : 2\alpha_{i2} \leq x_i \tag{39.11}$$

$$\forall i : x_i = x_{i1} + x_{i2} \tag{39.12}$$

$$\sum_{i=1}^{n} \text{Ct}\beta_i + \sum_{i=1}^{n} C_1 x_{i1} + C_2 x_{i2} \leq \text{PP} \tag{39.13}$$

$$\forall i, i'/H_i > H_{i'} : x_i \geq x_{i'} \tag{39.14}$$

$$\forall i : \sum_{i'=1}^{n} y_{ii'} = H_i \tag{39.15}$$

$$\forall i : \sum_{i'=1}^{n} y_{i'i} \leq x_i M \tag{39.16}$$

$$\forall i : y_{ii} \geq H_i \beta_i \tag{39.17}$$

$$\forall i, i'/D_{ii'} > 50 : y_{ii'} = 0 \tag{39.18}$$

$$\forall i : x_i \ \text{integers} \tag{39.1}$$

$$\forall i, i' : y_{ii'} \ \text{integers} \tag{39.2}$$

$$\forall i : \beta_i \ \text{binaries} \tag{39.5}$$

$$\forall i : x_{i1} \text{ integers}; x_{i2} \text{ integers} \tag{39.20}$$

$$\forall i : \alpha_{i1} \text{ binaries}; \alpha_{i2} \text{ binaries} \tag{39.21}$$

Implementation in LINGO

```
MODEL:
DATA:
 n=@OLE('C:\Files_\39_Excel.xlsx','n');
ENDDATA
SETS:
 Provinces/1..8/;
 Hospitals/1..n/:P,H,x,Beta,Alpha1,Alpha2,x1,x2;
 HH(Hospitals,Hospitals):D,y;
ENDSETS
DATA:
 E=20;
 P=@OLE('C:\Files_\39_Excel.xlsx','Province');
 H=@OLE('C:\Files_\39_Excel.xlsx','Citizens');
 D=@OLE('C:\Files_\39_Excel.xlsx','Distance');
 C1=@OLE('C:\Files_\39_Excel.xlsx','Cost1');
 C2=@OLE('C:\Files_\39_Excel.xlsx','Cost2');
 Ct=@OLE('C:\Files_\39_Excel.xlsx','Ct');
 PP=@OLE('C:\Files_\39_Excel.xlsx','Budget');
 M=@OLE('C:\Files_\39_Excel.xlsx','M');
ENDDATA
CALC:
@FOR(Hospitals(i):@FOR(Hospitals(ii)|i#LT#ii: D(ii,i)=D(i,ii)));
ENDCALC
!39.19;
MIN = @SUM(Hospitals(i):@SUM(Hospitals(ii):d(i,ii)*y(i,ii)));
!39.3;
@SUM(Hospitals(i):x(i))=E;
!39.4;
@FOR(Provinces(j):@SUM(Hospitals(i)|P(i)#EQ#j:x(i))>=1);
!39.6;
@FOR(Hospitals(i):x(i)>=Beta(i));
!39.7;
@FOR(Hospitals(i):x(i)<=13*Beta(i));
!39.8;
@FOR(Hospitals(i):Alpha1(i)+Alpha2(i)=1);

!39.9;
@FOR(Hospitals(i):x1(i)<=Alpha1(i));
!39.10;
@FOR(Hospitals(i):x2(i)<=13*Alpha2(i));
!39.11;
@FOR(Hospitals(i):2*Alpha2(i)<=x(i));
!39.12;
@FOR(Hospitals(i):x(i)=x1(i)+x2(i));
!39.13;
@SUM(Hospitals(i):Ct*Beta(i))+@SUM(Hospitals(i):C1*x1(i)+C2*x2
(i))<=PP;
!39.14;
@FOR(Hospitals(i):@FOR(Hospitals(ii)|H(i)#GT#H(ii):x(i)>=x(ii)));
!39.15;
```

```
Solution Report - 39_LingoE
Global optimal solution found.
Objective value:                                2759904.
Objective bound:                                2759904.
Infeasibilities:                                0.000000
Extended solver steps:                                 0
Total solver iterations:                              12
Elapsed runtime seconds:                            0.41

Model Class:                                        PILP

Total variables:                   737
Nonlinear variables:                 0
Integer variables:                 737

Total constraints:                 561
Nonlinear constraints:               0

Total nonzeros:                   2911
Nonlinear nonzeros:                  0
```

Fig. 7.4 Problem 39: LINGO solution report

```
@FOR(Hospitals(i):@SUM(Hospitals(ii):y(i,ii))=H(i));
!39.16;
@FOR(Hospitals(i):@SUM(Hospitals(ii):y(ii,i))<=x(i)*M);
!39.17;
@FOR(Hospitals(i):y(i,i)>=H(i)*Beta(i));
!39.18;
@FOR(Hospitals(i):@FOR(Hospitals(ii)|D(i,ii)#GT#50:y(i,ii)=0));
!39.1;
@FOR(Hospitals(i):@GIN(x(i)));
!39.2;
@FOR(Hospitals(i):@FOR(Hospitals(ii):@GIN(y(i,ii))));
!39.5;
@FOR(Hospitals(i):@BIN(Beta(i)));
!39.20;
@FOR(Hospitals(i):@GIN(x1(i)));@FOR(Hospitals(i):@GIN(x2(i)));
!39.21;
@FOR(Hospitals(i):@BIN(Alpha1(i)));@FOR(Hospitals(i):@BIN(Alpha2
(i)));
END
```

Solution

The solution is shown in Fig. 7.4.

7.4 Problem 40. Hiring of Employees

An international company is going to join the national market. The company will open eight new centers in four regions of the country. To do this, you need to hire workers. Workers can be fixed or with mobility. Each center has a demand for fixed workers, a demand for workers with mobility, and a remaining quantity that could be of any type. Fixed workers are specific to each center, while a mobility worker must be assigned to a minimum of two centers and a maximum of four centers.

A mobility worker cannot be assigned to centers in different regions or centers with distances greater than 100 km.

The cost of a fixed worker is a known cost that depends on the center. The cost of a mobility worker is calculated with the following formula:

Mobility worker cost = Z (fixed value) + (V * No. assigned centers) + (S * Maximum distance between assigned centers); Z, V, and S known values.

On the other hand, if the total number of permanent workers exceeds 100 people, we receive a work bonus of B \$.

We model the problem of minimizing the total costs of hiring.

Modeling

List of elements in Table 7.4.

Table 7.4 Problem 40: table of elements

			Data				
Elements	Set	QN	Name	Parameter	Type	Belonging	Value
Regions	$k = 1...4$	I_U					
Centers	$i, i' = 1...8$	I_U	Region	P_i	O	W	...
			Fixed worker demand	F_i	I	S	...
			Mob. worker demand	M_i	I	W	...
			Rest of demand	R_i	I	W	...
			Distance	$D_{ii'}$	C	S	...
			Cost	CF_i	C	S	...
Fixed workers	–	C_I	F_i; CF_i				...
Mobility workers	$j = 1...UB$	I_U					...
System	–	I_U	Constant 1	Z	C	W	
			Constant 2	V	C	W	
			Constant 3	S	C	W	
			Bonus	B	C	W	

It is necessary to establish an upper bound of the number of mobility workers, since they are individual elements with an indeterminate number.

$$UB = \left\lceil \frac{\sum_{i=1}^{8} M_i + R_i}{2} \right\rceil$$ A mobility worker must be assigned to two centers at least.

Decision Activities

Hire fixed workers for centers (*i*)

$$\forall i : x_i = \text{Number of fixed workers hired for the Center } i \tag{40.1}$$

Hire mobility workers (*j*) for centers (*i*)

$$\forall i,j : \alpha_{ij} = 1 \text{ if mobility workerjis hired for Center } i; 0 \text{ otherwise} \tag{40.2}$$

Specifications

- Quantitative selection specifications:

Activity	Elements selecting	Selectable elements	Selection	Constraints
$\forall i, j : \alpha_{ij}$	Mobility workers ($j = 1 \ldots$UB)	Centers ($i = 1 \ldots 8$)	≤ 4	$\forall j : \sum_{i=1}^{8} \alpha_{ij} \leq 4$ (40.3)
	Centers ($i = 1 \ldots 8$)	Mobility workers ($j = 1 \ldots$UB)	$\geq M_i$	$\forall i : \sum_{j=1}^{UB} \alpha_{ij} \geq M_i$ (40.4)

- Minimum of fixed workers (lower bound imposition):

$$\forall i : x_i \geq F_i \tag{40.5}$$

- Total demand for workers (demand specification):

$$\forall i : x_i + \sum_{j=1}^{UB} \alpha_{ij} = F_i + M_i + R_i \tag{40.6}$$

- Mobility worker hired in at least two centers (logical calculation + propositional logic specification):

By means of a logical calculation, we must find out if each worker with mobility has been hired:

Binary Logical Calculation

$$\forall j : \beta_j = 1 \text{ if mobility worker } j \text{ has been hired}; 0 \text{ otherwise.} \tag{40.7}$$

$$
\begin{aligned}
&\forall j : \beta_j = 1 \text{ IF AND ONLY IF } \sum_{i=1}^{8} \alpha_{ij} > 0 \\
&\Rightarrow f_{S_V} \Rightarrow \forall j : \text{IF } \sum_{i=1}^{8} \alpha_{ij} > 0 \text{ THEN } \beta_j = 1 \\
&\Rightarrow f_3, f_7, f_{14} \left(\text{UB}_{\sum_{i=1}^{8} \alpha_{ij}} = 4 \right) \Rightarrow \forall j : \sum_{i=1}^{8} \alpha_{ij} \leq 4\beta_j
\end{aligned}
\tag{40.8}
$$

Next, we impose that a hired mobility worker must be in at least two centers:

$$
\begin{aligned}
&\forall j : \text{IF } \beta_j = 1 \text{ THEN } \sum_{i=1}^{8} \alpha_{ij} \geq 2 \\
&\Rightarrow f_{15} \left(\text{LB}_{\sum_{i=1}^{8} \alpha_{ij}} = 0 \right) \Rightarrow \forall j : \sum_{i=1}^{8} \alpha_{ij} \geq 2\beta_j
\end{aligned}
\tag{40.9}
$$

- A worker with mobility cannot go to centers in different regions (propositional logic specification):

$$
\begin{aligned}
&\forall j, i, i' / i < i' \& P_i \neq P_{i'} : \text{NOT}\left(\alpha_{ij} = 1 \text{ AND } \alpha_{i'j} = 1\right) \\
&\Rightarrow f_{37}, f_{11} \Rightarrow \forall j, i, i' / i < i' \& P_i \neq P_{i'} : \alpha_{ij} + \alpha_{i'j} \leq 1
\end{aligned}
\tag{40.10}
$$

- A worker with mobility cannot go to centers with a distance greater than 100 km (propositional logic specification):

$$
\begin{aligned}
&\forall j, i, i' / i < i' \& D_{ii'} > 100 : \text{NOT}\left(\alpha_{ij} = 1 \text{ AND } \alpha_{i'j} = 1\right) \\
&\Rightarrow f_{37}, f_{11} \Rightarrow \forall j, i, i' / i < i' \& D_{ii'} > 100 : \alpha_{ij} + \alpha_{i'j} \leq 1
\end{aligned}
\tag{40.11}
$$

Objective Criterion

$$\text{Minimize Costs} - \text{Bonus}.$$

$$\text{Costs} = \text{Fixed workers' costs} + \text{Mobility workers' costs}.$$

$$\text{Fixed workers' costs} = \sum_{i=1}^{8} \text{CF}_i x_i$$

Mobility workers' costs = Z (fixed value) + (V * No. assigned centers) + (S * Maximum distance between assigned centers);

$$\text{Mobility workers' costs} = \sum_{j=1}^{UB} Z\beta_j + \sum_{j=1}^{UB} V \sum_{i=1}^{8} \alpha_{ij} + \sum_{j=1}^{UB} SD_j^{MAX}$$

It is necessary to calculate D_j^{MAX}: A binary logical calculation must be carried out to know if each j is assigned to centers i and i':

Binary Logical Calculation

$\forall j, i,$

$i'/i < i' : \delta_{jii'} = 1$ if mobility worker j has been assigned to Ccenter i and Center i'; 0 otherwise

(40.12)

$$\begin{aligned}
&\forall j, i, i'/i < i' : \delta_{jii'} = 1 \text{ IF AND ONLY IF } \alpha_{ij} = 1 \ \ Y \ \ \alpha_{i'j} = 1 \\
&\Rightarrow f_{S_V} \Rightarrow \forall j, i, i'/i < i' : \text{IF } \alpha_{ij} = 1 \text{ AND } \alpha_{i'j} = 1 \text{ THEN } \delta_{jii'} = 1 \\
&\Rightarrow f_3, f_7, f_{37}, f_{11} \Rightarrow \forall j, i, i'/i < i' : \text{IF } 1 - \delta_{jii'} = 1 \text{ THEN } \alpha_{ij} + \alpha_{i'j} \leq 1 \\
&\Rightarrow f_{14} \Rightarrow \forall j, i, i'/i < i' : \alpha_{ij} + \alpha_{i'j} \leq 1 + \delta_{jii'}
\end{aligned} \tag{40.13}$$

We define D_j^{MAX}as an upper bound calculation:

Upper Bound Calculation

$\forall j : D_j^{MAX}$

= Upper bound of the maximum distance between centers assigned to j.

$$\forall j, i, i'/i < i' : D_j^{MAX} \geq \delta_{jii'} D_{ii'} \tag{40.14}$$

Bonus: We define a logical calculation to know if we have had a bonus:

Binary Logical Calculation

$$\omega = 1 \text{ if a bonus is applied; 0 if not.} \tag{40.15}$$

$$\omega = 1 \text{ IF AND ONLY IF } \sum_{i=1}^{8} x_i \geq 101$$

$$\Rightarrow f_{S_V} \Rightarrow \text{IF } \omega = 1 \text{ THEN } \sum_{i=1}^{8} x_i \geq 101 \tag{40.16}$$

$$\Rightarrow f_{15}\left(\text{LB}_{\sum_{i=1}^{8} x_i} = 0\right) \Rightarrow \sum_{i=1}^{8} x_i \geq 101\omega$$

$$\text{Bonus} = B\omega$$

Objective Function

$$\text{Min } \sum_{i=1}^{8} \text{CF}_i x_i + \sum_{j=1}^{\text{UB}} Z\beta_j + \sum_{j=1}^{\text{UB}} V \sum_{i=1}^{8} \alpha_{ij} + \sum_{j=1}^{\text{UB}} \text{SD}_j^{\text{MAX}} - B\omega \tag{40.17}$$

Complete Model

$$\text{Min } \sum_{i=1}^{8} \text{CF}_i x_i + \sum_{j=1}^{\text{UB}} Z\beta_j + \sum_{j=1}^{\text{UB}} V \sum_{i=1}^{8} \alpha_{ij} + \sum_{j=1}^{\text{UB}} \text{SD}_j^{\text{MAX}} - B\omega \tag{40.17}$$

s.t.

$$\forall j : \sum_{i=1}^{8} \alpha_{ij} \leq 4 \tag{40.3}$$

$$\forall i : \sum_{j=1}^{\text{UB}} \alpha_{ij} \geq M_i \tag{40.4}$$

$$\forall i : x_i \geq F_i \tag{40.5}$$

$$\forall i : x_i + \sum_{j=1}^{\text{UB}} \alpha_{ij} = F_i + M_i + R_i \tag{40.6}$$

$$\forall j : \sum_{i=1}^{8} \alpha_{ij} \leq 4\beta_j \tag{40.8}$$

$$\forall j : \sum_{i=1}^{8} \alpha_{ij} \geq 2\beta_j \tag{40.9}$$

$$\forall j, i, i' / i < i' \& P_i \neq P_{i'} : \alpha_{ij} + \alpha_{i'j} \leq 1 \tag{40.10}$$

$$\forall j, i, i' / i < i' \& D_{ii'} > 100 : \alpha_{ij} + \alpha_{i'j} \leq 1 \tag{40.11}$$

$$\forall j, i, i'/i < i' : \alpha_{ij} + \alpha_{i'j} \leq 1 + \delta_{jii'} \quad (40.13)$$

$$\forall j, i, i'/i < i' : D_j^{\text{MAX}} \geq \delta_{jii'} D_{ii'} \quad (40.14)$$

$$\sum_{i=1}^{8} x_i \geq 101\omega \quad (40.16)$$

$$\forall i : x_i \text{ integers} \quad (40.1)$$

$$\forall i, j : \alpha_{ij} \text{ binaries} \quad (40.2)$$

$$\forall j : \beta_j \text{ binaries} \quad (40.7)$$

$$\forall j, i, i'/i < i' : \delta_{jii'} \text{ binaries} \quad (40.12)$$

$$\forall j : D_j^{\text{MAX}} \geq 0$$

$$\omega \text{ binarie} \quad (40.15)$$

Implementation in LINGO

```
MODEL:
DATA:
 n=@OLE('c:\Files_\40_Excel.xlsx','n');
 UB=@OLE('c:\Files_\40_Excel.xlsx','UB');
ENDDATA
SETS:
 Regions/1..4/;
 Centers/1..n/:P,F,M,R,CF,x;
 MobilityW/1..UB/:Beta,DMAX;
 CC(Centers,Centers):D;
 CM(Centers,MobilityW):Alpha;
 MCC(MobilityW,Centers,Centers):Delta;
ENDSETS
DATA:
 Z=@OLE('c:\Files_\40_Excel.xlsx','Z');
 V=@OLE('c:\Files_\40_Excel.xlsx','V');
 S=@OLE('c:\Files_\40_Excel.xlsx','S');
 B=@OLE('c:\Files_\40_Excel.xlsx','B');
 P=@OLE('c:\Files_\40_Excel.xlsx','Region');
 F=@OLE('c:\Files_\40_Excel.xlsx','FWD');
 M=@OLE('c:\Files_\40_Excel.xlsx','MWD');
 R=@OLE('c:\Files_\40_Excel.xlsx','Rest');
 CF=@OLE('c:\Files_\40_Excel.xlsx','Costs');
 D=@OLE('c:\Files_\40_Excel.xlsx','D');
ENDDATA
CALC:
@FOR(Centers(i):@FOR(Centers(ii)|i#LT#ii: D(ii,i)=D(i,ii)));
ENDCALC

!40.17;
MIN = @SUM(Centers(i):CF(i)*x(i))+@SUM(MobilityW(j):Z*Beta(j))+@SUM
(MobilityW(j):V*@SUM(Centers(i):Alpha(i,j)))+@SUM(MobilityW(j):
S*DMAX(j))-B*Omega;
!40.3;
```

```
@FOR(MobilityW(j):@SUM(Centers(i):Alpha(i,j))<=4);
!40.4;
@FOR(Centers(i):@SUM(MobilityW(j):Alpha(i,j))>=M(i));
!40.5;
@FOR(Centers(i):x(i)>=F(i));
!40.6;
@FOR(Centers(i):x(i)+@SUM(MobilityW(j):Alpha(i,j))=F(i)+M(i)+R
(i));
!40.8;
@FOR(MobilityW(j):@SUM(Centers(i):Alpha(i,j))<=4*Beta(j));
!40.9;
@FOR(MobilityW(j):@SUM(Centers(i):Alpha(i,j))>=2*Beta(j));
!40.10;
@FOR(MobilityW(j):@FOR(Centers(i):@FOR(Centers(ii)|i#LT#ii#AND#P
(i)#NE#P(ii): Alpha(i,j)+Alpha(ii,j)<=1)));
!40.11;
@FOR(MobilityW(j):@FOR(Centers(i):@FOR(Centers(ii)|i#LT#ii#AND#D
(i,ii)#GT#100: Alpha(i,j)+Alpha(ii,j)<=1)));
!40.13;
@FOR(MobilityW(j):@FOR(Centers(i):@FOR(Centers(ii)|i#LT#ii: Alpha
(i,j)+Alpha(ii,j)<=1+Delta(j,i,ii))));
!40.14;
@FOR(MobilityW(j):@FOR(Centers(i):@FOR(Centers(ii)|i#LT#ii: DMAX
(j)>=Delta(j,i,ii)*D(i,ii))));
!40.16;
@SUM(Centers(i):x(i))>=101*Omega;
!40.1;
@FOR(Centers(i):@GIN(x(i)));
!40.2;
@FOR(Centers(i):@FOR(MobilityW(j):@BIN(Alpha(i,j))));
!40.7;
@FOR(MobilityW(j):@BIN(Beta(j)));
!40.12;
@FOR(MobilityW(j):@FOR(Centers(i):@FOR(Centers(ii)|i#LT#ii:@BIN
(Delta(j,i,ii)))));
!40.15;
@BIN(Omega);
END
```

Solution

It is shown in Fig. 7.5.

7.5 Problem 41. Element Management

We have three groups of elements (whose significance is not necessary to include in the statement). The first group contains the primary elements. The second is for the secondary ones. The third are the auxiliary elements.

The first group consists of eight elements, the second one 20 elements, and the third 500 elements.

Primary elements have a value. Secondary elements have a suitability coefficient (continuous value within the interval [0.10]) for each primary element and a

```
Solution Report - 40_LingoE
Global optimal solution found.
Objective value:                              621200.0
Objective bound:                              621200.0
Infeasibilities:                              0.000000
Extended solver steps:                            5131
Total solver iterations:                         21908
Elapsed runtime seconds:                          0.64

Model Class:                                      MILP

Total variables:                  1045
Nonlinear variables:                 0
Integer variables:                 527

Total constraints:                1188
Nonlinear constraints:               0

Total nonzeros:                   3394
Nonlinear nonzeros:                  0
```

Fig. 7.5 Problem 40: LINGO solution report

compatibility coefficient between them of continuous value within the interval [0.1]. The auxiliary elements are made of one material (four different materials).

The problem involves associating each primary element with secondary and auxiliary elements, so that all elements are associated.

If a primary element has a value greater than 1, it must be associated with more than two secondary elements, and the compatibility coefficients between the associated secondary elements must be greater than 0.5.

Each secondary element is associated with a maximum of three primary ones. Each auxiliary element is associated with a primary element.

If a secondary element is associated with more than five auxiliary elements, there must be auxiliary elements of at least three materials.

Maximize total suitability between associated primary and secondary elements.

Modeling

Auxiliary elements can be grouped into four collective elements, and thus reduce the number of elements in the model and improve its efficiency. Allusions to collective elements can all be interpreted as numerals, except for one:

"Each auxiliary element is associated with a primary element"

This specification can also be redefined as follows: The total number of auxiliary elements assigned to the parent elements must be 500. That means that they are assigned only once. Taking this to each generated collective element, we must assign the total of the items of each one of them.

List of elements in Table 7.5.

$$\sum_{k=1}^{4} Q_k = 500$$

We could remove the materials as elements because there are no decisions or specifications on them.

Decision Activities

Table 7.5 Problem 41: table of elements

Elements	Set	QN	Data				
			Name	Parameter	Type	Belonging	Value
Primary elements	$i = 1\ldots8$	I_U	Value	P_i	C	W	...
			Suitability	II_{ij}	I	S	...
Secondary elements	$j, j' = 1\ldots20$	I_U	Compatibility	$\text{CC}_{jj'}$	C	S	
				I_{ij}			
Auxiliaries	$k = 1\ldots4$	C_D	Quantity	Q_k	I	W	...
			Material	m_k	O	W	...
Materials	$m = 1\ldots4$	I_U					

Associate secondary elements (*j*) with primary elements(*i*)

$\forall i,j : \alpha_{ij} = 1$ if we associate primary element i with secondary element j; 0 otherwise (41.1)

Associate auxiliaries (*k*) with primary elements (*i*)

In this case, we cannot use the primary elements as a direct object of the verb because we need to measure the number of auxiliaries that are associated with each primary element. Therefore:

$\forall i, k : x_{ik} =$ Number of auxiliary elements k associated with the primary element i. (41.2)

Specifications

– Quantitative selection specifications:

Activity	Elements selecting	Selectable elements	Selection	Constraints
$\forall i, j : \alpha_{ij}$	Primary ($i = 1\ldots8$)	Secondary ($j = 1\ldots20$)	≥ 1	$\forall i : \sum_{j=1}^{20} \alpha_{ij} \geq 1$ (41.3)
	Secondary ($j = 1\ldots20$)	Primary ($i = 1\ldots8$)	≥ 1	$\forall j : \sum_{i=1}^{8} \alpha_{ij} \geq 1$ (41.4)

– Associate all auxiliary elements (allocation specification)

$$\forall k : \sum_{i=1}^{8} x_{ik} = Q_k \tag{41.5}$$

– If a primary element has a value greater than 1, it must be associated with more than two secondary elements (propositional logic specification)

$$\begin{aligned}&\forall i: \text{IF } P_i > 1 \text{ THEN } \sum_{j=1}^{20} \alpha_{ij} > 2\\&\Rightarrow f_5, f_{45} \Rightarrow \forall i/P_i > 1: \sum_{j=1}^{20} \alpha_{ij} \geq 3\end{aligned} \tag{41.6}$$

– If a primary element has a value greater than 1, the compatibility coefficients between secondary elements must be greater than 0.5 between them (propositional logic specification).

$$\begin{aligned}&\forall i,j,j'/j < j': \text{IF } P_i > 1 \text{ AND } \alpha_{ij} = 1 \text{ AND } \alpha_{ij'} = 1 \text{ THEN } \mathrm{CC}_{jj'} > 0.5\\&\Rightarrow f_{45} \Rightarrow \forall i/P_i > 1, j, j'/j < j': \text{IF } \alpha_{ij} = 1 \text{ AND } \alpha_{ij'} = 1 \text{ THEN } \mathrm{CC}_{jj'} > 0.5\\&\Rightarrow f_3 \Rightarrow \forall i/P_i > 1, j, j'/j < j':\\&\text{IF } \mathrm{CC}_{jj'} \leq 0.5 \text{ THEN NOT}(\alpha_{ij} = 1 \text{ AND } \alpha_{ij'} = 1)\\&\Rightarrow f_{37}, f_{11}, f_{45} \Rightarrow \forall i/P_i > 1, j, j'/j < j' \& \mathrm{CC}_{jj'} \leq 0.5:\\&\alpha_{ij} + \alpha_{ij'} \leq 1\end{aligned} \tag{41.7}$$

– If a secondary element is associated with more than five auxiliary elements, there must be auxiliary elements of three materials (non-binary logical calculation + binary logical calculation + propositional logic specification)

First, we calculate the number of auxiliary elements with which each secondary element is associated. For this, we need a non-binary logical calculation that collects the auxiliary elements associated with each secondary element through each primary element.

Non-Binary Logical Calculation

$$\forall i,j,k: y_{ijk} = \text{Number of auxiliary elements of type } k \text{ associated with element } j \text{ through primary element } i. \tag{41.8}$$

$$\begin{aligned}
&\forall i,j,k: \text{IF } \ \alpha_{ij}=1 \ \text{ THEN } \ y_{ijk}=x_{ik}; \text{ELSE } \ y_{ijk}=0;\\
&\Rightarrow \forall i,j,k: \text{IF } \ \alpha_{ij}=1 \ \text{ THEN } \ y_{ijk}=x_{ik}; \text{ELSE } \ y_{ijk}=0;\\
&\Rightarrow \forall i,j,k: \text{IF } \ \alpha_{ij}=1 \ \text{ THEN } \ y_{ijk}=x_{ik};\\
&\Rightarrow f_{16}\left(\text{UB}_{y_{ijk}-x_{ik}}=0\right) \Rightarrow \forall i,j,k: y_{ijk}-x_{ik} \le \ 0
\end{aligned} \tag{41.9}$$

$$\Rightarrow f_{16}\left(\text{LB}_{y_{ijk}-x_{ik}}=-Q_k\right) \Rightarrow \forall i,j,k: y_{ijk}-x_{ik} \ge -Q_k\left(1-\alpha_{ij}\right) \tag{41.10}$$

$$\begin{aligned}
&\Rightarrow \forall i,j,k: \text{IF } \ \alpha_{ij}=0 \ \text{ THEN } \ y_{ijk}=0\\
&\Rightarrow f_{\text{LB}}, f_7 \Rightarrow \forall i,j: \text{IF } \ 1- \ \ \alpha_{ij}=1 \ \text{ THEN } \ y_{ijk} \le 0\\
&\Rightarrow f_{14}\left(\text{UB}_{y_{ijk}}=Q_k\right) \Rightarrow \forall i,j,k: y_{ijk} \le Q_k\alpha_{ij}
\end{aligned} \tag{41.11}$$

Auxiliary Calculation

$\forall j: z_j =$ Total number of auxiliary elements associated with secondary element j.

$$\forall j: z_j=\sum_{i=1}^{8}\sum_{k=1}^{4} y_{ijk} \tag{41.12}$$

Now we use a binary logical calculation to know if there are auxiliary elements of each category associated with each secondary element:
Binary Logical Calculation

$\forall j,k: \beta_{jk} = 1$ if there are auxiliary elements of category k associated with j;
0 otherwise.

(41.13)

$$\begin{aligned}
&\forall j,k: \beta_{jk}=1 \quad \text{IF AND ONLY IF } \sum_{i=1}^{8} y_{ijk} > 0\\
&\Rightarrow f_5 \Rightarrow \forall j,k: \beta_{jk}=1 \quad \text{IF AND ONLY IF } \ \sum_{i=1}^{8} y_{ijk} \ge 1
\end{aligned} \tag{41.14}$$

$$\begin{aligned}
&\Rightarrow f_{25}\left(\text{LB}_{\sum_{i=1}^{8} y_{ijk}}=0\right) \Rightarrow \forall j,k: \sum_{i=1}^{8} y_{ijk} \ge \beta_{jk}\\
&\Rightarrow f_{25}\left(\text{UB}_{\sum_{i=1}^{8} y_{ijk}}=Q_k\right) \Rightarrow \forall j,k: \sum_{i=1}^{8} y_{ijk} \le Q_k\beta_{jk}
\end{aligned} \tag{41.15}$$

Finally, we express the specification as a logical proposition:

$$\forall j: \text{IF } z_j > 5 \text{ THEN } \sum_{k=1}^{4} \beta_{jk} = 3$$

$$\Rightarrow f_5 \Rightarrow \forall j: \text{IF } z_j \geq 6 \text{ THEN } \sum_{k=1}^{4} \beta_{jk} = 3 \tag{41.16}$$

$$\Rightarrow f_{18}\left(\text{UB}_{z_j} = 500\right) \Rightarrow \forall j: \; z_j \leq 5\left(1 - \omega_j\right) + 500\omega_j$$

$$\Rightarrow f_{22}\left(\text{UB}_{\sum_{k=1}^{4} \beta_{jk}} = 4\right) \Rightarrow \forall j: \sum_{k=1}^{4} \beta_{jk} \leq 4 - \omega_j \tag{41.17}$$

$$\Rightarrow f_{22}\left(\text{LB}_{\sum_{k=1}^{4} \beta_{jk}} = 0\right) \Rightarrow \forall j: \sum_{k=1}^{4} \beta_{jk} \geq 3\omega_j \tag{41.18}$$

Objective Function
Maximize total suitability

$$\text{Max} \; \sum_{i=1}^{8} \sum_{j=1}^{20} \Pi_{ij} \alpha_{ij} \tag{41.19}$$

Complete Model

$$\text{Max} \; \sum_{i=1}^{8} \sum_{j=1}^{20} \Pi_{ij} \alpha_{ij} \tag{41.19}$$

s.t.

$$\forall i: \sum_{j=1}^{20} \alpha_{ij} \geq 1 \tag{41.3}$$

$$\forall j: \sum_{i=1}^{8} \alpha_{ij} \geq 1 \tag{41.4}$$

$$\forall k: \sum_{i=1}^{8} x_{ik} = Q_k \tag{41.5}$$

$$\forall i/P_i > 1: \; \sum_{j=1}^{20} \alpha_{ij} \geq 3 \tag{41.6}$$

$$\forall i/P_i > 1, j, j'/j < j' \& \text{CC}_{jj'} \leq 0.5 : \alpha_{ij} + \alpha_{ij'} \leq 1 \tag{41.7}$$

$$\forall i, j, k : y_{ijk} - x_{ik} \leq 0 \tag{41.9}$$

$$\forall i, j, k : y_{ijk} - x_{ik} \geq -Q_k(1 - \alpha_{ij}) \tag{41.10}$$

$$\forall i, j, k : y_{ijk} \leq Q_k \alpha_{ij} \tag{41.11}$$

$$\forall j : z_j = \sum_{i=1}^{8} \sum_{k=1}^{4} y_{ijk} \tag{41.12}$$

$$\forall j, k : \sum_{i=1}^{8} y_{ijk} \geq \beta_{jk} \tag{41.14}$$

$$\forall j, k : \sum_{i=1}^{8} y_{ijk} \leq Q_k \beta_{jk} \tag{41.15}$$

$$\forall j : \quad z_j \leq 5(1 - \omega_j) + 500\omega_j \tag{41.16}$$

$$\forall j : \sum_{k=1}^{4} \beta_{jk} \leq 4 - \omega_j \tag{41.17}$$

$$\forall j : \sum_{k=1}^{4} \beta_{jk} \geq 3\omega_j \tag{41.18}$$

$$\forall i, j : \alpha_{ij} \text{ binaries} \tag{41.1}$$

$$\forall i, k : x_{ik} \geq 0 \text{ integers} \tag{41.2}$$

$$\forall i, j, k : y_{ijk} \geq 0 \text{ integers} \tag{41.8}$$

$$\forall j : z_j \geq 0$$

$$\forall j, k : \beta_{jk} \text{ binaries} \tag{41.13}$$

$$\forall j : \omega_j \text{ binaries} \tag{41.20}$$

Implementation in LINGO

```
MODEL:
SETS:
 Primary/1..8/:P;
 Secondary/1..20/:z,Omega;
 Auxiliaries/1..4/:Q;
 PS(Primary,Secondary):II,Alpha;
 SS(Secondary,Secondary):CC;
 PA(Primary,Auxiliaries):x;
 SA(Secondary,Auxiliaries):Beta;
 PSA(Primary,Secondary,Auxiliaries):y;
ENDSETS
CALC:
@FOR(Secondary(j):@FOR(Secondary(jj)|j#LT#jj:CC(jj,j)=CC(j,jj)));
ENDCALC
```

```
DATA:
 P=@OLE('C:\Files_\41_Excel.xlsx','P');
 Q=@OLE('C:\Files_\41_Excel.xlsx','Q');
 II=@OLE('C:\Files_\41_Excel.xlsx','II');
 CC=@OLE('C:\Files_\41_Excel.xlsx','CC');
ENDDATA
!41.19;
MAX = @SUM(Primary(i):@SUM(Secondary(j):II(i,j)*Alpha(i,j)));
!41.3;
@FOR(Primary(i):@SUM(Secondary(j):Alpha(i,j))>=1);
!41.4;
@FOR(Secondary(j):@SUM(Primary(i):Alpha(i,j))>=1);
!41.5;
@FOR(Auxiliaries(k):@SUM(Primary(i):x(i,k))=Q(k));
!41.6;
@FOR(Primary(i)|P(i)#GT#1:@SUM(Secondary(j):ALpha(i,j))>=3);
!41.7;
@FOR(Primary(i)|P(i)#GT#1:@FOR(Secondary(j):@FOR(Secondary(jj)|
j#LT#jj#AND#CC(j,jj)#LE#0.5:ALpha(i,j)+ALpha(i,jj)<=1)));
!41.9;
@FOR(Primary(i):@FOR(Secondary(j):@FOR(Auxiliaries(k):y(i,j,k)-x
(i,k)<=0)));
!41.10;
@FOR(Primary(i):@FOR(Secondary(j):@FOR(Auxiliaries(k):y(i,j,k)-x
(i,k)>=-Q(k)*(1-Alpha(i,j)))));
!41.11;
@FOR(Primary(i):@FOR(Secondary(j):@FOR(Auxiliaries(k):y(i,j,k)<=Q
(k)*Alpha(i,j))));
!41.12;
@FOR(Secondary(j):z(j)=@SUM(Primary(i):@SUM(Auxiliaries(k):y(i,j,
k))));
!41.14;
@FOR(Secondary(j):@FOR(Auxiliaries(k):@SUM(Primary(i):y(i,j,k))
>=Beta(j,k)));
!41.15;
@FOR(Secondary(j):@FOR(Auxiliaries(k):@SUM(Primary(i):y(i,j,k))
<=Q(k)*Beta(j,k)));
!41.16;
@FOR(Secondary(j): z(j)<=5*(1-Omega(j))+500*Omega(j));
!41.17;
@FOR(Secondary(j): @SUM(Auxiliaries(k): Beta(j,k)) <=4 - Omega(j));
!41.18;
@FOR(Secondary(j): @SUM(Auxiliaries(k): Beta(j,k)) >=3*Omega(j));
! 41.1;
@FOR(Primary(i):@FOR(Secondary(j):@BIN(Alpha(i,j))));
!41.2;
@FOR(Primary(i):@FOR(Auxiliaries(k):@GIN(x(i,k))));
!41.8;
@FOR(Primary(i):@FOR(Secondary(j):@FOR(Auxiliaries(k):@GIN(y(i,j,
k)))));
!41.13;
@FOR(Secondary(j):@FOR(Auxiliaries(k):@BIN(Beta(j,k))));
!41.20;
@FOR(Secondary(j):@BIN(Omega(j)));
END
```

```
Solution Report - 41_LingoE
Global optimal solution found.
Objective value:                                 362.4000
Objective bound:                                 362.4000
Infeasibilities:                                 0.000000
Extended solver steps:                               1772
Total solver iterations:                           482197
Elapsed runtime seconds:                            55.23

Model Class:                                         MILP

Total variables:                    952
Nonlinear variables:                  0
Integer variables:                  932

Total constraints:                 2193
```

Fig. 7.6 Problem 41: LINGO solution report

Solution
It is shown in Fig. 7.6.

7.6 Problem 42. Creation of Sudokus

SUDOKU (Delahaye 2006): Grid made up by 81 squares arranged in 9 rows and 9 columns. We consider as a group of spaces each row, each column, and each square of nine spaces (made up of 3 × 3 spaces) as shown in Fig. 7.7.

The game consists of assigning a number 1–9 to each space without repeating any numbers within the row, column, or square. To facilitate the game, some boxes already filled out with numbers are shown.

The game ends when the entire grid is completed.

Knowing the sudoku game, we are going to create a model that generates sudoku puzzles. The specifications that are imposed to establish the level of difficulty of the game are the following:

- There must be a single row or column with no visible spaces.
- At least two rows with only two spaces visible.
- At least six sets (row, column, or square) with at least three visible spaces.
- A single square with five visible spaces.
- A single number that does not appear in any visible space.
- One of the numbers appears four times.

Objective: Minimize the number of visible spaces.

Fig. 7.7 Example of sudoku

5	3			7				
6			1	9	5			
	9	8					6	
8				6				3
4			8		3			1
7				2				6
	6					2	8	
			4	1	9			5
				8			7	9

Modelling

List of elements in Table 7.6.

Visible is a typology of spaces.

Table 7.6 Problem 42: table of elements

			DATA				
Elements	Set	QN	Name	Parameter	Type	Belonging	Value
Spaces	$i = 1 \ldots 81$	I_U	Row	F_i	O	W	–
			Column	C_i	O	W	–
			Square	R_i	O	W	–
Rows	$j = 1 \ldots 9$	I_U					
Columns	$k = 1 \ldots 9$	I_U					
Squares	$z = 1 \ldots 9$	I_U					
Numbers	$d = 1 \ldots 9$	I_U					
Visible	–	I_U					

Decision Variables

Set spaces (i) as visible

$$\forall i : \alpha_i = 1 \text{ if the space } i \text{ is set as visible; 0 otherwise.} \quad (42.1)$$

Assign numbers (d) to spaces (i)

$$\forall d, i : \beta_{di} = 1 \text{ if the number } d \text{ is assigned to space } i\text{; 0 otherwise.} \quad (42.2)$$

Specifications

– Quantitative selection specifications:

Activity	Elements selecting		Selectable elements	Selection	Constraints
$\forall d, i : \beta_{di}$	Spaces ($i = 1 \ldots 81$)		Numbers ($d = 1 \ldots 9$)	=1	$\forall i : \sum_{d=1}^{9} \beta_{di} = 1$ (42.3)
	Numbers ($d = 1 \ldots 9$)	Rows ($j = 1 \ldots 9$)	Spaces ($i/F_i = j$)	=1	$\forall j, d : \sum_{i/F_i=j} \beta_{di} = 1$ (42.4)
	Numbers ($d = 1 \ldots 9$)	Columns ($k = 1 \ldots 9$)	Spaces ($i/C_i = k$)	=1	$\forall k, d : \sum_{i/C_i=k} \beta_{di} = 1$ (42.5)
	Numbers ($d = 1 \ldots 9$)	Squares ($z = 1 \ldots 9$)	Spaces ($i/R_i = z$)	=1	$\forall z, d : \sum_{i/R_i=z} \beta_{di} = 1$ (42.6)

– There must be a single row or column without visible spaces (logical calculations + quantitative selection specification):

We must calculate if each row or column has zero visible spaces and then impose that the number of them is 1.

Binary Logical Calculations

$$\forall j : \delta_j^F = 1 \text{ if row } j \text{ has 0 visible spaces; 0 otherwise} \tag{42.7}$$

$$\forall k : \delta_k^C = 1 \text{ if column } k \text{ has 0 visible spaces; 0 otherwise} \tag{42.8}$$

$$\forall j : \delta_j^F = 1 \text{ IF AND ONLY IF } \sum_{i/F_i=j} \alpha_i =$$

$$\forall k : \delta_k^C = 1 \text{ IF AND ONLY IF } \sum_{i/C_i=k} \alpha_i = 0$$

$$\Rightarrow f_{\text{LB}} \Rightarrow \forall j : \delta_j^F = 1 \text{ IF AND ONLY IF } \sum_{i/F_i=j} \alpha_i \leq 0 \tag{42.9}$$

$$\Rightarrow f_{24}\left(\text{UB}_{\sum_{i/F_i=j} \alpha_i} = 9\right) \Rightarrow \forall j : \sum_{i/F_i=j} \alpha_i \leq 9\left(1 - \delta_j^F\right)$$

$$\Rightarrow f_{24}\left(\text{LB}_{\sum_{i/F_i=j} \alpha_i} = 0\right) \Rightarrow \forall j : \sum_{i/F_i=j} \alpha_i \geq 1 - \delta_j^F \tag{42.10}$$

$$\Rightarrow f_{\text{LB}} \Rightarrow \forall k : \delta_k^C = 1 \text{ IF AND ONLY IF } \sum_{i/C_i=k} \alpha_i \leq 0$$

$$\Rightarrow f_{24}\left(\text{UB}_{\sum_{i/C_i=k} \alpha_i} = 9\right) \Rightarrow \forall k : \sum_{i/C_i=k} \alpha_i \leq 9\left(1 - \delta_k^C\right) \tag{42.11}$$

$$\Rightarrow f_{24}\left(\mathrm{LB}_{\sum\limits_{i/C_i=k}\alpha_i}=0\right) \Rightarrow \forall k:\sum_{i/C_i=k}\alpha_i \geq 1-\ \delta_k^C \tag{42.12}$$

Quantitative selection specification:

$$\sum_{j=1}^{9}\delta_j^F+\sum_{k=1}^{9}\delta_k^C=1 \tag{42.13}$$

- At least two rows with only two visible spaces (logical calculation + quantitative selection specification):

We must calculate if each row has two visible spaces and then impose that the number of them be at least 2.

Binary Logical Calculation

$$\forall j:\varphi_j=1 \text{ if row } j \text{ has only two visible spaces; 0 otherwise} \tag{42.14}$$

$$\begin{aligned} &\forall j:\varphi_j=1 \text{ IF AND ONLY IF} \sum_{i/F_i=j}\alpha_i=2 \\ &\Rightarrow f_{S_V}\Rightarrow \forall j: \text{IF } \varphi_j=1 \text{ THEN} \sum_{i/F_i=j}\alpha_i=2 \\ &\Rightarrow f_{16}\left(\mathrm{UB}_{\sum\limits_{i/F_i=j}\alpha_i}=9\right) \Rightarrow \forall j:\sum_{i/F_i=j}\alpha_i \leq 2+7(1-\varphi_j) \end{aligned} \tag{42.15}$$

$$\Rightarrow f_{16}\left(\mathrm{LB}_{\sum\limits_{i/F_i=j}\alpha_i}=0\right) \Rightarrow \forall j:\sum_{i/F_i=j}\alpha_i \geq 2\varphi_j \tag{42.16}$$

Quantitative selection specification:

$$\sum_{j=1}^{9}\varphi_j \geq 2 \tag{42.17}$$

- At least six sets with at least three visible spaces (binary logical calculations + quantitative selection specification):

We must calculate if each row, column, or square has at least three visible spaces.

Binary Logical Calculations

$$\forall j:\mu_j^F=1 \text{ if the row } j \text{ has at least three visible spaces; 0 otherwise.} \tag{42.18}$$

$$\begin{aligned}
&\forall j : \mu_j^F = 1 \text{ IF AND ONLY IF } \sum_{i/F_i=j} \alpha_i \geq 3 \\
&\Rightarrow f_{S_V} \Rightarrow \forall j : \text{IF } \mu_j^F = 1 \text{ THEN } \sum_{i/F_i=j} \alpha_i \geq 3 \\
&\Rightarrow f_{15}\left(\text{LB}_{\sum_{i/F_i=j} \alpha_i} = 0\right) \Rightarrow \forall j : \sum\nolimits_{i/F_i=j} \alpha_i \geq 3\mu_j^F
\end{aligned} \tag{42.19}$$

$$\forall k : \mu_k^C = 1 \text{ if column } k \text{ has at least three visible spaces; 0 otherwise.} \tag{42.20}$$

$$\begin{aligned}
&\forall k : \mu_k^C = 1 \text{ IF AND ONLY IF } \sum_{i/C_i=k} \alpha_i \geq 3 \\
&\Rightarrow f_{S_V} \Rightarrow \forall k : \text{IF } \mu_k^C = 1 \text{ THEN } \sum_{i/C_i=k} \alpha_i \geq 3 \\
&\Rightarrow f_{15}\left(\text{LB}_{\sum_{i/C_i=k} \alpha_i} = 0\right) \Rightarrow \forall k : \sum_{i/C_i=k} \alpha_i \geq 3\mu_k^C
\end{aligned} \tag{42.21}$$

$$\forall z : \mu_z^R = 1 \text{ if column } k \text{ has at least three visible spaces; 0 otherwise.} \tag{42.22}$$

$$\begin{aligned}
&\forall z : \mu_z^R = 1 \text{ IF AND ONLY IF} \sum_{i/R_i=z} \alpha_i \geq 3 \\
&\Rightarrow f_{S_V} \Rightarrow \forall z : \text{IF } \mu_z^C = 1 \text{ THEN} \sum_{i/R_i=z} \alpha_i \geq 3 \\
&\Rightarrow f_{15}\left(\text{LB}_{\sum_{i/R_i=z} \alpha_i} = 0\right) \Rightarrow \forall z : \sum_{i/R_i=z} \alpha_i \geq 3\mu_z^R
\end{aligned} \tag{42.23}$$

Quantitative selection specification:

$$\sum_{j=1}^{9} \mu_j^F + \sum_{k=1}^{9} \mu_k^C + \sum_{z=1}^{9} \mu_z^R \geq 6 \tag{42.24}$$

– A single square with five visible spaces (logical calculation + quantitative selection specification).

Binary Logical Calculation

$$\forall z : \omega_z^R = 1 \text{ if square } z \text{ has exactly five visible spaces; 0 otherwise.} \tag{42.25}$$

$$\forall z : \omega_z^R = 1 \text{ IF AND ONLY IF } \sum_{i/R_i=z} \alpha_i = 5$$

$$\Rightarrow f_{26}\left(\text{UB}_{\sum_{i/R_i=z} \alpha_i} = 9\right) \Rightarrow \forall z : \sum_{i/R_i=z} \alpha_i \leq 5 + 4\left(1 - \omega_z^R\right) \tag{42.26}$$

$$\Rightarrow f_{26}\left(\text{LB}_{\sum_{i/R_i=z} \alpha_i} = 0\right) \Rightarrow \forall z : \sum_{i/R_i=z} \alpha_i \geq 5\omega_z^R \tag{42.27}$$

$$\Rightarrow f_{26} \Rightarrow \forall z : \sum_{i/R_i=z} \alpha_i \leq 4\lambda_{1z} + 9\omega_z^R + 9\lambda_{2z} \tag{42.28}$$

$$\Rightarrow f_{26} \Rightarrow \forall z : \sum_{i/R_i=z} \alpha_i \geq 6\lambda_{2z} \tag{42.29}$$

$$\Rightarrow f_{26} \Rightarrow \forall z : \lambda_{1z} + \lambda_{2z} = 1 - \omega_z^R \tag{42.30}$$

Quantitative selection specification:

$$\sum_{z=1}^{9} \omega_z^R = 1 \tag{42.31}$$

– A single number does not appear in any visible space (binary logical calculations + quantitative selection specification):

We first calculate whether each number appears in every visible space. Then we calculate from that calculation whether each number is visible, and finally, we impose the number of visible numbers to be 8.

Binary Logical Calculation

$$\forall i, d : \chi_{di} = 1 \text{ if the number } d \text{ is visible in space } i; 0 \text{ otherwise} \tag{42.32}$$

$$\forall i, d : \chi_{di} = 1 \text{ IF AND ONLY IF } \alpha_i = 1 \text{ AND } \beta_{di} = 1$$

$$\Rightarrow f_{37} \Rightarrow \forall i, d : \chi_{di} = 1 \text{ IF AND ONLY IF } \alpha_i + \beta_{di} \geq 2 \tag{42.33}$$

$$\Rightarrow f_{25}\left(\text{LB}_{\alpha_i + \beta_{di}} = 0\right) \Rightarrow \forall i, d : \alpha_i + \beta_{di} \geq 2\chi_{di}$$

$$\Rightarrow f_{25}\left(\text{UB}_{\alpha_i + \beta_{di}} = 2\right) \Rightarrow \forall i, d : \alpha_i + \beta_{di} \leq 1 + \chi_{di} \tag{42.34}$$

Binary Logical Calculation

$$\forall d : \pi_d = 1 \text{ if the number } d \text{ is visible}; 0 \text{ otherwise} \tag{42.35}$$

$$\forall d : \pi_d = 1 \text{ IF AND ONLY IF } \sum_{i=1}^{81} \chi_{di} \geq 1$$

$$\Rightarrow f_{25}\left(\text{LB}_{\sum_{i=1}^{81}\chi_{di}} = 0\right) \Rightarrow \forall d: \sum_{i=1}^{81} \chi_{di} \geq \pi_d \tag{42.36}$$

$$\Rightarrow f_{25}\left(\text{UB}_{\sum_{i=1}^{81}\chi_{di}} = 9\right) \Rightarrow \forall d: \sum_{i=1}^{81} \chi_{di} \leq 9\pi_d \tag{42.37}$$

Quantitative selection specification:

$$\sum_{d=1}^{9} \pi_d = 8 \tag{42.38}$$

- Only one of the numbers appears four times (binary logical calculation + quantitative selection specification):

I take advantage of the logical calculations from the previous specification and define a new logical calculation.

Binary Logical Calculation

$$\forall d : \theta_d = 1 \text{ if the numberdis visible 4 times.} \tag{42.39}$$

$$\forall d : \theta_d = 1 \text{ IF AND ONLY IF } \sum_{i=1}^{81} \chi_{di} = 4$$

$$\Rightarrow f_{26}\left(\text{UB}_{\sum_{i=1}^{81}\chi_{di}} = 9\right) \Rightarrow \forall d: \sum_{i=1}^{81} \chi_{di} \leq 4 + 5(1 - \theta_d) \tag{42.40}$$

$$\Rightarrow f_{26}\left(\text{LB}_{\sum_{i=1}^{81}\chi_{di}} = 0\right) \Rightarrow \forall d: \sum_{i=1}^{81} \chi_{di} \geq 4\theta_d \tag{42.41}$$

$$\Rightarrow f_{26} \Rightarrow \forall d: \sum_{i=1}^{81} \chi_{di} \leq 3\psi_{1d} + 9\theta_d + 9\psi_{2z} \tag{42.42}$$

$$\Rightarrow f_{26} \Rightarrow \forall d: \sum_{i=1}^{81} \chi_{di} \geq 5\psi_{2z} \tag{42.43}$$

$$\Rightarrow f_{26} \Rightarrow \forall d : \psi_{1d} + \psi_{2d} = 1 - \theta_d \tag{42.44}$$

Quantitative selection specification:

$$\sum_{d=1}^{9} \theta_d = 1 \tag{42.45}$$

Objective Function
Minimize the number of visible spaces.

$$\text{Min} \sum_{i=1}^{81} \alpha_i \tag{42.46}$$

We could have ignored the objective function if our objective had only been to build a sudoku puzzle with the requested specifications.
Complete Model

$$\text{Min} \sum_{i=1}^{81} \alpha_i \tag{42.46}$$

s.t.

$$\forall i : \sum_{d=1}^{9} \beta_{di} = 1 \tag{42.3}$$

$$\forall j, d : \sum_{i/F_i=j} \beta_{di} = 1 \tag{42.4}$$

$$\forall k, d : \sum_{i/C_i=k} \beta_{di} = 1 \tag{42.5}$$

$$\forall z, d : \sum_{i/R_i=z} \beta_{di} = 1 \tag{42.6}$$

$$\forall j : \sum_{i/F_i=j} \alpha_i \leq 9\left(1 - \delta_j^F\right) \tag{42.9}$$

$$\forall j : \sum_{i/F_i=j} \alpha_i \geq 1 - \delta_j^F \tag{42.10}$$

$$\forall k : \sum_{i/C_i=k} \alpha_i \leq 9(1 - \delta_k^C) \tag{42.11}$$

$$\forall k: \sum_{i/C_i=k} \alpha_i \geq 1 - \delta_k^C \tag{42.12}$$

$$\sum_{j=1}^{9} \delta_j^F + \sum_{k=1}^{9} \delta_k^C = 1 \tag{42.13}$$

$$\forall j: \sum_{i/F_i=j} \alpha_i \leq 2 + 7(1 - \varphi_j) \tag{42.15}$$

$$\forall j: \sum_{i/F_i=j} \alpha_i \geq 2\varphi_j \tag{42.16}$$

$$\sum_{j=1}^{9} \varphi_j \geq 2 \tag{42.17}$$

$$\forall j: \sum_{i/F_i=j} \alpha_i \geq 3\mu_j^F \tag{42.19}$$

$$\forall k: \sum_{i/C_i=k} \alpha_i \geq 3\mu_k^C \tag{42.21}$$

$$\forall z: \sum_{i/R_i=z} \alpha_i \geq 3\mu_z^R \tag{42.23}$$

$$\sum_{j=1}^{9} \mu_j^F + \sum_{k=1}^{9} \mu_k^C + \sum_{z=1}^{9} \mu_z^R \geq 4 \tag{42.24}$$

$$\forall z: \sum_{i/R_i=z} \alpha_i \leq 5 + 4(1 - \omega_z^R) \tag{42.26}$$

$$\forall z: \sum_{i/R_i=z} \alpha_i \geq 5\omega_z^R \tag{42.27}$$

$$\forall z: \sum_{i/R_i=z} \alpha_i \leq 4\lambda_{1z} + 9\omega_z^R + 9\lambda_{2z} \tag{42.28}$$

$$\forall z: \sum_{i/R_i=z} \alpha_i \geq 6\lambda_{2z} \tag{42.29}$$

$$\forall z: \lambda_{1z} + \lambda_{2z} = 1 - \omega_z^R \tag{42.30}$$

$$\sum_{z=1}^{9} \omega_z^R = 1 \tag{42.31}$$

$$\forall i, d: \quad \alpha_i + \beta_{di} \geq 2\chi_{di} \tag{42.33}$$

$$\forall i, d: \quad \alpha_i + \beta_{di} \leq 1 + \chi_{di} \tag{42.34}$$

$$\forall d: \sum_{i=1}^{81} \chi_{di} \geq \pi_d \tag{42.36}$$

$$\forall d: \sum_{i=1}^{81} \chi_{di} \leq 9\pi_d \tag{42.37}$$

$$\sum_{d=1}^{9} \pi_d = 8 \tag{42.38}$$

$$\forall d: \sum_{i=1}^{81} \chi_{di} \leq 4 + 5(1 - \theta_d) \tag{42.40}$$

$$\forall d: \sum_{i=1}^{81} \chi_{di} \geq 4\theta_d \tag{42.41}$$

$$\forall d: \sum_{i=1}^{81} \chi_{di} \leq 3\psi_{1d} + 9\theta_d + 9\psi_{2d} \tag{42.42}$$

$$\forall d: \sum_{i=1}^{81} \chi_{di} \geq 5\psi_{2d} \tag{42.43}$$

$$\forall d: \psi_{1d} + \psi_{2d} = 1 - \theta_d \tag{42.44}$$

$$\sum_{d=1}^{9} \theta_d = 1 \tag{42.45}$$

$$\forall i: \alpha_i \text{ binarias} \tag{42.1}$$

$$\forall d, i: \beta_{di} \text{ binarias} \tag{42.2}$$

$$\forall j: \delta_j^F \text{ binarias} \tag{42.7}$$

$$\forall k: \delta_k^C \text{ binarias} \tag{42.8}$$

$$\forall j: \varphi_j \text{ binarias} \tag{42.14}$$

$$\forall j: \mu_j^F \text{ binarias} \tag{42.18}$$

$$\forall k: \mu_k^C \text{ binarias} \tag{42.20}$$

$$\forall z: \mu_z^R \text{ binarias} \tag{42.22}$$

$$\forall z: \omega_z^R \text{ binarias} \tag{42.25}$$

$$\forall i, d: \chi_{di} \text{ binarias} \tag{42.32}$$

$$\forall d: \pi_d \text{ binarias} \tag{42.35}$$

$$\forall d : \theta_d \text{ binarias} \quad (42.39)$$

$$\forall z : \lambda_{1z}, \lambda_{2z} \text{ binarias} \quad (42.47)$$

$$\forall d : \psi_{1d}, \psi_{2d} \text{ binarias} \quad (42.48)$$

Implementation in LINGO

```
MODEL:
SETS:
 Spaces/1..81/:R,C,S,Alpha;
 Rows/1..9/:DeltaF,Ti,TitaF;
 Columns/1..9/:DeltaC,TitaC;
 Squares/1..9/:TitaR,OmegaR,y1,y2;
 Numbers/1..9/:Pi,O,Ro1,Ro2;
 DC(Numbers,Spaces):Beta,Sigma;
ENDSETS
DATA:
 R=@FILE('C:\Files_\42_Text.txt');
 C=@FILE('C:\Files_\42_Text.txt');
 S=@FILE('C:\Files_\42_Text.txt');
@OLE('C:\Files_\42_Excel.xlsx','Solution1')=@WRITEFOR(Spaces(i):
@IF(Alpha(i)#EQ#1,@SUM(Numbers(d):d*Beta(d,i)),' '));
@OLE('C:\Files_\42_Excel.xlsx','Solution2')=@WRITEFOR(Spaces(i):
@SUM(Numbers(d):d*Beta(d,i)));
ENDDATA
!42.46;
MIN= @SUM(Spaces(i):Alpha(i));
!42.3;
@FOR(Spaces(i):@SUM(Numbers(d):Beta(d,i))=1);
!42.4;
@FOR(Rows(j):@FOR(Numbers(d):@SUM(Spaces(i)|R(i)#EQ#j: Beta(d,i))
=1));
!42.5;
@FOR(Columns(k):@FOR(Numbers(d):@SUM(Spaces(i)|C(i)#EQ#k: Beta(d,
i))=1));
!42.6;
@FOR(Squares(z):@FOR(Numbers(d):@SUM(Spaces(i)|S(i)#EQ#z: Beta(d,
i))=1));
!42.9;
@FOR(Rows(j):@SUM(Spaces(i)|R(i)#EQ#j:Alpha(i))<=9*(1-DeltaF
(j)));
!42.10;
@FOR(Rows(j):@SUM(Spaces(i)|R(i)#EQ#j:Alpha(i))>=1-DeltaF(j));
!42.11;
@FOR(Columns(k):@SUM(Spaces(i)|C(i)#EQ#k:Alpha(i))<=9*(1-DeltaC
(k)));
!42.12;
@FOR(Columns(k):@SUM(Spaces(i)|C(i)#EQ#k:Alpha(i))>=1-DeltaC(k));
!42.13;
@SUM(Rows(j):DeltaF(j)) + @SUM(Columns(k):DeltaC(k))=1;
!42.15;
@FOR(Rows(j):@SUM(Spaces(i)|R(i)#EQ#j:Alpha(i))<=2+7*(1-Ti(j)));
!42.16;
```

```
@FOR(Rows(j):@SUM(Spaces(i)|R(i)#EQ#j:Alpha(i))>=2*Ti(j));
!42.17;
@SUM(Rows(j):Ti(j))>=2;
!42.19;
@FOR(Rows(j):@SUM(Spaces(i)|R(i)#EQ#j:Alpha(i))>=3*TitaF(j));
!42.21;
@FOR(Columns(k):@SUM(Spaces(i)|C(i)#EQ#k:Alpha(i))>=3*TitaC(k));
!42.23;
@FOR(Squares(z):@SUM(Spaces(i)|S(i)#EQ#z:Alpha(i))>=3*TitaR(z));
!42.24;
@SUM(Rows(j):TitaF(j))+@SUM(Columns(k):TitaC(k))+@SUM(Squares(z):
TitaR(z))>=6;
!42.26;
@FOR(Squares(z):@SUM(Spaces(i)|S(i)#EQ#z:Alpha(i))<=5 + 4*(1-OmegaR
(z)));
!42.27;
@FOR(Squares(z):@SUM(Spaces(i)|S(i)#EQ#z:Alpha(i))>=5*OmegaR(z));
!42.28;
@FOR(Squares(z):@SUM(Spaces(i)|S(i)#EQ#z:Alpha(i))<=4*y1(z)
+9*OmegaR(z)+9*y2(z));
!42.29;
@FOR(Squares(z):@SUM(Spaces(i)|S(i)#EQ#z:Alpha(i))>=6*y2(z));
!42.30;
@FOR(Squares(z):y1(z)+y2(z)=1-OmegaR(z));
!42.31;
@SUM(Squares(z):OmegaR(z))=1;
!42.33;
@FOR(Spaces(i):@FOR(Numbers(d): Alpha(i) + Beta(d,i) >=2*Sigma(d,
i)));
!42.34;
@FOR(Spaces(i):@FOR(Numbers(d): Alpha(i) + Beta(d,i) <=1 + Sigma(d,
i)));
!42.36;
@FOR(Numbers(d):@SUM(Spaces(i):Sigma(d,i))>=Pi(d));
!42.37;
@FOR(Numbers(d):@SUM(Spaces(i):Sigma(d,i))<=9*Pi(d));
!42.38;
@SUM(Numbers(d):Pi(d))=8;
!42.40;
@FOR(Numbers(d):@SUM(Spaces(i):Sigma(d,i))<=4 + 5*(1-O(d)));
!42.41;
@FOR(Numbers(d):@SUM(Spaces(i):Sigma(d,i))>=4*O(d));
!42.42;
@FOR(Numbers(d):@SUM(Spaces(i):Sigma(d,i))<=3*Ro1(d)+9*O(d)+9*Ro2
(d));
!42.43;
@FOR(Numbers(d):@SUM(Spaces(i):Sigma(d,i))>=5*Ro2(d));
!42.44;
@FOR(Numbers(d):Ro1(d)+Ro2(d)=1-O(d));
!42.45;
@SUM(Numbers(d):O(d))=1;
!42.1;
@FOR(Spaces(i):@BIN(Alpha(i)));
```

```
!42.2;
@FOR(Numbers(d):@FOR(Spaces(i):@BIN(Beta(d,i))));
!42.7;
@FOR(Rows(j):@BIN(DeltaF(j)));
!42.8;
@FOR(Columns(k):@BIN(DeltaC(k)));
!42.14;
@FOR(Rows(j):@BIN(Ti(j)));
!42.18;
@FOR(Rows(j):@BIN(TitaF(j)));
!42.20;
@FOR(Columns(k):@BIN(TitaC(k)));
!42.22;
@FOR(Squares(z):@BIN(TitaR(z)));
!42.25;
@FOR(Squares(z):@BIN(OmegaR(z)));
!42.32;
@FOR(Numbers(d):@FOR(Spaces(i):@BIN(Sigma(d,i))));
!42.35;
@FOR(Numbers(d):@BIN(Pi(d)));

!42.39;
@FOR(Numbers(d):@BIN(O(d)));
!42.47;
@FOR(Squares(z):@BIN(y1(z))); @FOR(Squares(z):@BIN(y2(z)));
!42.48;
@FOR(Numbers(d):@BIN(Ro1(d)));
@FOR(Numbers(d):@BIN(Ro2(d)));
END
```

Solution

To obtain results, we are simply going to look for a feasible solution (Fig. 7.8), since with it we can quickly build a sudoku puzzle.

Therefore, we eliminate the objective function in the resolution process.

In the file "42_Excel.xlsx" we have represented the solution (Fig. 7.9).

7.7 Problem 43. "Unblock Me"

Game description "Unblock Me" (Kira Games 2021): Blocks are positioned across the board in either horizontal or vertical positions; a horizontal block can only move right and left, while a vertical block can only move up and down. You must shift them around in a way that frees a path for the red block (Fig. 7.10).

It is a puzzle game that presents a board made up of several columns and rows. For the original game, the number of rows and columns is 6. In these rows and columns is placed a series of blocks, which we will call obstacles, and which can be horizontal or vertical. Horizontal blocks are placed in a row and can only be moved within that row (move right and left). Vertical obstacles can only move within the

```
Solution Report - 42_LingoE
Feasible solution found.
Infeasibilities:                                          0.000000
Extended solver steps:                                           0
Total solver iterations:                                     22364
Elapsed runtime seconds:                                      3.02


Export Summary Report
---------------------
Transfer Method:        OLE BASED
Workbook:               C:\Files_\42_Excel.xlsx
Ranges Specified:               1
    Solution1
Ranges Found:                   1
Range Size Mismatches:          0
Values Transferred:            81


Export Summary Report
---------------------
Transfer Method:        OLE BASED
Workbook:               C:\Files_\42_Excel.xlsx
Ranges Specified:               1
    Solution2
Ranges Found:                   1
Range Size Mismatches:          0
Values Transferred:            81

Model Class:                                              PILP

Total variables:                    1656
Nonlinear variables:                   0
Integer variables:                  1656

Total constraints:                  1977
Nonlinear constraints:                 0

Total nonzeros:                    13059
Nonlinear nonzeros:                    0
```

Fig. 7.8 Problem 42: LINGO solution report

Solution1

7	4		1		9	6		
	3		7					9
9		1						
3	9	5	2					1
								2
1								
4		3				5	9	
				7		2		

Solution2

7	4	8	1	2	9	6	5	3
2	3	6	7	5	4	1	8	9
9	5	1	6	8	3	4	2	7
3	9	5	2	6	7	8	4	1
8	2	7	4	3	1	9	6	5
6	1	4	5	9	8	7	3	2
1	6	2	9	4	5	3	7	8
4	7	3	8	1	2	5	9	6
5	8	9	3	7	6	2	1	4

Fig. 7.9 "42_Excel.xlsx"

column to which they belong (move up and down). On the other hand, we have the red block; it is a horizontal block located in a specific row (row 3 in the original game panel) and therefore only moves horizontally. The board has an exit at the end of that

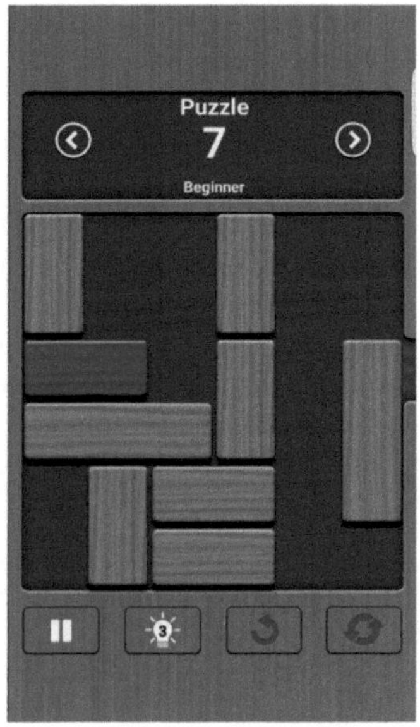

Fig. 7.10 Puzzle example

row. Our task is to get the red block out of the board. We must move obstacles around in a way that frees up a path for the red block.

We will create a model that mathematically represents the game with two scenarios:

- A scenario where it is imposed as a specification to get the red chip out of the board. In this scenario, no objective is imposed. Obtaining a feasible solution will mean that the red token has gone out of the board.
- A second scenario, as a complete optimization problem, in which the objective is to get the red chip out in the fewest possible steps (movements).

Modeling

We consider a valid model for any number of rows or columns, any number and size of obstacles, and any size and position of the red block.

Two versions are proposed. In the first version, it is proposed as a decision in each period to move blocks, and the positions of the blocks are updated once the movement has been made. In the second version, more efficient, we reduce the size of the model by integrating move and position in a single decision variable. Although the first version is more inefficient, we present the two versions to illustrate different ways of modeling.

In both versions, several movements will be allowed in each period, but within the same period, we will only allow movements of blocks of a single type (horizontal or vertical) and only one block per row or column. By allowing several movements, we are able to reduce the number of periods, and therefore the size of the models.

Both versions make use of the same table of elements.

For both versions, the two scenarios are considered.

Table of Elements

The red block can be considered as one more horizontal block. We incorporate a binary attribute to identify it.

The attributes Occupied Initial Column and Occupied Initial Row refer to the first occupied column or row by each block.

It is necessary to establish an upper bound on the number of periods of movements that could be carried out in the game, in principle, an unknown value that through a manual resolution we can obtain. In another case, it may be a value that we adjust after several resolutions. We will call M the upper bound. List of elements in Table 7.7.

Table 7.7 Problem 43: table of elements

			Data				
Elements	Set	QN	Name	Parameter	Type	Belonging	Value
Board	–	I_U	Number of rows	NF	I	W	–
			Number of columns	NC	I	W	–
Horizontal blocks	$i = 1...n_1$	I_U	Row	F_i	O	W	–
			Size	TH_i	I	W	–
			Occupied initial column	CI_i	O	W	–
			Red block	FR_i	B	W	–
Vertical blocks	$j = 1...n_2$	I_U	Column	C_j	O	W	–
			Size	TV_j	I	W	–
			Occupied initial row	FI_j	O	W	–
Rows	$f = 1...NF$	I_U					
Columns	$c = 1...NC$	I_U					
Periods	$t = 1...M$	I_U					

It is possible to simplify the definition of elements by grouping all blocks in the same set, incorporating a binary attribute (horizontal) to distinguish between horizontal and vertical blocks. The row or column number where the block is on the board will be collected in the Position attribute (Table 7.8).

We also include the range of possible positions in which the block can be moved within its location. This can be easily calculated by knowing the blocks that are in that same row or column and keeping in mind that you have to keep the entire block

Table 7.8 Problem 43: Reduced table of elements

Elements	Set	QN	Data				
			Name	Parameter	Type	Belonging	Value
Board	–	I_U	Number of rows	NF	I	W	–
			Number of columns	NC	I	W	–
Blocks	$i = 1 \ldots n$	I_U	Horizontal	H_i	B	W	–
			Position (row of columns)	P_i	O	W	–
			Size	TT_i	I	W	–
			Occupied initial position	PI_i	O	W	–
			Move interval starting	II_i	O	W	–
			Move interval ending	FI_i	O	W	–
			Red block	FR_i	B	W	–
Rows	$f = 1 \ldots NF$	I_U					
Columns	$c = 1 \ldots NC$	I_U					
Positions	$j = 1 \ldots \mathrm{Max}$ (NF, NC)	I_U					
Periods	$t = 1 \ldots M$	I_U					

within the location (row or column) to which it belongs. A horizontal block of size 3 (occupies 3 columns) alone in its row can only move to positions 1, 2, 3… NC − 3 + 1. We define this movement limit for each block i as:

$$\mathrm{NP}_i = (H_i \cdot \mathrm{NC}) + ((1 - H_i) \cdot \mathrm{NF}) - \mathrm{TT}_i + 1$$

Version 1: Scenario 1

Decision Activities

Move blocks(*i*) to positions(*j*) in periods (*t*)

And the events considered would be as follows.

$$i = 1 \ldots n; j = I_i \ldots F_i; t = 1 \ldots M :$$

$\alpha_{ijt} = 1$ if we move block i to position j in period t; 0 otherwise.

Specifications

– Quantitative selection specifications: In each period, at most one block is moved from each position (column or row):

$$\forall t, j : \sum_{i/P_i=j} \sum_{j'=\text{II}_i}^{\text{FI}_i} \alpha_{ij't} \leq 1$$

- Only horizontal or vertical blocks can be moved in the same t, but not both (propositional logic specification):

$$\forall t, i/H_i = 1 : \text{IF} \sum_{j=\text{II}_i}^{\text{FI}_i} \alpha_{ijt} = 1 \text{ THEN} \sum_{i'/H_{i'} \neq H_i} \sum_{j=II_{i'}}^{FI_{i'}} \alpha_{i'jt} = 0$$

$$\Rightarrow f_{\text{LB}} \Rightarrow \text{IF} \sum_{j=\text{II}_i}^{\text{FI}_i} \alpha_{ijt} = 1 \text{ THEN} \sum_{i'/H_{i'} \neq H_i} \sum_{j=II_{i'}}^{FI_{i'}} \alpha_{i'jt} \leq 0$$

$$\Rightarrow f_{14}\left(\text{UB}_{\sum_{i'/H_{i'} \neq H_i} \sum_{j=II_{i'}}^{FI_{i'}} \alpha_{i'jt}} = \text{NC}\right) \Rightarrow \sum_{i'/H_{i'} \neq H_i} \sum_{j=\text{II}_{i'}}^{FI_{i'}} \alpha_{i'jt} \leq \text{NC}\left(1 - \sum_{j=\text{II}_i}^{FI_i} \alpha_{ijt}\right)$$

- Valid movements I: In the movement of a block, it cannot overlap with any block of the row or column to which it belongs (binary logical calculation + propositional logic specification).

For this, we need the position of the blocks each time we are going to make a move. We define a logical calculation to obtain the position of each block based on its position in the previous period or the movement made in the previous period.

Binary Logical Calculation: Position of i in t

$\forall i,\ j \in [\text{II}_i, \text{FI}_i],\ t > 1 : \lambda_{ijt} = 1$ if in period t ($t > 1$) block i has position j; 0 otherwise.

For $t = 1$, we assign to the variables the value of the initial position.

$$\forall i, j/\text{PI}_i = j, t = 1 : \lambda_{ijt} = 1$$

$$\forall i, j/\text{PI}_i \neq j, t = 1 : \lambda_{ijt} = 0$$

For $t > 1$: block i will have position j if it moved in the period before j or if it had it in the previous period and did not move.

$$\forall i,j \in [\mathrm{II}_i, \mathrm{FI}_i], t/t > 1 : \lambda_{ijt} = 1 \text{ IF AND ONLY IF}$$
$$\alpha_{ijt-1} = 1 \text{ OR } \left(\lambda_{ijt-1} = 1 \text{ AND } \sum_{j'/j' \neq j \& j' \in [\mathrm{II}_i, \mathrm{FI}_i]} \alpha_{ij't-1} = 0\right)$$
$$\Rightarrow f_{37} \Rightarrow \forall i,j \in [\mathrm{II}_i, \mathrm{FI}_i], t/t > 1 : \lambda_{ijt} = 1 \text{ IF AND ONLY IF}$$
$$\alpha_{ijt-1} = 1 \text{ OR } \left(\lambda_{ijt-1} + 1 - \sum_{j'/j' \neq j \& j' \in [\mathrm{II}_i, \mathrm{FI}_i]} \alpha_{ij't-1} \geq 2\right)$$
$$\forall i,j \in [\mathrm{II}_i, \mathrm{FI}_i], t : \omega_{ijt} = 1 \text{ IF AND ONLY IF } \lambda_{ijt} - \sum_{j'/j' \neq j \& j' \in [\mathrm{II}_i, \mathrm{FI}_i]} \alpha_{ij't} \geq 1$$
$$\Rightarrow f_{25}\left(\mathrm{LB}_{\lambda_{ijt} - \sum_{j'/j' \neq j \& j' \in [\mathrm{II}_i, \mathrm{FI}_i]} \alpha_{ij't}} = -1\right) \Rightarrow$$
$$\Rightarrow \forall i,j \in [\mathrm{II}_i, \mathrm{FI}_i], t : \lambda_{ijt} - \sum_{j'/j' \neq j \& j' \in [\mathrm{II}_i, \mathrm{FI}_i]} \alpha_{ij't} \geq 2\omega_{ijt} - 1$$
$$\Rightarrow f_{25}\left(\mathrm{UB}_{\lambda_{ijt} - \sum_{j'/j' \neq j \& j' \in [\mathrm{II}_i, \mathrm{FI}_i]} \alpha_{ij't}} = 1\right) \Rightarrow$$
$$\Rightarrow \forall i,j \in [\mathrm{II}_i, \mathrm{FI}_i], t : \lambda_{ijt} - \sum_{j'/j' \neq j \& j' \in [\mathrm{II}_i, \mathrm{FI}_i]} \alpha_{ij't} \leq \omega_{ijt}$$
$$\Rightarrow \forall i,j \in [\mathrm{II}_i, \mathrm{FI}_i], t/t > 1 : \lambda_{ijt} = 1 \text{ IF AND ONLY IF } \alpha_{ijt-1} = 1 \text{ OR } \omega_{ijt-1} = 1$$
$$\Rightarrow f_{34} \Rightarrow \forall i,j \in [\mathrm{II}_i, \mathrm{FI}_i], t/t > 1 : \lambda_{ijt} = 1 \text{ IF AND ONLY IF } \alpha_{ijt-1} + \omega_{ijt-1} \geq 1$$
$$\Rightarrow f_{25}\left(\mathrm{LB}_{\alpha_{ijt-1} + \omega_{ijt-1}} = 0\right) \Rightarrow \forall i,j \in [\mathrm{II}_i, \mathrm{FI}_i], t/t > 1 : \alpha_{ijt-1} + \omega_{ijt-1} \geq \lambda_{ijt}$$
$$\Rightarrow f_{25}\left(\mathrm{UB}_{\alpha_{ijt-1} + \omega_{ijt-1}} = 1\right) \Rightarrow \forall i,j \in [\mathrm{II}_i, \mathrm{FI}_i], t/t > 1 : \alpha_{ijt-1} + \omega_{ijt-1} \leq \lambda_{ijt}$$

Specification modeling: Prevent overlapping with blocks of the same position (row or column) during block movement:

$$\forall i, i'/P_i = P_{i'} \& H_i = H_{i'} \& \mathrm{PI}_{i'} > \mathrm{PI}_i, j \in [\mathrm{II}_i, \mathrm{FI}_i], t :$$
$$\text{IF } \alpha_{ijt} = 1 \text{ THEN } \sum_{j'=\mathrm{II}_{i'}}^{\mathrm{FI}_{i'}} j' \cdot \lambda_{i'j't} \geq j + \mathrm{TT}_i$$
$$\Rightarrow f_{15}\left(\mathrm{LB}_{\sum_{j'=\mathrm{II}_{i'}}^{\mathrm{FI}_{i'}} j' \cdot \lambda_{i'j't}} = 1\right) \Rightarrow \sum_{j'=\mathrm{II}_{i'}}^{\mathrm{FI}_{i'}} j' \cdot \lambda_{i'j't} \geq (j + \mathrm{TT}_i)\alpha_{ijt} + (1 - \alpha_{ijt})$$
$$\Rightarrow \sum_{j'=\mathrm{II}_{i'}}^{\mathrm{FI}_{i'}} j' \cdot \lambda_{i'j't} \geq 1 + (j + \mathrm{TT}_i - 1)\alpha_{ijt}$$

$$\forall i, i'/P_i = P_{i'} \& H_i = H_{i'} \& \mathrm{PI}_{i'} < \mathrm{PI}_i, j \in [\mathrm{II}_i, \mathrm{FI}_i], t:$$

$$\text{IF } \alpha_{ijt} = 1 \text{ THEN } \sum_{j'=\mathrm{II}_{i'}}^{\mathrm{FI}_{i'}} j' \cdot \lambda_{i'j't} + \mathrm{TT}_{i'} \le j$$

$$\Rightarrow \text{IF } \alpha_{ijt} = 1 \text{ THEN } \sum_{j'=\mathrm{II}_{i'}}^{\mathrm{FI}_{i'}} j' \cdot \lambda_{i'j't} \le j - \mathrm{TT}_{i'}$$

$$\Rightarrow f_{14}\left(\mathrm{UB}_{\sum_{j'=\mathrm{II}_{i'}}^{\mathrm{FI}_{i'}} j' \cdot \lambda_{i'j't}} = 1\right) \Rightarrow$$

$$\Rightarrow \sum_{j'=\mathrm{II}_{i'}}^{\mathrm{FI}_{i'}} j' \cdot \lambda_{i'j't} \le j - \mathrm{TT}_{i'} + (1 - (j - \mathrm{TT}_{i'}))(1 - \alpha_{ijt})$$

$$\Rightarrow \sum_{j'=\mathrm{II}_{i'}}^{\mathrm{FI}_{i'}} j' \cdot \lambda_{i'j't} \le 1 + (j - \mathrm{TT}_{i'} - 1)\alpha_{ijt}$$

- Valid movements II: In the displacement of a block, it cannot be crossed with any block. We already have the position of all the blocks in each movement t (propositional logic specification).

$$\forall i, j/j \in [\mathrm{II}_i \mathrm{FI}_i], j'/j' \in [\mathrm{II}_i \mathrm{FI}_i] \& j < j', t:$$

$$\text{IF } \lambda_{ijt} = 1 \text{ AND}$$

$$\alpha_{ij't} = 1 \text{ THEN } \sum_{i'/H_{i'} \ne H_i \& j + \mathrm{TT}_i \le P_{i'} \le j' + \mathrm{TT}_i - 1} \sum_{k=P_i - \mathrm{TT}_{i'} + 1}^{P_i} \lambda_{i'kt}$$

$$= 0 \Rightarrow f_{\mathrm{LB}} \Rightarrow \text{IF } \lambda_{ijt} = 1 \text{ AND}$$

$$\alpha_{ij't} = 1 \text{ THEN } \sum_{i'/H_{i'} \ne H_i \& j + \mathrm{TT}_i \le P_{i'} \le j' + \mathrm{TT}_i - 1} \sum_{k=P_i - \mathrm{TT}_{i'} + 1}^{P_i} \lambda_{i'kt}$$

$$\le 0 \Rightarrow f_{37} \Rightarrow \text{IF } \lambda_{ijt} + \alpha_{ij't} \ge 2 \text{ THEN } \sum_{i'/H_{i'} \ne H_i \& j + \mathrm{TT}_i \le P_{i'} \le j' + \mathrm{TT}_i - 1} \sum_{k=P_i - \mathrm{TT}_{i'} + 1}^{P_i} \lambda_{i'kt}$$

$$\le 0 \Rightarrow f_{18}\left(\mathrm{UB}_{\alpha_{ijt} + \lambda_{ij't}} = 2\right) \Rightarrow \lambda_{ijt} + \alpha_{ij't} \le 1$$

$$+ \theta_{ijj't} \Rightarrow f_{20}\left(\mathrm{UB}_{ijj'} = \sum_{i'} H_{i'} \ne H_i \& j + \mathrm{TT}_i \le P_{i'} \le j' + \mathrm{TT}_i - 1\right) \Rightarrow$$

$$\sum_{i'/H_{i'} \ne H_i \& j + \mathrm{TT}_i \le P_{i'} \le j' + \mathrm{TT}_i - 1} \sum_{k=P_i - \mathrm{TT}_{i'} + 1}^{P_i} \lambda_{i'kt}$$

$$\le U_{ijj'}(1 - \theta_{ijj't})$$

$\forall i, j/j \in [\mathrm{II}_i, \mathrm{FI}_i], j'/j' \in [\mathrm{II}_i, \mathrm{FI}_i] \& j > j', t : \text{IF } \lambda_{ijt} = 1 \text{ AND } \alpha_{ij't} = 1$

$$\text{THEN} \sum_{i'/H_{i'} \neq H_i \& j' \leq P_{i'} \leq j-1} \sum_{k=P_i-\mathrm{TT}_{i'}+1}^{P_i} \lambda_{i'kt} = 0$$

$$\Rightarrow f_{\mathrm{LB}} \Rightarrow \text{IF } \lambda_{ijt} = 1 \text{ AND } \alpha_{ij't} = 1 \text{ THEN} \sum_{i'/H_{i'} \neq H_i \& j' \leq P_{i'} \leq j-1} \sum_{k=P_i-\mathrm{TT}_{i'}+1}^{P_i} \lambda_{i'kt} \leq 0$$

$$\Rightarrow f_{37} \Rightarrow \text{IF } \lambda_{ijt} + \alpha_{ij't} \geq 2 \text{ THEN} \sum_{i'/H_{i'} \neq H_i \& j' \leq P_{i'} \leq j-1} \sum_{k=P_i-\mathrm{TT}_{i'}+1}^{P_i} \lambda_{i'kt} \leq 0$$

$$\Rightarrow f_{18}\left(\mathrm{UB}_{\alpha_{ijt}+\lambda_{ij't}} = 2\right) \Rightarrow \lambda_{ijt} + \alpha_{ij't} \leq 1 + \theta_{ijj't}$$

$$\Rightarrow f_{20}\left(\mathrm{UB}_{ijj'} = \sum_{i'} H_{i'} \neq H_i \& j' \leq P_{i'} \leq j-1\right) \Rightarrow$$

$$\Rightarrow \sum_{i'/H_{i'} \neq H_i \& j' \leq P_{i'} \leq j-1} \sum_{k=P_i-\mathrm{TT}_{i'}+1}^{P_i} \lambda_{i'kt} \leq \mathrm{UB}_{ijj'}\left(1 - \theta_{ijj't}\right)$$

– Getting the red block out (quantitative selection specification):

Getting the red block out of the board is equivalent to getting the block to the end of the row it is in, that is, reaching the column FI_i, $i/\mathrm{FR}_i = 1$.

$$\forall i/\mathrm{FR}_i = 1 : \sum_{t=1}^{M} \lambda_{i\mathrm{FI}_i t} \geq 1$$

Version 1: Scenario 2
Objective Function

Minimize the number of moves until you get the red chip out.

In the second scenario, we must take into account all the movements of each period in the objective function. We must maintain the set of restrictions that Scenario 1 supposes and incorporate the minimization of movements as objective function.

$$\text{Min} \sum_{i=1}^{n} \sum_{j=\mathrm{II}_i}^{\mathrm{FI}_i} \sum_{t=1}^{M} \alpha_{ijt}$$

Version 2: Scenario 1

In this second version, we slightly changed the decision activity. We go from Moving Blocks to Positioning Blocks. The difference with version 1 is that we integrate into a single decision variable, the decision variable, and the logical calculation of the first version. Therefore, in each period the position of each block is decided, and we maintain the same specifications.

Decision Activities

Position blocks(*i*) in positions (*j*) in periods (*t*)

And the events considered would be as follows:

$$i = 1 \ldots n; j = \text{II}_i \ldots \text{FI}_i; t = 1 \ldots M :$$
$$\lambda_{ijt} = 1 \text{ if we position block } i \text{ to position } j \text{ in period } t; 0 \text{ otherwise.} \tag{43.1}$$

Specifications

- Quantitative selection specifications:

  In each period, each block must be in one position:

$$\forall t, \forall i : \sum_{j=\text{II}_i}^{\text{FI}_i} \lambda_{ijt} = 1 \tag{43.2}$$

- Initial position (allocation specification):

  For $t = 1$, we assign to the variables the value of the initial position.

$$\forall i, j/\text{PI}_i = j, t = 1 : \lambda_{ijt} = 1 \tag{43.3}$$
$$\forall i, j/\text{PI}_i \neq j, t = 1 : \lambda_{ijt} = 0 \tag{43.4}$$

- Blocks that move (change position) in each t: From each row or column, only one block can be moved in each t (propositional logic specification):

$$\begin{aligned}
&\forall t/t < M, i, j \in [\text{II}_i, \text{FI}_i], i'/i \neq i' \& H_{i'} = H_i \& P_i = P_{i'}, j' \in [\text{II}_{i'}, \text{FI}_{i'}] : \\
&\text{IF } \lambda_{ijt+1} - \lambda_{ijt} = 1 \text{ THEN } \lambda_{i'j't+1} - \lambda_{i'j't} = 0 \\
&\Rightarrow f_{\text{UB}}, f_{\text{LB}} \Rightarrow \forall t/t < M, i, j \in [\text{II}_i, \text{FI}_i], \\
&i'/i \neq i' \& H_{i'} = H_i \& P_i = P_{i'}, j' \in [\text{II}_{i'}, \text{FI}_{i'}] : \\
&\text{IF } \lambda_{ijt+1} - \lambda_{ijt} \geq 1 \text{ THEN } \lambda_{i'j't+1} - \lambda_{i'j't} \leq 0 \\
&\Rightarrow f_{18}\left(\text{UB}_{\lambda_{ijt+1} - \lambda_{ijt}} = 1\right) \Rightarrow \forall t/t < M, i, j \in [\text{II}_i, \text{FI}_i] : \\
&\lambda_{ijt+1} - \lambda_{ijt} \leq \pi_{ijt}
\end{aligned} \tag{43.5}$$

$$\Rightarrow f_{20}\left(\text{UB}_{\lambda_{i'j't+1}-\lambda_{i'j't}}=1\right)\Rightarrow \forall t/t<M,i,j\in[\text{II}_i,\text{FI}_i],$$
$$i'/i\neq i'\&H_{i'}=H_i\&P_i=P_{i'},j'\in[\text{II}_{i'},\text{FI}_{i'}]: \tag{43.6}$$
$$\lambda_{i'j't+1}-\lambda_{i'j't}\leq 1-\pi_{ijt}$$

– Blocks that move (change position) in each t: Either horizontal or vertical tokens can be moved in the same t, but not both (propositional logic specification):

$$\forall t/t<M,i/H_i=1,j\in[\text{II}_i,\text{FI}_i],i'/H_{i'}\neq H_i,j'\in[\text{II}_{i'},\text{FI}_{i'}]:$$
$$\text{IF } \lambda_{ijt+1}-\lambda_{ijt}=1 \text{ THEN } \lambda_{i'j't+1}-\lambda_{i'j't}=0$$
$$\Rightarrow f_{\text{UB}},f_{\text{LB}}\Rightarrow \forall t/t<M,i/H_i=1,$$
$$j\in[\text{II}_i,\text{FI}_i],i'/H_{i'}\neq H_i,j'\in[\text{II}_{i'},\text{FI}_{i'}]: \tag{43.5}$$
$$\text{IF } \lambda_{ijt+1}-\lambda_{ijt}\geq 1 \text{ THEN } \lambda_{i'j't+1}-\lambda_{i'j't}\leq 0$$
$$\Rightarrow f_{18}\left(\text{UB}_{\lambda_{ijt+1}-\lambda_{ijt}}=1\right)\Rightarrow$$
$$\Rightarrow \forall t/t<M,i/H_i=1,j\in[\text{II}_i,\text{FI}_i]:$$
$$\lambda_{ijt+1}-\lambda_{ijt}\leq \pi_{ijt}$$

$$\Rightarrow f_{20}\left(\text{UB}_{\lambda_{i'j't+1}-\lambda_{i'j't}}=1\right)\Rightarrow$$
$$\Rightarrow \forall t/t<M,i/H_i=1,j\in[\text{II}_i,\text{FI}_i], \tag{43.7}$$
$$i'/H_{i'}\neq H_i,j'\in[\text{II}_{i'},\text{FI}_{i'}]:$$
$$\lambda_{i'j't+1}-\lambda_{i'j't}\leq 1-\pi_{ijt}$$

– Valid movements I: Prevent overlapping with blocks of the same position (row or column) during the movement of the block (propositional logic specification):

$$\forall i,i'/P_i=P_{i'}\&H_i=H_{i'}\&\text{PI}_{i'}>\text{PI}_i,j\in[\text{II}_i,\text{FI}_i],t:$$
$$\text{IF } \lambda_{ijt}=1 \text{ THEN } \sum_{j'=\text{II}_{i'}}^{\text{FI}_{i'}} j'\cdot\lambda_{i'j't}\geq j+\text{TT}_i$$
$$\Rightarrow f_{15}\left(\text{LB}_{\sum_{j'=\text{II}_{i'}}^{\text{FI}_{i'}} j'\cdot\lambda_{i'j't}}=1\right)\Rightarrow \sum_{j'=\text{II}_{i'}}^{\text{FI}_{i'}} j'\cdot\lambda_{i'j't}\geq (j+\text{TT}_i)\lambda_{ijt}+(1-\lambda_{ijt})$$
$$\Rightarrow \sum_{j'=\text{II}_{i'}}^{\text{FI}_{i'}} j'\cdot\lambda_{i'j't}\geq 1+(j+\text{TT}_i-1)\lambda_{ijt} \tag{43.8}$$

$$\forall i, i'/P_i = P_{i'} \& H_i = H_{i'} \& \text{PI}_{i'} < \text{PI}_i, j \in [\text{II}_i, \text{FI}_i], t:$$

$$\text{IF } \lambda_{ijt} = 1 \text{ THEN } \sum_{j'=\text{II}_{i'}}^{\text{FI}_{i'}} j' \cdot \lambda_{i'j't} + \text{TT}_{i'} \le j$$

$$\Rightarrow \text{IF } \lambda_{ijt} = 1 \text{ THEN } \sum_{j'=\text{II}_{i'}}^{\text{FI}_{i'}} j' \cdot \lambda_{i'j't} \le j - \text{TT}_{i'}$$

$$\Rightarrow f_{14}\left(\text{UB}_{\sum_{j'=\text{II}_{i'}}^{\text{FI}_{i'}} j' \cdot \lambda_{i'j't}} = 1\right) \Rightarrow \tag{43.9}$$

$$\Rightarrow \sum_{j'=\text{II}_{i'}}^{\text{FI}_{i'}} j' \cdot \lambda_{i'j't} \le j - \text{TT}_{i'} + (1 - (j - \text{TT}_{i'}))(1 - \lambda_{ijt})$$

$$\Rightarrow \sum_{j'=\text{II}_{i'}}^{\text{FI}_{i'}} j' \cdot \lambda_{i'j't} \le 1 + (j - \text{TT}_{i'} - 1)\lambda_{ijt}$$

– Valid movements II: In the movement of a block, it cannot be crossed with any block (propositional logic specification).

$$\forall t < M, i, j/j \in [\text{II}_i, \text{FI}_i], j'/j' \in [\text{II}_i, \text{FI}_i] \& j < j':$$

$$\text{IF } \lambda_{ijt} = 1 \text{ AND } \lambda_{ij't+1} = 1 \text{ THEN } \sum_{i'/H_{i'} \ne H_i \& j + \text{TT}_i \le P_{i'} \le j' + \text{TT}_i - 1} \ \sum_{k = P_i - \text{TT}_{i'} + 1}^{P_i} \lambda_{i'kt} = 0$$

$$\Rightarrow f_{\text{LB}} \Rightarrow \text{IF } \lambda_{ijt} = 1 \text{ AND } \lambda_{ij't+1} = 1$$

$$\text{THEN } \sum_{i'/H_{i'} \ne H_i \& j + \text{TT}_i \le P_{i'} \le j' + \text{TT}_i - 1} \ \sum_{k = P_i - \text{TT}_{i'} + 1}^{P_i} \lambda_{i'kt} \le 0$$

$$\Rightarrow f_{37} \Rightarrow \text{IF } \lambda_{ijt} + \lambda_{ij't+1} \ge 2$$

$$\text{THEN } \sum_{i'/H_{i'} \ne H_i \& j + \text{TT}_i \le P_{i'} \le j' + \text{TT}_i - 1} \ \sum_{k = P_i - \text{TT}_{i'} + 1}^{P_i} \lambda_{i'kt} \le 0$$

$$\Rightarrow f_{18}\left(\text{UB}_{\lambda_{ijt} + \lambda_{ij't+1}} = 2\right) \Rightarrow \forall t < M, i, j/j \in [\text{II}_i, \text{FI}_i],$$

$$j'/j' \in [\text{II}_i, \text{FI}_i] \& j < j':$$

$$\lambda_{ijt} + \lambda_{ij't+1} \le 1 + \theta_{ijj't} \tag{43.10}$$

$$\Rightarrow f_{20}\left(\mathrm{UB}_{ijj'} = \sum_{i'} H_{i'} \neq H_i \& j + \mathrm{TT}_i \leq P_{i'} \leq j' + \mathrm{TT}_i - 1\right) \Rightarrow$$
$$\Rightarrow \forall t < M, i, j/j \in [\mathrm{II}_i, \mathrm{FI}_i], j'/j' \in [\mathrm{II}_i, \mathrm{FI}_i] \& j < j' : \quad (43.11)$$
$$\sum_{i'/H_{i'} \neq H_i \& j + \mathrm{TT}_i \leq P_{i'} \leq j' + \mathrm{TT}_i - 1} \sum_{k = P_i - \mathrm{TT}_{i'} + 1}^{P_i} \lambda_{i'kt} \leq UB_{ijj'}\left(1 - \theta_{ijj't}\right)$$

$$\forall t < M, i, j/j \in [\mathrm{II}_i, \mathrm{FI}_i], j'/j' \in [\mathrm{II}_i, \mathrm{FI}_i] \& j > j' :$$
$$\text{IF } \lambda_{ijt} = 1 \text{ AND } \lambda_{ij't+1} = 1 \text{ THEN } \sum_{i'/H_{i'} \neq H_i \& j' \leq P_{i'} \leq j-1} \sum_{k = P_i - \mathrm{TT}_{i'} + 1}^{P_i} \lambda_{i'kt} = 0$$
$$\Rightarrow f_{\mathrm{LB}} \Rightarrow \text{IF } \lambda_{ijt} = 1 \text{ AND } \lambda_{ij't+1} = 1$$
$$\text{THEN } \sum_{i'/H_{i'} \neq H_i \& j' \leq P_{i'} \leq j-1} \sum_{k = P_i - \mathrm{TT}_{i'} + 1}^{P_i} \lambda_{i'kt} \leq 0$$
$$\Rightarrow f_{37} \Rightarrow \text{IF } \lambda_{ijt} + \lambda_{ij't+1} \geq 2$$
$$\text{THEN } \sum_{i'/H_{i'} \neq H_i \& j' \leq P_{i'} \leq j-1} \sum_{k = P_i - \mathrm{TT}_{i'} + 1}^{P_i} \lambda_{i'kt} \leq 0$$
$$\Rightarrow f_{18}\left(\mathrm{UB}_{\lambda_{ijt} + \lambda_{ij't+1}} = 2\right) \Rightarrow$$
$$\Rightarrow \forall t < M, i, j/j \in [\mathrm{II}_i, \mathrm{FI}_i], j'/j' \in [\mathrm{II}_i, \mathrm{FI}_i] \& j > j' :$$
$$\lambda_{ijt} + \lambda_{ij't+1} \leq 1 + \theta_{ijj't}$$
$$(43.12)$$

$$\Rightarrow f_{20}\left(\mathrm{UB}_{ijj'} = \sum_{i'} H_{i'} \neq H_i \& j' \leq P_{i'} \leq j - 1\right) \Rightarrow$$
$$\Rightarrow \forall t < M, i, j/j \in [\mathrm{II}_i, \mathrm{FI}_i], j'/j' \in [\mathrm{II}_i, \mathrm{FI}_i] \& j > j' : \quad (43.13)$$
$$\sum_{i'/H_{i'} \neq H_i \& j' \leq P_{i'} \leq j-1} \sum_{i'/H_{i'} \neq H_i \& j' \leq P_{i'} \leq j-1}^{P_i} \lambda_{i'kt} \leq \mathrm{UB}_{ijj'}\left(1 - \theta_{ijj't}\right)$$

- Getting the red block out as specification (quantitative selection specification):

  Identical to version 1:

$$\forall i/\mathrm{FR}_i = 1 : \sum_{t=1}^{M} \lambda_{i\mathrm{FI}_i t} \geq 1 \quad (43.14)$$

Version 2: Scenario 2

Objective Criterion

Minimize the number of moves (position changes) until you get the red block out.

To minimize the number of movements, we must use an upper bound calculation for each block i in each period t:

Upper Bound Calculation

x_{it}^{MAX} = Upper bound of the position change of a block in each period t

$$\forall i, j \in [\mathrm{II}_i, \mathrm{FI}_i], t/t < M : x_{it}^{\mathrm{MAX}} \geq \lambda_{ijt+1} - \lambda_{ijt} \tag{43.15}$$

Objective Function

$$\mathrm{Min} \sum_{i=1}^{n} \sum_{t=1}^{M-1} x_{it}^{\mathrm{MAX}} \tag{43.16}$$

Complete Model

$$\mathrm{Min} \sum_{i=1}^{n} \sum_{t=1}^{M-1} x_{it}^{\mathrm{MAX}} \quad \text{(NOT FOR SCENARIO 1)} \tag{43.16}$$

s.t.

$$\forall t, \forall i : \sum_{j=\mathrm{II}_i}^{\mathrm{FI}_i} \lambda_{ijt} = 1 \tag{43.2}$$

$$\forall i, j/\mathrm{PI}_i = j, t = 1 : \lambda_{ijt} = 1 \tag{43.3}$$

$$\forall i, j/\mathrm{PI}_i \neq j, t = 1 : \lambda_{ijt} = 0 \tag{43.4}$$

$$\forall t/t < M, j \in [\mathrm{II}_i, \mathrm{FI}_i] : \lambda_{ijt+1} - \lambda_{ijt} \leq \pi_{ijt} \tag{43.5}$$

$$\forall t/t < M, i, j \in [\mathrm{II}_i, \mathrm{FI}_i], i'/i \neq i' \& H_{i'} = H_i \& P_i = P_{i'}, j' \in [\mathrm{II}_{i'}, \mathrm{FI}_{i'}] : \lambda_{i'j't+1} - \lambda_{i'j't} \leq 1 - \pi_{ijt} \tag{43.6}$$

$$\forall t/t < M, i/H_i = 1, j \in [\mathrm{II}_i, \mathrm{FI}_i], i'/H_{i'} \neq H_i, j' \in [\mathrm{II}_{i'}, \mathrm{FI}_{i'}] : \lambda_{i'j't+1} - \lambda_{i'j't} \leq 1 - \pi_{ijt} \tag{43.7}$$

$$\forall i, i'/P_i = P_{i'} \& H_i = H_{i'} \& \mathrm{PI}_{i'} > \mathrm{PI}_i, j \in [\mathrm{II}_i, \mathrm{FI}_i], t : \sum_{j'=\mathrm{II}_{i'}}^{\mathrm{FI}_{i'}} j' \cdot \lambda_{i'j't} \geq 1 + (j + \mathrm{TT}_i - 1)\lambda_{ijt} \tag{43.8}$$

$$\forall i,i'/P_i = P_{i'} \& H_i = H_{i'} \& \mathrm{PI}_{i'} < \mathrm{PI}_i, j \in [\mathrm{II}_i, \mathrm{FI}_i], t : \sum_{j'=\mathrm{II}_{i'}}^{\mathrm{FI}_{i'}} j' \cdot \lambda_{i'j't} \leq 1 + (j - \mathrm{TT}_{i'} - 1)\lambda_{ijt} \tag{43.9}$$

$$\forall i,j/j \in [\mathrm{II}_i, \mathrm{FI}_i], j'/j' \in [\mathrm{II}_i, \mathrm{FI}_i] \& j \neq j', t < M : \lambda_{ijt} + \lambda_{ij't+1} \leq 1 + \theta_{ijj't} \tag{43.10, 43.12}$$

$$\forall i,j/j \in [\mathrm{II}_i, \mathrm{FI}_i], j'/j' \in [\mathrm{II}_i, \mathrm{FI}_i] \& j < j', t < M : \sum_{i'/H_{i'} \neq H_i \& j+\mathrm{TT}_i \leq P_{i'} \leq j'+\mathrm{TT}_i - 1} \sum_{k=P_i - \mathrm{TT}_{i'}+1}^{P_i} \lambda_{i'kt} \leq \mathrm{UB}_{ijj'}\left(1 - \theta_{ijj't}\right)$$
$$\mathrm{UB}_{ijj'} = \sum_{i'} H_{i'} \neq H_i \& j + \mathrm{TT}_i \leq P_{i'} \leq j' + \mathrm{TT}_i - 1 \tag{43.11}$$

$$\forall i,j/j \in [\mathrm{II}_i, \mathrm{FI}_i], j'/j' \in [\mathrm{II}_i, \mathrm{FI}_i] \& j > j', t < M : \sum_{i'/H_{i'} \neq H_i \& j' \leq P_{i'} \leq j-1} \sum_{k=P_i - \mathrm{TT}_{i'}+1}^{P_i} \lambda_{i'kt} \leq \mathrm{UB}_{ijj'}\left(1 - \theta_{ijj't}\right)$$
$$\mathrm{UB}_{ijj'} = \sum_{i'} H_{i'} \neq H_i \& j' \leq P_{i'} \leq j - 1 \tag{43.13}$$

$$\forall i/\mathrm{FR}_i = 1 : \sum_{t=1}^{M} \lambda_{i\mathrm{FI}_i t} \geq 1 \tag{43.14}$$

$$\forall i,j \in [\mathrm{II}_i, \mathrm{FI}_i], t/t < M : x_{it}^{\mathrm{MAX}} \geq \lambda_{ijt+1} - \lambda_{ijt} \quad \text{(NOT FOR SCENARIO 1)} \tag{43.15}$$

$$i = 1 \ldots n; j = \mathrm{II}_i \ldots \mathrm{FI}_i; t = 1 \ldots M : \lambda_{ijt} \text{ binaries} \tag{43.1}$$

$$\forall i,j \in [\mathrm{II}_i, \mathrm{FI}_i], \forall t < M : \pi_{ijt} \text{ binaries} \tag{43.17}$$

$$\forall i,j/j \in [\mathrm{II}_i, \mathrm{FI}_i], j'/j' \in [\mathrm{II}_i, \mathrm{FI}_i] \& j \neq j', t : \theta_{ijj't} \text{ binaries} \tag{43.18}$$

Implementation in LINGO

```
MODEL:
DATA:
 n=@FILE('C:\Files_\43_Text.txt');
 NF=@FILE('C:\Files_\43_Text.txt');
 NC=@FILE('C:\Files_\43_Text.txt');
 NMAX=@FILE('C:\Files_\43_Text.txt');
 M=@FILE('C:\Files_\43_Text.txt');
ENDDATA
SETS:
 Blocks/1..n/: H,P,TT,PI,II,FI,FR;
 Rows/1..NF/;
```

```
 Columns/1..NC/;
 Positions/1..NMAX/;
 Periods/1..M/;
 FP(Blocks,Periods):xMAX;
 FPP(Blocks,Positions,Periods):Lamda,PiG;
 FPPP(Blocks,Positions,Positions,Periods):Thita;
ENDSETS
DATA:
 H=@FILE('C:\Files_\43_Text.txt');
 P=@FILE('C:\Files_\43_Text.txt');
 TT=@FILE('C:\Files_\43_Text.txt');
 PI=@FILE('C:\Files_\43_Text.txt');
 II=@FILE('C:\Files_\43_Text.txt');
 FI=@FILE('C:\Files_\43_Text.txt');
 FR=@FILE('C:\Files_\43_Text.txt');
@OLE('C:\Files_\43_Excel.xlsx','Solution')=@WRITEFOR(Blocks(i):
@WRITEFOR(Periods(t):@SUM(Positions(j):j*Lamda(i,j,t))));
ENDDATA
!43.16;
Min= @SUM(Blocks(i):@SUM(Periods(t)|t#LT#M:xMAX(i,t)));
!43.2;
@FOR(Periods(t):@FOR(Blocks(i):@SUM(Positions(j)|j#GE#II(i)
#AND#j#LE#FI(i):Lamda(i,j,t))=1));
!43.3;
@FOR(Blocks(i):@FOR(Positions(j)|PI(i)#EQ#j:Lamda(i,j,1)=1));
!43.4;
@FOR(Blocks(i):@FOR(Positions(j)|PI(i)#NE#j:Lamda(i,j,1)=0));
!43.5;
@FOR(Periods(t)|t#LT#M:@FOR(Blocks(i):@FOR(Positions(j)|j#GE#II
(i)#AND#j#LE#FI(i):
Lamda(i,j,t+1)-Lamda(i,j,t)<=PiG(i,j,t))));
!43.6;
@FOR(Periods(t)|t#LT#M:@FOR(Blocks(i):@FOR(Positions(j)|j#GE#II
(i)#AND#j#LE#FI(i):@FOR(Blocks(ip)|i#NE#ip #AND# H(ip)#EQ#H(i)
#AND#P(i)#EQ#P(ip):@FOR(Positions(jj)|jj#GE#II(ip)#AND#jj#LE#FI
(ip):
Lamda(ip,jj,t+1)-Lamda(ip,jj,t)<=1-PiG(i,j,t))))));
!43.7;
@FOR(Periods(t)|t#LT#M:@FOR(Blocks(i)|H(i)#EQ#1:@FOR(Positions
(j)|j#GE#II(i)#AND#j#LE#FI(i):@FOR(Blocks(ip)|i#NE#ip#AND#H(ip)
#NE#H(i):@FOR(Positions(jj)|jj#GE#II(ip)#AND#jj#LE#FI(ip):
Lamda(ip,jj,t+1)-Lamda(ip,jj,t)<=1-PiG(i,j,t))))));
!43.8;
@FOR(Blocks(i):@FOR(Blocks(ip)|P(i)#EQ#P(ip)#AND#H(i)#EQ#H(ip)
#AND#PI(ip)#GT#PI(i):@FOR(Positions(j)|j#GE#II(i)#AND#j#LE#FI(i):
@FOR(Periods(t):@SUM(Positions(jj)|jj#GE#II(ip)#AND#jj#LE#FI(ip):
jj*Lamda(ip,jj,t))>=1+(j+TT(i)-1)*Lamda(i,j,t)))));
!43.9;
@FOR(Blocks(i):@FOR(Blocks(ip)|P(i)#EQ#P(ip)#AND#H(i)#EQ#H(ip)
#AND#PI(ip)#LT#PI(i):@FOR(Positions(j)|j#GE#II(i)#AND#j#LE#FI(i):
@FOR(Periods(t):@SUM(Positions(jj)|jj#GE#II(ip)#AND#jj#LE#FI(ip):
jj*Lamda(ip,jj,t))<=1+(j-TT(ip)-1)*Lamda(i,j,t)))));
!43.10;!43.12;
```

```
@FOR(Blocks(i):@FOR(Positions(j)|j#GE#II(i)#AND#j#LE#FI(i):@FOR
(Positions(jj)|jj#GE#II(i)#AND#jj#LE#FI(i)#AND#j#NE#jj:@FOR
(Periods(t)|t#LT#M:Lamda(i,j,t)+Lamda(i,jj,t+1)<=1+Thita(i,j,jj,
t)))));
!43.11;
@FOR(Blocks(i):@FOR(Positions(j)|j#GE#II(i)#AND#j#LE#FI(i):@FOR
(Positions(jj)|jj#GE#II(i)#AND#jj#LE#FI(i)#AND#j#LT#jj:@FOR
(Periods(t)|t#LT#M:
@SUM(Blocks(ip)|H(ip)#NE#H(i)#AND#j+TT(i)#LE#P(ip)#AND#P(ip)
#LE#jj+TT(i)-1:@SUM(Positions(k)|k#GE#P(i)-TT(ip)+1#AND#k#LE#P
(i):Lamda(ip,k,t)))<=@SUM(Blocks(ip)|H(ip)#NE#H(i)#AND#j+TT(i)
#LE#P(ip)#AND#P(ip)#LE#jj+TT(i)-1:1)*(1-Thita(i,j,jj,t))))));
!43.13;
@FOR(Blocks(i):@FOR(Positions(j)|j#GE#II(i)#AND#j#LE#FI(i):@FOR
(Positions(jj)|jj#GE#II(i)#AND#jj#LE#FI(i)#AND#j#GT#jj:@FOR
(Periods(t)|t#LT#M:
@SUM(Blocks(ip)|H(ip)#NE#H(i)#AND#jj#LE#P(ip)#AND#P(ip)#LE#j-1:
@SUM(Positions(k)|k#GE#P(i)-TT(ip)+1#AND#k#LE#P(i):Lamda(ip,k,
t)))<=@SUM(Blocks(ip)|H(ip)#NE#H(i)#AND#jj#LE#P(ip)#AND#P(ip)
#LE#j-1:1)*(1-Thita(i,j,jj,t))))));
!43.14;
@FOR(Blocks(i)|FR(i)#EQ#1:@SUM(Periods(t):Lamda(i,FI(i),t))>=1);
!43.15;
@FOR(Blocks(i):@FOR(Positions(j)|j#GE#II(i)#AND#j#LE#FI(i):@FOR
(Periods(t)|t#LT#M:xMAX(i,t)>=Lamda(i,j,t+1)-Lamda(i,j,t))));
!43.1;
@FOR(Blocks(i):@FOR(Positions(j)|j#GE#II(i)#AND#j#LE#FI(i):@FOR
(Periods(t):@BIN(Lamda(i,j,t)))));
!43.17;
@FOR(Blocks(i):@FOR(Positions(j)|j#GE#II(i)#AND#j#LE#FI(i):@FOR
(Periods(t):@BIN(PiG(i,j,t)))));
!43.18;
@FOR(Blocks(i):@FOR(Positions(j)|j#GE#II(i)#AND#j#LE#FI(i):@FOR
(Positions(jj)|jj#GE#II(i)#AND#jj#LE#FI(i)#AND#j#NE#jj:@FOR
(Periods(t):@BIN(Thita(i,j,jj,t))))));
END
```

Solution

Like in the previous problem, the creation of sudokus, to improve the resolution time, we are only going to look for a feasible solution, since with it we manage to get the red block out of the board. The solution is shown in Fig. 7.11.

In the file "43_Excel.xlsx" (Fig. 7.12) we have represented the position of the nine blocks during the 14 periods considered. The values in the first range of cells express the position number of the block within its row or column.

Block 2 is the red block.

In the second range of cells, we have calculated the blocks that have changed positions from one period to another.

```
Solution Report - 43_LingoE
Feasible solution found.
Infeasibilities:                                    0.000000
Extended solver steps:                                  2958
Total solver iterations:                               47202
Elapsed runtime seconds:                                9.11



Export Summary Report
---------------------
Transfer Method:         OLE BASED
Workbook:                C:\Files_\43_Excel.xlsx
Ranges Specified:                1
   Solution
Ranges Found:                    1
Range Size Mismatches:           0
Values Transferred:            126

Model Class:                                            MILP

Total variables:                      4491
Nonlinear variables:                     0
Integer variables:                    2957

Total constraints:                    9428
Nonlinear constraints:                   0

Total nonzeros:                      28274
Nonlinear nonzeros:                      0
```

Fig. 7.11 Problem 43: LINGO solution report

	t	1	2	3	4	5	6	7	8	9	10	11	12	13	14
i	1	1	1	1	1	1	1	1	1	1	1	1	1	1	1
	2	1	1	1	1	4	4	4	3	3	2	2	1	1	5
	3	1	1	1	1	3	3	2	1	1	1	1	1	1	1
	4	5	5	5	5	5	1	1	1	1	1	1	1	1	1
	5	3	3	5	5	5	5	5	5	5	5	5	2	2	1
	6	3	3	5	5	5	5	5	5	5	4	4	1	1	4
	7	1	1	1	1	1	1	1	1	1	1	1	1	1	1
	8	3	3	3	5	5	5	5	5	4	4	3	3	4	4
	9	3	1	1	1	1	1	1	1	1	1	1	1	4	4

1														
2					4			3		2		1		5
3					3		2	1						
4						1								
5			5									2		1
6			5							4		1		4
7														
8				5					4		3		4	
9		1											4	

Fig. 7.12 "43_Excel.xlsx"

7.8 Problem 44. Distribution of State Aid

We are going to study the distribution of state aid that has a budget of S $ for the rural development of small shops in a region. Each shop has an area (in m^2), a number of workers, and the annual turnover (previous year) and belongs to one or more sectors. For new shops without a previous invoice, these data are calculated based on the sectors to which it belongs and the size it owns. There are the following sectors in small shops:

Textile, food, footwear and leather, drug stores and perfumeries, orthopedics and herbalists, flowers and plants, toys and sports, construction and DIY, furniture, paper and graphic arts, bazaars ($i = 1 \ldots 11$).

In allocating the budget to each shop, the following factors must be considered:

When a company belongs to several sectors, for the allocation, the square meters of the sector are distributed proportionally. Companies that belong to more than one sector with an area of less than 100 m^2 may only receive the help of one of them, the one with the highest amount. Stores over 100 m^2 can receive subsidies from up to two sectors, those with the highest amounts.

If an establishment has the ratio (annual turnover/number of workers) above TT$/worker, it will receive a discount on grants of 20%.

The previous discounts for invoicing plus the quantities that could not be assigned due to belonging to more sectors will be distributed additionally in a second phase according to a percentage of importance for the allocation (PI$_i$), calculated from a study of the degree of prejudice with respect to the great trade. This second distribution applies to all stores and all their sectors.

Objective: Maximize the allocation to the sector with the lowest allocation per m^2.

Modeling

The following calculated data have been included in the table of elements (Table 7.9):

The total of sectors of a shop:

$$\forall j : N_j = \sum_{i=1}^{11} A_{ij}$$

The total m^2 of each sector:

$$\forall i : m_i = \sum_{j/A_{ij}=1} \frac{V_j}{N_j}$$

The difficulty of this modeling lies fundamentally in the establishment of decisions. According to the description, it seems that the distribution of aid to the sectors in the first phase is imposed as a decision, and from there, we will distribute that

Table 7.9 Problem 44: table of elements

Elements	Set	QN	Data				
			Name	Parameter	Type	Belonging	Value
Shops	$j = 1\ldots n$	I_U	Size	V_j	C(m^2)	W	...
			Sector	Aij	B	S	...
			Turnover	Fi	C($)	W	...
			N° sectors	NS$_j$	I	W	...
			Workers	ti	I	W	...
Sectors	$i = 1\ldots 11$	I_U	Percentage	PI$_i$	C(%)	W	...
			Meters	m_i	C(m^2)	W	...
State aid	–	I_M	Amount	S	C($)	W	...

money between the shops based on calculations and not on decisions. The second phase can also be obtained by auxiliary calculations.

Decision Activities

Distribute state aid to sectors (*i*) [in a first phase].

$$\forall i : x_i = \text{Amount of state aid distributed to sector } i$$

Specifications

– Distribution of aid money (allocation specification based on availability attribute)

$$\sum_{i=1}^{11} x_i = S \tag{44.1}$$

Objective Criterion

Maximize the allocation to the sector with the lowest allocation per m^2.

We start by calculating the distribution between shops:

First, we make the gross initial distribution:

$\forall i,\ \forall\ j/A_{ij} = 1 : y_{ij} =$ Gross initial distribution. Theoretical distribution corresponding to shop j with respect to sector i.

To each shop in a sector, a value is obtained by a simple rule of three:

$$\begin{aligned} &x_i \rightarrow m_i \\ &y_{ij} \rightarrow \frac{V_j}{\mathrm{NS}_j} \\ &\forall i, \forall j/A_{ij} = 1 : y_{ij} = \frac{x_i}{m_i} \frac{V_j}{\mathrm{NS}_j} \end{aligned} \tag{44.2}$$

As we say, it is a gross distribution, since if it belongs to more than one sector, the criterion indicates that:

- If it is less than 100 m^2, only the sector with the highest amount corresponds to it;
- If it is more than 100 m^2, only the two best sectors correspond to it.

The problem with this modeling is that there can be a tie. Without ties, it would be easier. The modeling should pick up the sector with the highest income in the first case (less than 100 m^2) and the two sectors with the highest amount in the second case. In the event of a tie, we will opt for the one with the lowest index.

In addition to this, we must store the discarded sectors to accumulate that money for distribution in the second phase.

It is necessary to define binary logical calculations to collect the result of pairwise comparisons.

Binary Logical Calculation

$$\forall i, i'/i < i', \forall j/A_{ij} = 1\&A_{i'j} = 1 : \alpha_{ii'j} = 1 \text{ IF AND ONLY IF } y_{ij} \geq y_{i'j} \quad (44.3)$$

$$\Rightarrow f_{S_V} \Rightarrow \forall i, i'/i < i', \forall j/A_{ij} = 1\&A_{i'j} = 1 : \text{IF } \ y_{ij} \geq y_{i'j} \ \text{ THEN } \ \alpha_{ii'y} = 1$$
$$\Rightarrow f_3, f_7, f_4 \Rightarrow \forall i, i'/i < i', \forall j/A_{ij} = 1\&A_{i'j} = 1 :$$
$$\text{IF } \ 1 - \alpha_{ii'y} = 1 \ \text{ THEN } \ y_{ij} - y_{i'j} \leq \xi \quad (44.4)$$
$$\Rightarrow f_{14} \Rightarrow \forall i, i'/i < i', \forall j/A_{ij} = 1\&A_{i'j} = 1 :$$
$$y_{ij} - y_{i'j} \leq \xi \ + (S - \xi)\alpha_{ii'y}$$

Auxiliary Calculation

We define an auxiliary calculation to obtain the value of the opposite comparisons:

$$\forall i, i'/i > i', \forall j/A_{ij} = 1\&A_{i'j} = 1 : \alpha_{ii'j} = 1 - \alpha_{i'ij} \quad (44.5)$$

Finally, we define a non-binary logical calculation to collect the granted value in each sector. Here, I differentiate the case of premises smaller than 100 m^2 or from 100 m^2:

Non-Binary Logical Calculation

$$\forall i, j/A_{ij} = 1\&\text{NS}_j > 1 : \ z_{ij} = \text{Amount granted from sector } i \text{ for shop } j.$$

$$\forall i, j/A_{ij} = 1\&\text{NS}_j > 1\&V_j < 100 : \text{IF} \sum_{i' \neq i} \alpha_{ii'j} = \text{NS}_j - 1 \ \text{ THEN } \ z_{ij} = y_{ij};$$

$$\text{OTHERWISE } \ z_{ij} = 0;$$

Modeling:

$$\Rightarrow f_{\text{UB}} \Rightarrow \forall i,j/A_{ij} = 1\&\text{NS}_j > 1\&V_j < 100 :$$
$$\text{IF } \sum_{i' \neq i} \alpha_{ii'j} \geq \text{NS}_j - 1 \text{ THEN } z_{ij} = y_{ij}; \text{OTHERWISE } z_{ij} = 0;$$
$$\Rightarrow f_{\text{UB}}(z_{ij} \leq y_{ij}) \Rightarrow \forall i,j/A_{ij} = 1\&\text{NS}_j > 1\&V_j < 100 : \tag{44.6}$$
$$\text{IF} \sum_{i' \neq i} \alpha_{ii'j} \geq \text{NS}_j - 1 \text{ THEN } z_{ij} \geq y_{ij}; \text{OTHERWISE } z_{ij} = 0;$$
$$\Rightarrow z_{ij} \leq y_{ij}$$

(44.6) has been introduced to allow for the previous simplification of the upper bound f_{UB}.

$$\Rightarrow f_{18}\left(\text{UB}_{\sum_{i' \neq i} \alpha_{ii'j}} = \text{NS}_j - 1\right) \Rightarrow$$
$$\Rightarrow \sum_{i' \neq i} \alpha_{ii'j} \leq (\text{NS}_j - 2)(1 - \omega_{ij}) + (\text{NS}_j - 1)\omega_{ij} \tag{44.7}$$

$$\Rightarrow f_{21}\left(\text{LB}_{z_{ij} - y_{ij}} = -S\right) \Rightarrow z_{ij} - y_{ij} \geq -S(1 - \omega_{ij}) \tag{44.8}$$

On the other hand, it is necessary to model the adjustment to the value 0 when the conditions are not satisfied:

\scale90%

$$\Rightarrow \forall i,j/A_{ij} = 1\&\text{NS}_j > 1\&V_j < 100 : \text{IF} \sum_{i' \neq i} \alpha_{ii'j} < \text{NS}_j - 1 \text{ THEN } z_{ij} = 0;$$

$$\Rightarrow f_{\text{LB}} \Rightarrow \forall i,j/A_{ij} = 1\&\text{NS}_j > 1\&V_j < 100 : \text{IF} \sum_{i' \neq i} \alpha_{ii'j} < \text{NS}_j - 1 \text{ THEN } z_{ij} \leq 0;$$

$$\Rightarrow f_4 \Rightarrow \forall i,j/A_{ij} = 1\&\text{NS}_j > 1\&V_j < 100 : \text{IF} \sum_{i' \neq i} \alpha_{ii'j} \leq \text{NS}_j - 2 \text{ THEN } z_{ij} \leq 0;$$

$$\Rightarrow f_{17}\left(\text{LB}_{\sum_{i' \neq i} \alpha_{ii'j}} = 0\right) \Rightarrow \sum_{i' \neq i} \alpha_{ii'j} \geq (\text{NS}_j - 1)(1 - \pi_{ij}) \tag{44.9}$$

$$\Rightarrow f_{20}(\text{UB}_{z_{ij}} = S) \Rightarrow z_{ij} \leq S(1 - \pi_{ij}) \tag{44.10}$$

In the case of choosing the two best sectors (>100 m^2):

$$\forall i,j/A_{ij}=1\&\text{NS}_j>1\&V_j\geq 100:$$
$$\text{IF}\sum_{i'\neq i}\alpha_{ii'j}\geq \text{NS}_j-2\ \text{ THEN }\ z_{ij}=y_{ij};\text{OTHERWISE }\ z_{ij}=0;$$

Modeling: (Equivalent to the previous)

$$\begin{aligned}&\Rightarrow f_{\text{UB}}\Rightarrow \forall i,j/A_{ij}=1\&\text{NS}_j>1\&V_j\geq 100:\\&IF\sum_{i'\neq i}\alpha_{ii'j}\geq \text{NS}_j-2\ \text{ THEN }\ z_{ij}\geq y_{ij};\text{OTHERWISE }\ z_{ij}=0;\\&\Rightarrow z_{ij}\leq y_{ij}\end{aligned}\tag{44.11}$$

$$\begin{aligned}&\Rightarrow f_{18}\left(\text{UB}_{\sum_{i'\neq i}\alpha_{ii'j}}=\text{NS}_j-1\right)\Rightarrow\\&\Rightarrow\ \sum_{i'\neq i}\alpha_{ii'j}\leq(\text{NS}_j-3)(1-\omega_{ij})+(\text{NS}_j-1)\omega_{ij}\end{aligned}\tag{44.12}$$

$$\Rightarrow f_{21}\left(\text{LB}_{z_{ij}-y_{ij}}=-S\right)\Rightarrow z_{ij}-y_{ij}\geq -S(1-\omega_{ij})\tag{44.13}$$

\scale90%

$$\Rightarrow \forall i,j/A_{ij}=1\&\text{NS}_j>1\&V_j\geq 100:\text{IF}\sum_{i'\neq i}\alpha_{ii'j}<\text{NS}_j-2\ \text{ THEN }\ z_{ij}=0;$$

$$\Rightarrow f_{\text{LB}}\Rightarrow \forall i,j/A_{ij}=1\&\text{NS}_j>1\&V_j\geq 100:\text{IF}\sum_{i'\neq i}\alpha_{ii'j}<\text{NS}_j-2\ \text{ THEN }\ z_{ij}\leq 0;$$

$$\Rightarrow f_4\Rightarrow \forall i,j/A_{ij}=1\&\text{NS}_j>1\&V_j\geq 100:\text{IF }\sum_{i'\neq i}\alpha_{ii'j}\leq\text{NS}_j-3\ \text{ THEN }\ z_{ij}\leq 0;$$

$$\Rightarrow f_{17}\left(\text{LB}_{\sum_{i'\neq i}\alpha_{ii'j}}=0\right)\Rightarrow\ \sum_{i'\neq i}\alpha_{ii'j}\geq(\text{NS}_j-2)(1-\pi_{ij})\tag{44.14}$$

$$\Rightarrow f_{20}(\text{UB}_{z_{ij}}=S)\Rightarrow z_{ij}\leq S(1-\pi_{ij})\tag{44.15}$$

Finally, it would be necessary to model that the real allocation to the shops of a single sector coincides with the gross one in an auxiliary calculation:

$$\forall i,j/A_{ij}=1\&\text{NS}_j=1:\ \ z_{ij}=\text{Amount granted from sector } i \text{ for shop } j.$$

$$\forall i,j/A_{ij}=1\&\text{NS}_j=1:\ \ z_{ij}=y_{ij}\tag{44.16}$$

20% discount on the allocation if an establishment has the ratio (annual turnover/number of workers) above *TT* \$/worker:

Auxiliary Calculation:

$$\forall j/\frac{F_j}{t_j} > \text{TT}, i/A_{ij} = 1 : r_{ij} = \text{Discounted value for shops } j \text{ in sectors } i.$$

$$\forall j/\frac{F_j}{t_j} > \text{TT}, i/A_{ij} = 1 : r_{ij} = 0.2z_{ij} \tag{44.17}$$

Distribution of the remaining amounts:

We must distribute the remaining quantities among all the stores, according to the PI_i index. We collect the remaining amount:

Auxiliary Calculation:

$$q = \text{The remaining amount.}$$

$$q = \sum_{j/\frac{F_j}{t_j} > \text{TT}} \sum_{i/A_{ij}=1} r_{ij} + \sum_{j/\text{NS}_j > 1} \sum_{i/A_{ij}=1} \left(y_{ij} - z_{ij}\right) \tag{44.18}$$

We make the second distribution:

Auxiliary Calculation

$$\forall i : u_i = \text{Amount received by sector } i$$

$$\forall i : u_i = \frac{\text{PI}_i}{100} q \frac{m_i}{\sum_{i'=1}^{11} m_{i'}} \tag{44.19}$$

Auxiliary Calculation

$\forall i, j/A_{ij} = 1 : d_{ij} =$ Amount in the second phase for shop j with respect to sector i.

To each shop in a sector, a value obtained by a rule of three:

$$\begin{gathered} u_i \rightarrow m_i \\ v_{ij} \rightarrow \frac{V_j}{\text{NS}_j} \\ \forall i, \forall j/A_{ij} = 1 : d_{ij} = \frac{u_i}{m_i} \frac{V_j}{\text{NS}_j} \end{gathered} \tag{44.20}$$

Therefore, the amount that each shop j has received will be:

$$\forall j/j/\frac{F_j}{t_j} > TT : \sum_{i/A_{ij}=1} (z_{ij} - r_{ij} + d_{ij})$$

$$\forall j/j/\frac{F_j}{t_j} \leq TT : \sum_{i/A_{ij}=1} (z_{ij} + d_{ij})$$

We calculate the allocation of each sector per m^2:
Auxiliary Calculation

$$\forall i : b_i = \text{Total amount received by sector } i \text{ per m}^2$$

$$\forall i : b_i = \frac{1}{m_i}\left(\sum_{j/A_{ij}=1\&\frac{F_j}{t_j}>\mathrm{TT}} z_{ij} - r_{ij} + d_{ij} + \sum_{j/A_{ij}=1\&\frac{F_j}{t_j}\leq\mathrm{TT}} z_{ij} + d_{ij} \right) \quad (44.21)$$

Lower Bound Calculation
b_{MIN}: Lower bound of the allocation per m^2 to the sectors

$$\forall i : b_{\mathrm{MIN}} \leq b_i \quad (44.22)$$

Objective Function:

$$\text{Max } b_{\mathrm{MIN}} \quad (44.23)$$

Complete Model

$$\text{Max } b_{\mathrm{MIN}} \quad (44.23)$$

s.t.

$$\sum_{i=1}^{11} x_i = S \quad (44.1)$$

$$\forall i, \forall j/A_{ij} = 1 : y_{ij} = \frac{x_i}{m_i}\frac{V_j}{\mathrm{NS}_j} \quad (44.2)$$

$$\forall i, i'/i < i', \forall j/A_{ij} = 1\&A_{i'j} = 1 : \quad y_{ij} - y_{i'j} \leq \xi \ + (S - \xi)\alpha_{ii'y} \quad (44.4)$$

$$\forall i, i'/i > i', \forall j/A_{ij} = 1\&A_{i'j} = 1 : \alpha_{ii'j} = 1 - \alpha_{i'ij} \quad (44.5)$$

$$\forall i,j/A_{ij} = 1\&\mathrm{NS}_j > 1\&V_j < 100 : z_{ij} \leq y_{ij} \quad (44.6)$$

$$\forall i,j/A_{ij}=1\&\mathrm{NS}_j>1\&V_j<100:\sum_{i'\neq i}\alpha_{ii'j}\leq\left(\mathrm{NS}_j-2\right)\left(1-\omega_{ij}\right)+\left(\mathrm{NS}_j-1\right)\omega_{ij} \tag{44.7}$$

$$\forall i,j/A_{ij}=1\&\mathrm{NS}_j>1\&V_j<100:z_{ij}-y_{ij}\geq-S\left(1-\omega_{ij}\right) \tag{44.8}$$

$$\forall i,j/A_{ij}=1\&\mathrm{NS}_j>1\&V_j<100:\ \sum_{i'\neq i}\alpha_{ii'j}\geq\left(\mathrm{NS}_j-1\right)\left(1-\pi_{ij}\right) \tag{44.9}$$

$$\forall i,j/A_{ij}=1\&\mathrm{NS}_j>1\&V_j<100:z_{ij}\leq S\left(1-\pi_{ij}\right) \tag{44.10}$$

$$\forall i,j/A_{ij}=1\&\mathrm{NS}_j>1\&V_j\geq 100:z_{ij}\leq y_{ij} \tag{44.11}$$

$$\forall i,j/A_{ij}=1\&\mathrm{NS}_j>1\&V_j\geq 100:\ \sum_{i'\neq i}\alpha_{ii'j}\leq\left(\mathrm{NS}_j-3\right)\left(1-\omega_{ij}\right)+\left(\mathrm{NS}_j-1\right)\omega_{ij} \tag{44.12}$$

$$\forall i,j/A_{ij}=1\&\mathrm{NS}_j>1\&V_j\geq 100:z_{ij}-y_{ij}\geq-S\left(1-\omega_{ij}\right) \tag{44.13}$$

$$\forall i,j/A_{ij}=1\&\mathrm{NS}_j>1\&V_j\geq 100:\sum_{i'\neq i}\alpha_{ii'j}\geq\left(\mathrm{NS}_j-2\right)\left(1-\pi_{ij}\right) \tag{44.14}$$

$$\forall i,j/A_{ij}=1\&\mathrm{NS}_j>1\&V_j\geq 100:z_{ij}\leq S\left(1-\pi_{ij}\right) \tag{44.15}$$

$$\forall i,j/A_{ij}=1\&\mathrm{NS}_j=1:\ \ z_{ij}=y_{ij} \tag{44.16}$$

$$\forall j/\frac{F_j}{t_j}>\mathrm{TT},i/A_{ij}=1:r_{ij}=0.2z_{ij} \tag{44.17}$$

$$q=\sum_{j/\frac{F_j}{t_j}>\mathrm{TT}}\sum_{i/A_{ij}=1}r_{ij}+\sum_{j/\mathrm{NS}_j>1}\sum_{i/A_{ij}=1}\left(y_{ij}-z_{ij}\right) \tag{44.18}$$

$$\forall i:u_i=\frac{\mathrm{PI}_i}{100}q\frac{m_i}{\sum_{i'=1}^{11}m_{i'}} \tag{44.19}$$

$$\forall i,\forall j/A_{ij}=1:d_{ij}=\frac{u_i}{m_i}\frac{V_j}{\mathrm{NS}_j} \tag{44.20}$$

$$\forall i:b_i=\frac{1}{m_i}\left(\sum_{j/A_{ij}=1\&\frac{F_j}{t_j}>\mathrm{TT}}z_{ij}-r_{ij}+d_{ij}+\sum_{j/A_{ij}=1\&\frac{F_j}{t_j}\leq\mathrm{TT}}z_{ij}+d_{ij}\right) \tag{44.21}$$

$$\forall i:b_{\mathrm{MIN}}\leq b_i \tag{44.22}$$

$$\forall i : x_i \geq 0, u_i \geq 0, b_i \geq 0$$

$$\forall i, i'/i \neq i', \forall j/A_{ij} = 1 \& A_{i'j} = 1 : \alpha_{ii'j} \text{ binaries} \tag{44.3}$$

$$\forall i,j/A_{ij} = 1 : y_{ij} \geq 0, z_{ij} \geq 0, d_{ij} \geq 0$$

$$\forall j/\frac{F_j}{t_j} > \text{TT}, i/A_{ij} = 1 : r_{ij} \geq 0$$

$$q \geq 0$$

$$b_{\text{MIN}} \geq 0$$

$$\forall i,j/A_{ij} = 1 \& \text{NS}_j > 1 : \omega_{ij}, \pi_{ij} \text{ binaries} \tag{44.24}$$

Implementation in LINGO

```
MODEL:
DATA:
 N=@ODBC('44_ODBC','System','N');
ENDDATA
SETS:
 Shops/1..N/:V,F,t,NS;
 Sectors/1..11/:PI,m,x,u,b;
CS(Sectors,Shops)/@ODBC('44_ODBC','CS','S','C')/:y,z,Omega,PiG,r,
d;
 SSC(Sectors,Sectors,Shops):Alpha;
ENDSETS
DATA:
 V=@ODBC('44_ODBC','Shops','Tamano');
 F=@ODBC('44_ODBC','Shops','Facturacion');
 t=@ODBC('44_ODBC','Shops','Trabajadores');
 NS=@ODBC('44_ODBC','Shops','NumSectores');
 PI =@ODBC('44_ODBC','Sectors','PI');
 S=@ODBC('44_ODBC','System','Subvencion');
 TT=@ODBC('44_ODBC','System','TT');
 Epsilon=0.01;
ENDDATA
@FOR(Sectors(i):m(i)=@SUM(CS(i,j):V(j)));
!44.23;
Max = bMIN;
!44.1;
@SUM(Sectors(i):x(i))= S;
!44.2;
@FOR(CS(i,j):y(i,j)=(x(i)/m(i))*(v(j)/NS(j)));
!44.4;
@FOR(Sectors(i)|i#LT#11:@FOR(CS(i,j):@FOR(Sectors(ii)|i#LT#ii:
@FOR(CS(ii,j):y(i,j)- y(ii,j)<=Epsilon + (S-Epsilon)*Alpha(i,ii,
j)))));
!44.5;
@FOR(Sectors(i):@FOR(Sectors(ii)|i#GT#ii:@FOR(CS(i,j):@FOR(CS(ii,
j):Alpha(i,ii,j)=1-Alpha(ii,i,j)))));
!44.6;
@FOR(Sectors(i):@FOR(CS(i,j)|NS(j)#GT#1#AND#V(j)#LT#100: z(i,j)<=y
(i,j)));
```

```
!44.7;
@FOR(Sectors(i):@FOR(CS(i,j)|NS(j)#GT#1#AND#V(j)#LT#100:@SUM
(Sectors(ii)|ii#NE#i: Alpha(i,ii,j))<=(NS(j)-2)*(1-Omega(i,j))+(NS
(j)-1)*Omega(i,j)));
!44.8;
@FOR(Sectors(i):@FOR(CS(i,j)|NS(j)#GT#1#AND#V(j)#LT#100: z(i,j)-y
(i,j)>=-S*(1-Omega(i,j))));
!44.9;
@FOR(Sectors(i):@FOR(CS(i,j)|NS(j)#GT#1#AND#V(j)#LT#100: @SUM
(Sectors(ii)|ii#NE#i: Alpha(i,ii,j))>=(NS(j)-1)*(1-PiG(i,j))));
!44.10;
@FOR(Sectors(i):@FOR(CS(i,j)|NS(j)#GT#1#AND#V(j)#LT#100: z(i,j)
<=S*(1-PiG(i,j))));
!44.11;
@FOR(Sectors(i):@FOR(CS(i,j)|NS(j)#GT#1#AND#V(j)#GE#100: z(i,j)<=y
(i,j)));
!44.12;
@FOR(Sectors(i):@FOR(CS(i,j)|NS(j)#GT#1#AND#V(j)#GE#100:@SUM
(Sectors(ii)|ii#NE#i: Alpha(i,ii,j))<=(NS(j)-3)*(1-Omega(i,j))+(NS
(j)-1)*Omega(i,j)));
!44.13;
@FOR(Sectors(i):@FOR(CS(i,j)|NS(j)#GT#1#AND#V(j)#GE#100: z(i,j)-y
(i,j)>=-S*(1-Omega(i,j))));
!44.14;
@FOR(Sectors(i):@FOR(CS(i,j)|NS(j)#GT#1#AND#V(j)#GE#100:@SUM
(Sectors(ii)|ii#NE#i: Alpha(i,ii,j))>=(NS(j)-2)*(1-PiG(i,j))));
!44.15;
@FOR(Sectors(i):@FOR(CS(i,j)|NS(j)#GT#1#AND#V(j)#GE#100:z(i,j)
<=S*(1-PiG(i,j))));
!44.16;
@FOR(Sectors(i):@FOR(CS(i,j)|NS(j)#EQ#1:z(i,j)=y(i,j)));
!44.17;
@FOR(CS(i,j)|(F(j)/t(j))#GT#TT: r(i,j)=0.2*z(i,j));
!44.18;
q=@SUM(CS(i,j)|(F(j)/t(j))#GT#TT:r(i,j))+@SUM(CS(i,j)|NS(j)#GT#1:
(y(i,j)-z(i,j)));
!44.19;
@FOR(Sectors(i):u(i)=(PI(i)/100)*q*(m(i)/@SUM(Sectors(ii):m
(ii))));
!44.20;
@FOR(CS(i,j):d(i,j)=(u(i)/m(i))*(V(j)/NS(j)));
!44.21;
@FOR(Sectors(i):b(i)=(1/m(i))*(@SUM(CS(i,j)|(F(j)/t(j))#GT#TT:z
(i,j)-r(i,j)+d(i,j))+@SUM(CS(i,j)|(F(j)/t(j))#LE#TT:z(i,j)+d(i,
j))));
!44.22;
@FOR(Sectors(i):bMIN<=b(i));
!44.3;
@FOR(Sectors(i):@FOR(Sectors(ii)|i#NE#ii:@FOR(CS(i,j):@FOR(CS(ii,
j):@BIN(Alpha(i,ii,j))))));
!44.24;
@FOR(CS(i,j)|NS(j)#GT#1: @BIN(Omega(i,j)));
```

```
Solution Report - 44_LingoE
Global optimal solution found.
Objective value:                                    33.07652
Objective bound:                                    33.07652
Infeasibilities:                                    0.000000
Extended solver steps:                                     0
Total solver iterations:                                 828
Elapsed runtime seconds:                                0.19

Model Class:                                            MILP

Total variables:                      5319
Nonlinear variables:                     0
Integer variables:                     222

Total constraints:                     597
Nonlinear constraints:                   0

Total nonzeros:                       2527
Nonlinear nonzeros:                      0
```

Fig. 7.13 Problem 44: LINGO solution report

```
@FOR(CS(i,j)|NS(j)#GT#1: @BIN(PiG(i,j)));
END
```

Solution

It is shown in Fig. 7.13.

7.9 Problem 45. Planning of a Formula One Race

We are going to optimize the planning of a Formula One race with tire changes and fueling in the race.

Our car has a 100-l fuel tank. The consumption per lap is 4.5 l of fuel.

There are two types of tires that the car can use:

- Hard tires
- Soft tires

Hard tires have less wear than soft tires and hold more laps, although the car speed is lower than with soft tires.

Hard tires can withstand up to 50 laps on the circuit, but soft tires can withstand up to 20 laps.

The FIA (International Automobile Federation) only allows the use of two sets of soft tires during the race and a maximum of three tire changes.

The team establishes that at fuel stops, tires are also changed.

The race is scheduled to last 60 laps.

The speed and, therefore, the time that the car can take to make a turn depend on the type of tires, the wear of the tires, and the weight of the car. All this, for simplification, is going to be estimated as follows:

If the car takes hard tires: Time per lap (in seconds) = $TB_1 + F$ * Real weight of the car.

If the car has soft tires: Time per lap (in seconds) = $TB_2 + F$ * Real weight of the car

Where TB_1 and TB_2 are known time data and F is an increment factor that is also known.

The real weight of the car is calculated as the weight of the car (605 k: vehicle and driver) plus the weight of the fuel carried (1 l of fuel weighs 800 g).

At each pit stop, 30 s are lost if fuel is refueled (whether or not tires are changed) and 25 s are lost if only tires are changed.

The initial fuel and tires initially placed are considered to be placed in lap 1.

Modeling (Table 7.10)

Table 7.10 Problem 45: table of elements

			Data				
Elements	Set	QN	Name	Parameter	Type	Belonging	Value
Laps	$t = 1...60$	I_U					
Tires	$j = 1, 2$ (hard, soft)	I_U	Base time	TB_j	C(s)	W	–
			Endurance laps	V_j	I	W	{50,20}
			Time for change	tc_j (tc)	C(s)	W	25
Car	–	I_U	Weight	P	C (Kg)	W	605
			Factor	F	C	W	
			Tank capacity	K	C(l)	W	100
			Consumption per lap	C	C(l)	W	4.5
Fuel	–	I_M	Weight	PG	C (Kg)	W	0.8
			Refueling time	tg	C(s)	W	30

Decision Activities
Refuel [fuel] in laps (*t*)

$$\forall t: \; x_t = \text{Amount of fuel refuelled on lap } t$$

Put tires (*j*) on laps (*t*)

$$\forall j, t: \; \alpha_{jt} = 1 \text{ if we put tiresjon lap } t \tag{45.1}$$

Specifications

- If the car stops to refuel in a lap, the tires must be changed (propositional logic specification).

$$\begin{aligned} &\forall t : \text{IF } x_t > 0 \text{ THEN } \sum_{j=1}^{2} \alpha_{jt} = 1 \\ &\Rightarrow \sum_{j=1}^{2} \alpha_{jt} \text{ is binary expression} \\ &\Rightarrow f_3, f_7, f_{14}(\text{UB}_{x_t} = K) \Rightarrow \forall t : x_t \le K \sum_{j=1}^{2} \alpha_{jt} \end{aligned} \tag{45.2}$$

- Endurance of the tires. We must change tires before their endurance time runs out (propositional logic specification).

$$\begin{aligned} &\forall j, t/t + V_j \le 60 : \text{IF } \alpha_{jt} = 1 \text{ THEN } \sum_{t'=t+1}^{t+V_j} \alpha_{jt'} > 0 \\ &\Rightarrow f_5, f_{15} \left(\text{LB}_{\sum_{t'=t+1}^{t+V_j} \alpha_{jt'}} = 0 \right) \Rightarrow \forall j, t/t + V_j \le 60 : \sum_{t'=t+1}^{t+V_j} \alpha_{jt'} \ge \alpha_{jt} \end{aligned} \tag{45.3}$$

- Maximum three tire changes, not including the initial set of tires (quantitative selection rule).

$$\sum_{t=2}^{60} \sum_{j=1}^{2} \alpha_{jt} \le 3 \tag{45.4}$$

- Maximum two sets of soft tires (quantitative selection specification):

$$\sum_{t=1}^{60} \alpha_{2t} \leq 2 \tag{45.5}$$

- We must have fuel in all laps (lower bound imposition)

  We calculate the fuel at the beginning of each lap:
  Auxiliary Calculation

$$\forall t : y_t = \text{Fuel at the beginning of lap } t.$$

$$\forall t : y_t = \sum_{t'=1}^{t} x_{t'} - C(t-1) \tag{45.6}$$

  In each lap, we need at least the fuel necessary for one lap:

$$\forall t : y_t \geq C \tag{45.7}$$

  Objective Criterion
  Minimize the total time of the race

$$\text{Total time} = \sum_{t=1}^{60} \text{Time of the lap } t$$

For the time of each lap we need to know the type of tire that the car has on each lap. Stops and fueling in each lap are already collected in the decision variables.
Binary Logical Calculation
Type of tire in each lap

$$\forall j, t : \beta_{jt} = 1 \text{ if the car has type } j \text{ tires on lap } t; 0 \text{ otherwise.} \tag{45.8}$$

The calculation is different for $t = 1$ than for $t > 1$:

$$\forall j, t = 1 : \beta_{jt} = 1 \text{ IF AND ONLY IF } \alpha_{jt} = 1$$

$$\forall j, t > 1 : \beta_{jt} = 1 \text{ IF AND ONLY IF } \alpha_{jt} = 1 \text{ OR } \left(\beta_{jt-1} = 1 \text{ AND } \sum_{j=1}^{2} \alpha_{jt} = 0 \right)$$

Modeling:

$$\begin{aligned} &\forall j, t = 1 : \beta_{jt} = 1 \text{ IF AND ONLY IF } \alpha_{jt} = 1 \\ &\Rightarrow f_{23} \Rightarrow \forall j, t = 1 : \beta_{jt} = \alpha_{jt} \end{aligned} \tag{45.9}$$

$$\begin{aligned}
&\forall j, t > 1 : \beta_{jt} = 1 \text{ IF AND ONLY IF } \alpha_{jt} = 1 \text{ OR} \\
&\left(\beta_{jt-1} = 1 \text{ AND } \sum_{j'=1}^{2} \alpha_{j't} = 0\right) \\
&\Rightarrow f_7, f_{37}\left(\sum_{j=1}^{2} \alpha_{jt} \text{ is binary expression}\right) \Rightarrow \\
&\Rightarrow \forall j, t > 1 : \beta_{jt} = 1 \text{ IF AND ONLY IF } \alpha_{jt} = 1 \text{ OR} \\
&\left(\beta_{jt-1} + 1 - \sum_{j'=1}^{2} \alpha_{j't} \geq 2\right)
\end{aligned} \tag{45.10}$$

$$\omega_{jt} = 1 \text{ IF AND ONLY IF}\left(\beta_{jt-1} - \sum_{j'=1}^{2} \alpha_{j't} \geq 1\right)$$

$$\Rightarrow f_{25}\left(\text{LB}_{\beta_{jt-1} - \sum_{j'=1}^{2} \alpha_{j't}} = -1\right) \Rightarrow \beta_{jt-1} - \sum_{j'=1}^{2} \alpha_{j't} \geq 2\omega_{jt} - 1 \tag{45.11}$$

$$\Rightarrow f_{25}\left(\text{UB}_{\beta_{jt-1} - \sum_{j'=1}^{2} \alpha_{j't}} = 1\right) \Rightarrow\Rightarrow \beta_{jt-1} - \sum_{j'=1}^{2} \alpha_{j't} \leq \omega_{jt} \tag{45.12}$$

$$\begin{aligned}
&\forall j, t > 1 : \beta_{jt} = 1 \text{ IF AND ONLY IF } \alpha_{jt} = 1 \text{ OR } \omega_{jt} = 1 \\
&\Rightarrow f_{34} \Rightarrow \forall j, t > 1 : \beta_{jt} = 1 \text{ IF AND ONLY IF } \alpha_{jt} + \omega_{jt} \geq 1 \\
&\Rightarrow (\alpha_{jt} + \omega_{jt}) \text{ is binary expression} \\
&\Rightarrow f_{23} \Rightarrow \forall j, t > 1 : \beta_{jt} = \alpha_{jt} + \omega_{jt}
\end{aligned} \tag{45.13}$$

On the other hand, we need to accumulate the time to refuel and the time of the tire change when it is done without refueling (when refueling, only the time of refueling is imputed). This is achieved by making a logical calculation for each fact.

Binary Logical Calculation: Refuel on each lap t ($t > 1$)

In the first lap, the time to refuel and put tires is not counted.

$$\forall t, t > 1 : \quad \pi_t = 1 \text{ if refuel on lap } t; 0 \text{ otherwise} \tag{45.14}$$

$$\begin{aligned}
&\forall t/t > 1 : \quad \pi_t = 1 \text{ IF AND ONLY IF } x_t > 0 \\
&\Rightarrow f_{S_V} \Rightarrow \forall t/t > 1 : \quad \text{IF } x_t > 0 \text{ THEN } \pi_t = 1 \\
&\Rightarrow f_3, f_7, f_{14}(\text{UB}_{x_t} = K) \Rightarrow \forall t/t > 1 : \quad x_t \leq K\pi_t
\end{aligned} \tag{45.15}$$

Binary Logical Calculation: Change tires in lap t ($t > 1$)

$$\forall t, t > 1: \ \delta_t = 1 \text{ if we change tires and we do not refuel on lap } t; \text{ 0 otherwise.} \tag{45.16}$$

$$\forall t, t > 1: \ \delta_t = 1 \text{ IF AND ONLY IF } \pi_t = 0 \text{ AND } \sum_{j=1}^{2} \alpha_{jt} = 1$$

$$\Rightarrow f_{S_v} \Rightarrow \forall t, t > 1: \ \text{IF } \pi_t = 0 \text{ AND } \sum_{j=1}^{2} \alpha_{jt} = 1 \text{ THEN } \delta_t = 1$$

$$\Rightarrow f_7, f_{37} \Rightarrow \forall t, t > 1: \ \text{IF } 1 - \pi_t + \sum_{j=1}^{2} \alpha_{jt} \geq 2 \text{ THEN } \delta_t = 1 \tag{45.17}$$

$$\Rightarrow \forall t, t > 1: \ \text{IF } \sum_{j=1}^{2} \alpha_{jt} - \pi_t \geq 1 \text{ THEN } \delta_t = 1$$

$$\Rightarrow f_{3,}f_7, f_{14} \left(\text{UB}_{\sum_{j=1}^{2} \alpha_{jt} - \pi_t} = 1 \right) \Rightarrow \forall t, t > 1: \sum_{j=1}^{2} \alpha_{jt} - \pi_t \leq \delta_t$$

Objective Function

$$\text{Min} \quad \sum_{t=1}^{60} \sum_{\text{j}=1}^{2} \left(\text{TB}_\text{j} \beta_{jt} \right) + F(P + \text{PG}y_t) + +tg\pi_t + tc\delta_t \tag{45.18}$$

Complete Model

$$\text{Min} \quad \sum_{t=1}^{60} \sum_{\text{j}=1}^{2} \left(\text{TB}_\text{j} \beta_{jt} \right) + F(P + \text{PG}y_t) + tg\pi_t + tc\delta_t \tag{45.18}$$

s.t.

$$\forall t: x_t \leq \ K \sum_{j=1}^{2} \alpha_{jt} \tag{45.2}$$

$$\forall j, t/t + V_j \leq 60: \sum_{t'=t+1}^{t+V_j} \alpha_{\ jt'} \geq \alpha_{jt} \tag{45.3}$$

$$\sum_{t=2}^{60} \sum_{j=1}^{2} \alpha_{jt} \leq 3 \tag{45.4}$$

$$\sum_{t=1}^{60} \alpha_{2t} \leq 2 \quad (45.5)$$

$$\forall t : y_t = \sum_{t'=1}^{t} x_{t'} - C(t-1) \quad (45.6)$$

$$\forall t : y_t \geq C \quad (45.7)$$

$$\forall j, t = 1 : \beta_{jt} = \alpha_{jt} \quad (45.9)$$

$$\forall j, t > 1 : \beta_{jt-1} - \sum_{j'=1}^{2} \alpha_{j't} \geq 2\omega_{jt} - 1 \quad (45.11)$$

$$\forall j, t > 1 : \beta_{jt-1} - \sum_{j'=1}^{2} \alpha_{j't} \leq \omega_{jt} \quad (45.12)$$

$$\forall j, t > 1 : \beta_{jt} = \alpha_{jt} + \omega_{jt} \quad (45.13)$$

$$\forall t / t > 1 : x_t \leq K\pi_t \quad (45.15)$$

$$\forall t, t > 1 : \sum_{j=1}^{2} \alpha_{jt} - \pi_t \leq \delta_t \quad (45.17)$$

$$\forall t : x_t \geq 0$$

$$\forall j, t : \alpha_{jt} \text{ binaries} \quad (45.1)$$

$$\forall t : y_t \geq 0$$

$$\forall j, t : \beta_{jt} \text{ binaries} \quad (45.8)$$

$$\forall j, t / t > 1 : \omega_{jt} \text{ binaries} \quad (45.10)$$

$$\forall t, t > 1 : \pi_t \text{ binaries} \quad (45.14)$$

$$\forall t, t > 1 : \delta_t \text{ binaries} \quad (45.16)$$

Implementation in LINGO

```
MODEL:
SETS:
 Laps/1..50/:y,Pi,Delta,x;
 Tires/1..2/:TB,V,Tc;
 TL(Tires,Laps):ALpha,Beta,Omega;
ENDSETS
DATA:
 TB=68 80;
 V= 50 20;
 Tc= 25 25;
 P= 605;
 F= 0.2;
 K= 100;
 C= 4.5;
```

```
 PG= 0.8;
 tg= 30;
ENDDATA
!45.18;
MIN = @SUM(Laps(t):@SUM(Tires(j):TB(j)*Beta(j,t))+F*(P+PG*y(t))
+tg*Pi(t)+ @SUM(Tires(j):Tc(j)*Delta(t)));
!45.2;
@FOR(Laps(t):x(t)<=K*@SUM(Tires(j):Alpha(j,t)));
!45.3;
@FOR(Tires(j):@FOR(Laps(t)|t+V(j)#LE#60:@SUM(Laps(tt)|tt#GE#t
+1#AND#tt#LE#t+V(j):Alpha(j,tt))>=Alpha(j,t)));
!45.4;
@SUM(Laps(t)|t#GT#1:@SUM(Tires(j):ALpha(j,t))) <=3;
!45.5;
@SUM(Laps(t):Alpha(2,t))<=2;
!45.6;
@FOR(Laps(t):y(t)=@SUM(Laps(tt)|tt#LE#t:x(tt))-C*(t-1));
!45.7;
@FOR(Laps(t):y(t)>=C);
!45.9;
@FOR(Tires(j):Beta(j,1)=Alpha(j,1));
!45.11;
@FOR(Tires(j):@FOR(Laps(t)|t#GT#1:Beta(j,t-1)-@SUM(Tires(jj):
Alpha(jj,t))>=2*Omega(j,t)-1));
!45.12;
@FOR(Tires(j):@FOR(Laps(t)|t#GT#1:Beta(j,t-1)-@SUM(Tires(jj):
Alpha(jj,t))<=Omega(j,t)));
!45.13;
@FOR(Tires(j):@FOR(Laps(t)|t#GT#1:Beta(j,t)=Alpha(j,t)+Omega(j,
t)));
!45.15;
@FOR(Laps(t)|t#GT#1:x(t)<=K*Pi(t));
!45.17;
@FOR(Laps(t)|t#GT#1:@SUM(Tires(j):ALpha(j,t))-Pi(t)<=Delta(t));
!45.1;
@FOR(Tires(j):@FOR(Laps(t):@BIN(Alpha(j,t))));
!45.8;
@FOR(Tires(j):@FOR(Laps(t):@BIN(Beta(j,t))));
!45.10;
@FOR(Tires(j):@FOR(Laps(t)|t#GT#1:@BIN(Omega(j,t))));
!45.14;
@FOR(Laps(t):@BIN(Pi(t)));

!45.16;
@FOR(Laps(t):@BIN(Delta(t)));
END
```

Solution

It is shown in Fig. 7.14.

```
Solution Report - 45_LingoE
Global optimal solution found.
Objective value:                                    9783.360
Objective bound:                                    9783.360
Infeasibilities:                                    0.000000
Extended solver steps:                                187975
Total solver iterations:                             3550631
Elapsed runtime seconds:                              239.15

Model Class:                                            MILP

Total variables:                    500
Nonlinear variables:                  0
Integer variables:                  398

Total constraints:                  597
Nonlinear constraints:                0

Total nonzeros:                    4539
Nonlinear nonzeros:                   0
```

Fig. 7.14 Problem 45: LINGO solution report

7.10 Problem 46. Ready Mixed Concrete Delivery to Construction Sites

In modern construction, ready mixed concrete (RMC) is one of the most popular building materials in the construction industry, and concrete demand keeps an increasing rate in recent years (Albayrak and Albayrak 2015).

A concrete delivery company wants to propose a vehicle allocation system to attend construction sites planned for each day, taking into account the capacity of the company and the priority of the construction sites.

The company owns a plant where ready mixed concrete is loaded into vehicles. The company has a fleet of $i = 1...8$ vehicles, each with a K_i capacity of m^3 of concrete. The geographical position of the construction sites is known, and therefore, the time to go from the plant to the place and the time to return are known. The order download time is also known for each construction site. The charging time of the vehicle in the central area is negligible.

Each construction site has a demand for ready mixed concrete, which can be covered with the number of vehicles required. If a work is served, the full demanded order must be served. Each construction site schedules a time interval in which the concrete must arrive.

The company wants to establish a program that maximizes the total weight of the service. The program decides which vehicles go to a construction site and at what moment the delivery leaves the central. Each construction site is assigned a weight or priority and the objective is to maximize the total weight of the construction site served. It is assumed that a vehicle will go to a construction site at most once. If we allowed more than one trip for a vehicle to a construction site, the trips in the problem would have to be considered as individual elements to identify the time

Table 7.11 Problem 46: table of elements

Elements	Set	QN	Data				
			Name	Parameter	Type	Belonging	Value
Construction sites	$j = 1 \ldots n$	I_U	Order	D_j	C (m^3)	S	–
			Time from central	TI_j	C (min)	S	–
			Time to central	TV_j	C (min)	S	–
			Download time	TD_j	C (min)	W	–
			Weight (priority)	P_j	C	W	–
			Interval start	II_j	C (min)	W	–
			Interval end	FI_j	C (min)	W	–
Vehicles	$i = 1 \ldots 8$	I_M	Capacity	K_i	C (m^3)	S	–
Central	–	I_U		TI_j; TV_j			
Ready mixed Concrete	–	I_M		D_j; K_i			

of departure and the load with each trip, perfectly feasible, although it is ruled out due to the characteristics of the unloading process.

Modeling

The list of elements is shown in Table 7.11.

We declare the vehicles as measurable because they have to be scheduled in the horizon time.

Decision Activities

Deliver ready mixed concrete to the construction sites (*j*) with vehicles (*i*)

$\forall i,j : x_{ij} =$ Amount of ready mixed concrete that the vehicle i delivers on the construction site j

Schedule vehicles (*i*) to go to the construction sites (*j*)

$\forall i,j : y_{ij} =$ Instant in which vehicle i is scheduled to go to the construction site j

Specifications

– Vehicle capacity (capacity specification)

$$\forall i,j : x_{ij} \leq K_i \tag{46.1}$$

– If a vehicle serves a construction site, it is unloaded within the preset interval (propositional logic specification).

$$\forall i,j : \text{IF } x_{ij} > 0 \text{ THEN } y_{ij} + \text{TI}_j \geq \text{II}_j \text{ AND } y_{ij} + \text{TI}_j \leq \text{FI}_j$$

Binary Logical Calculation

$$\forall i,j : \alpha_{ij} = 1 \text{ if vehicle } i \text{ serves construction site } j; \tag{46.2}$$

$$\begin{aligned} &\forall i,j : \alpha_{ij} = 1 \text{ IF AND ONLY IF } x_{ij} > 0 \\ &\Rightarrow f_{S_V} \Rightarrow \forall i,j : \text{IF } x_{ij} > 0 \text{ THEN } \alpha_{ij} = 1 \\ &\Rightarrow f_3, f_7, f_{14}\left(\text{UB}_{x_{ij}} = D_j\right) \Rightarrow \forall i,j : x_{ij} \leq D_j \alpha_{ij} \end{aligned} \tag{46.3}$$

$$\begin{aligned} &\forall i,j : \text{IF } \alpha_{ij} = 1 \text{ THEN } y_{ij} + \text{TI}_j \geq \text{II}_j \text{ AND } y_{ij} + \text{TI}_j \leq \text{FI}_j \\ &\Rightarrow f_{41} \Rightarrow \forall i,j : \text{IF } \alpha_{ij} = 1 \text{ THEN } y_{ij} + \text{TI}_j \geq \text{II}_j \\ &\Rightarrow \forall i,j : \text{IF } \alpha_{ij} = 1 \text{ THEN } y_{ij} \geq \text{II}_j - \text{TI}_j \\ &\Rightarrow f_{15}\left(\text{LB}_{y_{ij}} = 0\right) \Rightarrow \forall i,j : y_{ij} \geq \left(\text{II}_j - \text{TI}_j\right)\alpha_{ij} \end{aligned} \tag{46.4}$$

$$\begin{aligned} &\Rightarrow f_{41} \Rightarrow \forall i,j : \text{IF } \alpha_{ij} = 1 \text{ THEN } y_{ij} + \text{TI}_j \leq \text{FI}_j \\ &\Rightarrow \forall i,j : \text{IF } \alpha_{ij} = 1 \text{ THEN } y_{ij} \leq \text{FI}_j - \text{TI}_j \\ &\Rightarrow f_{14}\left(\text{UB}_{y_{ij}} = \text{FI}_j - \text{TI}_j\right) \Rightarrow \forall i,j : y_{ij} \leq \left(\text{FI}_j - \text{TI}_j\right) \end{aligned} \tag{46.5}$$

– A vehicle cannot be in two construction sites at the same time (propositional logic specification).

$$\begin{aligned} &\forall i,j,j'/j < j' : \text{IF } \alpha_{ij} = 1 \text{ AND } \alpha_{ij'} = 1 \text{ THEN} \\ &y_{ij} + \text{TI}_j + \text{TD}_j + \text{TV}_j \leq y_{ij'} \text{ OR } y_{ij'} + \text{TI}_{j'} + TD_{j'} + TV_{j'} \leq y_{ij} \\ &\Rightarrow \forall i,j,j'/j < j' : \\ &\delta_{1ijj'} = 1 \text{ IF AND ONLY IF } y_{ij} + \text{TI}_j + \text{TD}_j + \text{TV}_j \leq y_{ij'} \end{aligned} \tag{46.6}$$

$$\begin{aligned} &\Rightarrow f_{S_V} \Rightarrow \forall i,j,j'/j < j' : \text{IF } \delta_{1ijj'} = 1 \text{ THEN } y_{ij} + \text{TI}_j + \text{TD}_j + \text{TV}_j \leq y_{ij'} \\ &\Rightarrow f_{14}\left(\text{UB}_{y_{ij} - y_{ij'}} = \text{FI}_j\right) \Rightarrow \forall i,j,j'/j < j' : \\ &y_{ij} - y_{ij'} \leq -\left(\text{TI}_j + \text{TD}_j + \text{TV}_j\right) + \left(\text{FI}_j + \text{TI}_j + \text{TD}_j + \text{TV}_j\right)\left(1 - \delta_{1ijj'}\right) \end{aligned} \tag{46.7}$$

$$\begin{aligned} &\Rightarrow \forall i,j,j'/j < j' : \\ &\delta_{2ijj'} = 1 \text{ IF AND ONLY IF } y_{ij'} + \text{TI}_{j'} + \text{TD}_{j'} + \text{TV}_{j'} \leq y_{ij} \end{aligned} \tag{46.8}$$

$$\Rightarrow f_{S_V} \Rightarrow \forall i,j,j'/j<j' : \text{IF } \delta_{2ijj'}=1 \text{ THEN } y_{ij'}+\text{TI}_{j'}+\text{TD}_{j'}+\text{TV}_{j'} \le y_{ij}$$

$$\Rightarrow f_{14}\left(\text{UB}_{y_{ij'}-y_{ij}}=\text{FI}_{j'}\right) \Rightarrow \forall i,j,j'/j<j' :$$

$$y_{ij'}-y_{ij} \le -\left(\text{TI}_{j'}+\text{TD}_{j'}+\text{TV}_{j'}\right)+\left(\text{FI}_{j'}+\text{TI}_{j'}+\text{TD}_{j'}+\text{TV}_{j'}\right)\left(1-\delta_{2ijj'}\right) \tag{46.9}$$

$$\forall i,j,j'/j<j' : \text{IF } \alpha_{ij}=1 \text{ AND } \alpha_{ij'}=1 \text{ THEN } \delta_{1ijj'}=1 \text{ OR } \delta_{2ijj'}=1$$

$$\Rightarrow f_{37}, f_{35} \Rightarrow \forall i,j,j'/j<j' : \text{IF } \alpha_{ij}+\alpha_{ij'} \ge 2 \text{ THEN } \delta_{1ijj'}+\delta_{2ijj'} \ge 1$$

$$\delta_{1ijj'}+\delta_{2ijj'} \text{ is binary expression :}$$

$$\Rightarrow f_3, f_7 \Rightarrow \forall i,j,j'/j<j' : \text{IF } 1-\left(\delta_{1ijj'}+\delta_{2ijj'}\right)=1 \text{ THEN } \alpha_{ij}+\alpha_{ij'} \le 1$$

$$\Rightarrow f_{14}\left(\text{UB}_{\alpha_{ij}+\alpha_{ij'}}=2\right) \Rightarrow$$

$$\Rightarrow \forall i,j,j'/j<j' : \alpha_{ij}+\alpha_{ij'} \le 1+\left(\delta_{1ijj'}+\delta_{2ijj'}\right) \tag{46.10}$$

– If concrete is served to a construction site, the entire order must be served (propositional logic specification).

$$\forall i,j : \text{IF } \alpha_{ij}=1 \text{ THEN } \sum_{i'=1}^{8} x_{i'j}=D_j$$

$$\Rightarrow f_{\text{UB}} \Rightarrow \forall i,j : \text{IF } \alpha_{ij}=1 \text{ THEN } \sum_{i'=1}^{8} x_{i'j} \ge D_j \tag{46.11}$$

$$\Rightarrow f_{15}\left(\text{LB}_{\sum_{i'=1}^{8} x_{i'j}}=0\right) \Rightarrow \forall i,j : \sum_{i'=1}^{8} x_{i'j} \ge D_j\alpha_{ij}$$

Objective Criterion
Maximize the total weight of the construction sites served.
We must collect in a logical calculation the construction sites served:
Binary Logical Calculation

$$\forall j : \pi_j=1 \text{ IF AND ONLY IF } \sum_{i=1}^{8} \alpha_{ij}>0 \tag{46.12}$$

$$\Rightarrow f_{S_V} \Rightarrow \forall j: \text{IF } \pi_j = 1 \text{ THEN } \sum_{i=1}^{8} \alpha_{ij} > 0$$

$$\Rightarrow f_5 \Rightarrow \forall j: \text{IF } \pi_j = 1 \text{ THEN } \sum_{i=1}^{8} \alpha_{ij} \geq 1 \tag{46.13}$$

$$\Rightarrow f_{15}\left(\text{LB}_{\sum_{i=1}^{8} \alpha_{ij}} = 0\right) \Rightarrow \forall j: \sum_{i=1}^{8} \alpha_{ij} \geq \pi_j$$

Objective Function

$$\text{Max} \sum_{j=1}^{n} P_j \pi_j \tag{46.14}$$

Complete Model

$$\text{Max} \sum_{j=1}^{n} P_j \pi_j \tag{46.14}$$

s.t.

$$\forall i,j: x_{ij} \leq K_i \tag{46.1}$$

$$\forall i,j: x_{ij} \leq D_j \alpha_{ij} \tag{46.3}$$

$$\forall i,j: y_{ij} \geq (\text{II}_j - \text{TI}_j)\alpha_{ij} \tag{46.4}$$

$$\forall i,j: y_{ij} \leq (\text{FI}_j - \text{TI}_j) \tag{46.5}$$

$$\forall i,j,j'/j < j':$$
$$y_{ij} - y_{ij'} \leq -(\text{TI}_j + \text{TD}_j + \text{TV}_j) + (\text{FI}_j + \text{TI}_j + \text{TD}_j + \text{TV}_j)(1 - \delta_{1ijj'}) \tag{46.7}$$

$$\forall i,j,j'/j < j':$$
$$y_{ij'} - y_{ij} \leq -(\text{TI}_{j'} + \text{TD}_{j'} + \text{TV}_{j'}) + (\text{FI}_{j'} + \text{TI}_{j'} + \text{TD}_{j'} + \text{TV}_{j'})(1 - \delta_{2ijj'}) \tag{46.9}$$

$$\forall i,j,j'/j < j': \alpha_{ij} + \alpha_{ij'} \leq 1 + (\delta_{1ijj'} + \delta_{2ijj'}) \tag{46.10}$$

$$\forall i,j: \sum_{i'=1}^{8} x_{i'j} \geq D_j \alpha_{ij} \tag{46.11}$$

$$\forall j: \sum_{i=1}^{8} \alpha_{ij} \geq \pi_j \qquad (46.13)$$

$$\forall i,j: x_{ij} \geq 0$$

$$\forall i,j: y_{ij} \geq 0$$

$$\forall i,j: \alpha_{ij} \text{ binaries} \qquad (46.2)$$

$$\forall i,j,j'/j < j': \quad \delta_{1ijj'}, \delta_{2ijj'} \text{ binaries} \qquad (46.6, 46.8)$$

$$\forall j: \pi_j \text{ binaries} \qquad (46.12)$$

Implementation in LINGO

```
MODEL:
DATA:
 n=@ODBC('46_ODBC','System','n');
ENDDATA
SETS:
 ConstructionSites/1..n/:D,TI,TV,TD,P,II,FI,Pi;
 Vehicles/1..8/:K;
 VO(Vehicles,ConstructionSites):x,y,Alpha;
 VOO(Vehicles,ConstructionSites,ConstructionSites):Delta1,Delta2;
ENDSETS
DATA:
 D=@ODBC('46_ODBC','ConstructionSites','D');
 TI=@ODBC('46_ODBC','ConstructionSites','TI');
 TV=@ODBC('46_ODBC','ConstructionSites','TV');
 TD=@ODBC('46_ODBC','ConstructionSites','TD');
 P=@ODBC('46_ODBC','ConstructionSites','Priority');
 II=@ODBC('46_ODBC','ConstructionSites','II');
 FI=@ODBC('46_ODBC','ConstructionSites','FI');
 K=@ODBC('46_ODBC','Vehicles','Capacity');
ENDDATA
!46.14;
MAX= @SUM(ConstructionSites(j):P(j)*Pi(j));
!46.1;
@FOR(Vehicles(i):@FOR(ConstructionSites(j):x(i,j)<=K(i)));
!46.3;
@FOR(Vehicles(i):@FOR(ConstructionSites(j):x(i,j)<=D(j)*Alpha(i,
j)));
!46.4;
@FOR(Vehicles(i):@FOR(ConstructionSites(j):y(i,j)>=(II(j)-TI(j))
*Alpha(i,j)));
!46.5;
@FOR(Vehicles(i):@FOR(ConstructionSites(j):y(i,j)<=FI(j)-TI(j)));
!46.7;
@FOR(Vehicles(i):@FOR(ConstructionSites(j):@FOR(ConstructionSites
(jj)|j#LT#jj:y(i,j)-y(i,jj)<=-(TI(j)+TD(j)+TV(j))+(FI(j)+TI(j)+TD
(j)+TV(j))*(1-Delta1(i,j,jj)))));
!46.9;
@FOR(Vehicles(i):@FOR(ConstructionSites(j):@FOR(ConstructionSites
(jj)|j#LT#jj:y(i,jj)-y(i,j)<=-(TI(jj)+TD(jj)+TV(jj))+(FI(jj)+TI
```

```
(jj)+TD(jj)+TV(jj))*(1-Delta2(i,j,jj))))));
!46.10;
@FOR(Vehicles(i):@FOR(ConstructionSites(j):@FOR(ConstructionSites
(jj)|j#LT#jj:Alpha(i,j)+Alpha(i,jj)<=1+(Delta1(i,j,jj)+Delta2(i,
j,jj)))));
!46.11;
@FOR(Vehicles(i):@FOR(ConstructionSites(j):@SUM(Vehicles(ip):x
(ip,j))>=D(j)*Alpha(i,j)));
!46.13;
@FOR(ConstructionSites(j):@SUM(Vehicles(i):Alpha(i,j))>=Pi(j));
!46.2;
@FOR(Vehicles(i):@FOR(ConstructionSites(j):@BIN(Alpha(i,j))));
!46.6;
@FOR(Vehicles(i):@FOR(ConstructionSites(j):@FOR(ConstructionSites
(jj):@BIN(Delta1(i,j,jj)))));
!46.8;
@FOR(Vehicles(i):@FOR(ConstructionSites(j):@FOR(ConstructionSites
(jj):@BIN(Delta2(i,j,jj)))));
!46.12;
@FOR(ConstructionSites(j):@BIN(Pi(j)));
END
```

Solution

The solution is shown in Fig. 7.15.

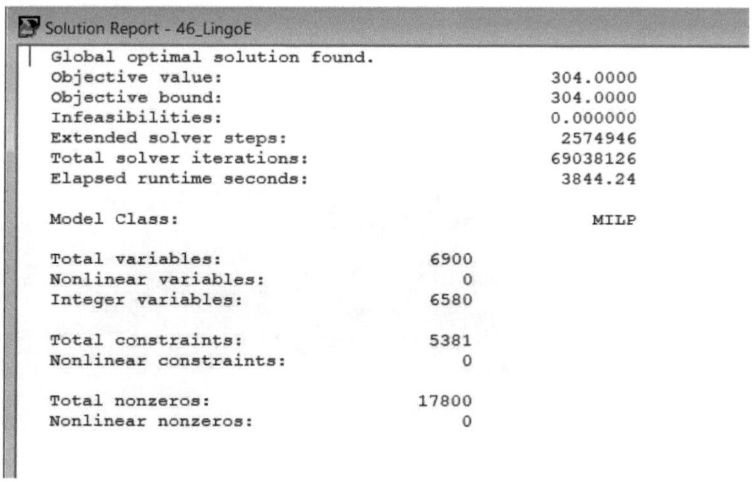

Solution Report - 46_LingoE

Global optimal solution found.		
Objective value:		304.0000
Objective bound:		304.0000
Infeasibilities:		0.000000
Extended solver steps:		2574946
Total solver iterations:		69038126
Elapsed runtime seconds:		3844.24
Model Class:		MILP
Total variables:	6900	
Nonlinear variables:	0	
Integer variables:	6580	
Total constraints:	5381	
Nonlinear constraints:	0	
Total nonzeros:	17800	
Nonlinear nonzeros:	0	

Fig. 7.15 Problem 46: LINGO solution report

7.11 Problem 47. Planning of a Supermarket Car Park

In the design of the parking of a new establishment of a supermarket chain, a system has been established for the allocation of parking places for shopping carts. The parking lot is divided into individual parking, a total of 80, which must be used to park vehicles or shopping carts.

For the parking places that are decided to be for shopping carts, a structure, as in the photo of Fig. 7.16, will be installed that distinguishes them from the parking places for vehicles.

For the allocation of cart parking places, the distance between each pair of car parks and the distance from each parking place to the entrance of the establishment have been collected.

The parking has three zones. Each parking space belongs to a zone. The company wants there to be at least one parking lot for shopping carts in each area.

The company has calculated that it should place R1 large shopping carts and R2 small shopping carts. There could be a maximum of 30 large carts and 40 small carts in a parking place. The company does not allow there to be any stall with less than 20 large carts and 30 small carts.

The company must decide where to locate shopping cart parking places, raising the objective of minimizing the sum of the greatest distance from each vehicle parking place lot to a shopping cart parking lot in its area, plus the sum of the distances of all shopping cart parking places to the entrance of the establishment.

Fig. 7.16 Photo of a parking place for shopping carts (Source: Internet)

Modeling

The list of elements is shown in Table 7.12.

Table 7.12 Problem 47: table of elements

Elements	Set	QN	Data				
			Name	Parameter	Type	Belonging	Value
Car parks	$i, i' = 1\ldots80$	I_U	Distance between car parks	$D_{ii'}$	C (m)	S	–
			Distance to the establishment	DE_i	C (m)	W	–
			Zone	Z_i	O	W	–
Types of parking places	$j = 1, 2$ {for vehicles, for shopping carts}	I_U					
Zones	$k = 1\ldots3$	I_U					
Shopping carts	r=1, 2 {large, small}	C_D	Stocks	E_r	I	W	{R_1, R_2}
			Maximum units[a]	M_r	I	W	{30,40}
			Minimum units[a]	N_r	I	W	{20,30}

[a]It is not necessary to share these attributes with the car parks, since it is the same value for all. It would have been necessary if the car parks had different dimensions and therefore different capacities

Decision Activities

Assign shopping carts (*r*) in car parks (*i*)

$$\forall r, i : x_{ri} = \text{Number of shopping carts } r \text{ assigned to car park } i \qquad (47.1)$$

Assign types of parking places (*j*) to car parks (*i*)

As there is a positive minimum number of shopping carts in a shopping cart parking, the allocation of the parking type to each parking place becomes a binary logical calculation based on the previous activity. If we assign a positive number of shopping carts, we define the car park as parking for shopping carts, and if we do not assign carts then the parking is for vehicles.

Binary Logical Calculation

$\forall i, j : \alpha_{ij} = 1$ if type j is assigned to car park i; 0 otherwise.

For convenience, we can dispense with the two types of parking as elements and only consider the type parking for shopping carts:

$$\forall i : \alpha_i = 1 \text{ if we select parking } i \text{ for shopping cart parking; 0 otherwise.} \quad (47.2)$$

(It would be equivalent to the variable α_{i2}). Variables α_{i1} have no decision capacity since $\alpha_{i1} = 1 - \alpha_{i2.}$

Definition:

$$\forall i : \alpha_i = 1 \text{ IF AND ONLY IF } \sum_{r=1}^{2} x_{ri} > 0$$

In principle, it is only necessary to define the value 1:

$$\Rightarrow f_{S_V} \Rightarrow \forall i : \text{IF } \sum_{r=1}^{2} x_{ri} > 0 \text{ THEN } \alpha_i = 1$$

In the case of not placing shopping carts in a car park, the variable would be free to value 0 or 1. Since there are equilibrium terms in the objective function referring to the relationship between parking spaces for carts and for vehicles, it is not advisable to simplify values of α_i. However, there is a later specification (47.6) that allows us to simplify the definition. In that specification, the value 0 of α_i is indirectly defined.

Modeling of the value $\alpha_i = 1$:

$$\Rightarrow f_3, f_7, f_{14} \left(\text{UB}_{\sum_{r=1}^{2} x_{ri}} = M_1 + M_2 \right) \Rightarrow \forall i : \sum_{r=1}^{2} x_{ri} \leq (M_1 + M_2)\alpha_i \quad (47.3)$$

Note: If α_i had been considered as a decision activity, it would be necessary to consider the relationship between the variables α_i (or α_{i2}) and x_{ri} as an implicit specification.

Specifications

- Placement of all carts (allocation specification)

$$\forall r : \sum_{i=1}^{80} x_{ri} = E_r \quad (47.4)$$

- Maximum number of carts in a car park (capacity specification or upper bound imposition):

$$\forall i, r : x_{ri} \leq M_r \quad (47.5)$$

- Minimum number of carts in a cart parking (propositional logic specification):

$$\forall i, r: \text{IF } \alpha_i = 1 \text{ THEN } x_{ri} \geq N_r \\ \Rightarrow f_{15}(\text{LB}_{x_{ri}} = 0) \Rightarrow \forall i, r: \quad x_{ri} \geq N_r \alpha_i \tag{47.6}$$

- At least one shopping cart parking place in each area (quantitative selection specification):

$$\forall k: \sum_{i/Z_i = k} \alpha_i \geq 1 \tag{47.7}$$

Objective Criterion

Minimize the sum of the greatest distance from each vehicle parking lot to a shopping cart park in your zone (*Sum*1) plus the sum of the distances from all car parks for shopping carts to the entrance of the establishment (*Sum*2).

$$\text{Min } Sum1 + Sum2$$

*Sum*1

We define a calculation to obtain an upper bound of the distance from each car park to a shopping cart parking in its area. Later, we will only take into account the car parks for vehicles.

Upper Bound Calculation

$\forall i: x_i^{\text{MAX}}$ = Greater distance from each car park i to a shopping cart parking in its zone.

$$\forall i, i'/i \neq i' \& Z_i = Z_{i'}: \quad x_i^{\text{MAX}} \geq D_{ii'} \alpha_{i'}$$

Now we only have to add those x_i^{MAX} that correspond to car parks for vehicles. It would be necessary to define a non-binary logical calculation that only took into account the car parks for vehicles.

Non-Binary Logical Calculation

$$\forall i: y_i = \text{Distance from car park } i.$$

$$\forall i: \text{IF } \alpha_i = 0 \text{ THEN } y_i = x_i^{\text{MAX}}; \text{OTHERWISE } y_i = 0$$

$$\forall i: \text{IF } \alpha_i = 0 \text{ THEN } y_i = x_i^{\text{MAX}}; \text{OTHERWISE } y_i = 0$$

$$Sum1 = \sum_{i=1}^{80} y_i$$

All of this could have been modeled with a conditional upper bound calculation:

Conditional Upper Bound Calculation

$\forall i : xc_i^{\text{MAX}} =$ Greater distance from each car park i to a shopping cart parking in its zone.

$$\begin{aligned}
&\forall i, i'/i \neq i' \& Z_i = Z_{i'} : \text{IF } \alpha_i = 0 \text{ AND } \alpha_{i'} = 1 \text{ THEN } xc_i^{\text{MAX}} \geq D_{ii'}\alpha_{i'} \\
&\Rightarrow \forall i, i'/i \neq i' \& Z_i = Z_{i'} : \text{IF } \alpha_i = 0 \text{ AND } \alpha_{i'} = 1 \text{ THEN } xc_i^{\text{MAX}} \rightarrow D_{ii'} \\
&\Rightarrow f_7, f_{37} \Rightarrow \forall i, i'/i \neq i' \& Z_i = Z_{i'} : \text{IF } \alpha_{i'} - \alpha_i \geq 1 \text{ THEN } xc_i^{\text{MAX}} \geq D_{ii'} \\
&\Rightarrow f_{18}\left(\text{UB}_{\alpha_{i'} - \alpha_i} = 1\right) \Rightarrow \forall i, i'/i \neq i' \& Z_i = Z_{i'} : \alpha_{i'} - \alpha_i \leq \omega_{ii'}
\end{aligned} \tag{47.8}$$

$$\Rightarrow f_{21}\left(\text{LB}_{x_i^{\text{MAX}}} = 0\right) \Rightarrow \forall i, i'/i \neq i' \& Z_i = Z_{i'} : xc_i^{\text{MAX}} \geq D_{ii'}\omega_{ii'} \tag{47.9}$$

$$Sum1 = \sum_{i=1}^{80} xc_i^{\text{MAX}}$$

Sum2

Distance from all shopping cart parking places to the entrance of the establishment.

$$Sum2 = \sum_{i=1}^{80} \text{DE}_i \alpha_i$$

Objective Function

$$\text{Min} \sum_{i=1}^{80} xc_i^{\text{MAX}} + \sum_{i=1}^{80} \text{DE}_i \alpha_i \tag{47.10}$$

Complete Model

$$\text{Min} \sum_{i=1}^{80} xc_i^{\text{MAX}} + \sum_{i=1}^{80} \text{DE}_i \alpha_i \tag{47.10}$$

s.t.

$$\forall i : \sum_{r=1}^{2} x_{ri} \leq (M_1 + M_2)\alpha_i \tag{47.3}$$

$$\forall r : \sum_{i=1}^{80} x_{ri} = E_r \tag{47.4}$$

$$\forall i, r : x_{ri} \leq M_r \tag{47.5}$$

$$\forall i, r : x_{ri} \geq N_r \alpha_i \tag{47.6}$$

$$\forall k : \sum_{i/Z_i=k} \alpha_i \geq 1 \tag{47.7}$$

$$\forall i, i'/i \neq i' \& Z_i = Z_{i'} : \alpha_{i'} - \alpha_i \leq \omega_{ii'} \tag{47.8}$$

$$\forall i, i'/i \neq i' \& Z_i = Z_{i'} : xc_i^{\text{MAX}} \geq D_{ii'} \omega_{ii'} \tag{47.9}$$

$$\forall r, i : x_{ri} \geq 0 \text{ integers} \tag{47.1}$$

$$\forall i : \alpha_i \text{ binaries} \tag{47.2}$$

$$\forall i, i'/i \neq i' \& Z_i = Z_{i'} : \omega_{ii'} \text{ binaries} \tag{47.11}$$

Implementation in LINGO

```
MODEL:
SETS:
 Carparks/1..80/:DE,Z,Alpha,xcMAX;
 ShoppingCarts/1..2/:E,N,M;
 Zones/1..3/;
 AA(Carparks,Carparks):D,Omega;
 CA( ShoppingCarts,Carparks):x;
ENDSETS
DATA:
 DE=@OLE('C:\Files_\47_Excel.xlsx','DE');
 Z=@OLE('C:\Files_\47_Excel.xlsx','Z');
 D=@OLE('C:\Files_\47_Excel.xlsx','D');
 E=@OLE('C:\Files_\47_Excel.xlsx','E');
 M=@OLE('C:\Files_\47_Excel.xlsx','M');
 N=@OLE('C:\Files_\47_Excel.xlsx','N');
ENDDATA
CALC:
@FOR(Carparks(i):@FOR(Carparks(ii)|i#LT#ii:D(ii,i)=D(i,ii)));
ENDCALC
!47.10;
MIN = @SUM(Carparks(i):xcMAX(i))+@SUM(Carparks(i):DE(i)*Alpha(i));
!47.3;
@FOR(Carparks(i):@SUM(ShoppingCarts(r):x(r,i))<=(M(1)+M(2))*Alpha
(i));
!47.4;
@FOR(ShoppingCarts(r):@SUM(Carparks(i):x(r,i))=E(r));
!47.5;
@FOR(Carparks(i):@FOR(ShoppingCarts(r):x(r,i)<=M(r)));
!47.6;
@FOR(Carparks(i):@FOR(ShoppingCarts(r):x(r,i)>=N(r)*Alpha(i)));
!47.7;
@FOR(Zones(k):@SUM(Carparks(i)|Z(i)#EQ#k:Alpha(i))>=1);
!47.8;
@FOR(Carparks(i):@FOR(Carparks(ii)|i#NE#ii#AND#Z(i)#EQ#Z(ii):
Alpha(ii)-Alpha(i)<=Omega(i,ii)));
!47.9;
@FOR(Carparks(i):@FOR(Carparks(ii)|i#NE#ii#AND#Z(i)#EQ#Z(ii):
xcMAX(i)>=D(i,ii)*Omega(i,ii)));
!47.1;
```

```
Solution Report - 47_LingoE
Global optimal solution found.
Objective value:                                  1107.900
Objective bound:                                  1107.900
Infeasibilities:                                  0.000000
Extended solver steps:                               66572
Total solver iterations:                          11278183
Elapsed runtime seconds:                           1824.07

Model Class:                                          MILP

Total variables:                     6720
Nonlinear variables:                    0
Integer variables:                   2496

Total constraints:                   4918
Nonlinear constraints:                  0

Total nonzeros:                     12400
Nonlinear nonzeros:                     0
```

Fig. 7.17 Problem 47: LINGO solution report

```
@FOR(Carparks(i):@FOR(ShoppingCarts(r):@GIN(x(r,i))));
!47.2;
@FOR(Carparks(i):@BIN(Alpha(i)));
!47.11;
@FOR(Carparks(i):@FOR(Carparks(ii)|i#NE#ii#AND#Z(i)#EQ#Z(ii):@BIN
(Omega(i,ii))));
END
```

Solution
The solution is shown in Fig. 7.17.

7.12 Problem 48. Assignment of Teacher Posts

The public administration wants to change the system for allocating posts to secondary school teachers who pass a competition; m posts are offered (each vacant place belongs to a teaching institute). Since m posts are offered, there are m candidate teachers (the m best marks of the competition).

To these places must be added the r posts currently covered, but whose teachers have requested a change of destination. These teachers apply for a specific post.

These teachers also have a grade, which takes into account both the years of work and the initial grade that the teacher had in their corresponding competition. The election system must be based on that grade, but with an important difference: the system either gives them the posts they have applied for or they keep the one they currently have.

Regarding the new teachers, each can request up to three posts and the system will try to give him one of those posts, or another as close as possible (we know the distances between posts). The places requested by the teacher are in priority order.

Adding to the m vacant positions the r change request positions, the allocation system has a total of $n = m + r$ positions that enter the contest. The m new teachers can apply for any of the n posts.

The following rules must always be respected when allocating:

- We cannot assign the post requested by a teacher with a position to a teacher with a lower grade than him.
- If a new teacher has a higher grade than another teacher, new or with a position, and they have applied for the same position in the same order of choice or in any other, we will never be able to assign it to the lowest grade unless the teacher with the highest grade is assigned a requested position from a higher election order.

The objective is to maximize the overall satisfaction of the allocation, both for new teachers and for teachers with a position requesting change. The way to measure satisfaction will be as follows:

Regarding new teachers, if they are assigned one of the requested positions, satisfaction corresponds to the constant values S_1, S_2, or S_3, depending on the order in its list of priority of the allocated position ($S_1 > S_2 > S_3$). If you do not receive one of the requested places, satisfaction will be measured as the F factor minus the shortest distance from the assigned place with respect to those requested. F will take the value of S_1, S_2, or S_3, depending on the requested position with a smaller distance from the one allocated.

Regarding teachers with position, the change to the requested place is valued as an S_4 satisfaction, and null satisfaction if they stay in the current place.

The S_1, S_2, S_3, and S_4 values are established by the administration based on an analytical study.

Modeling

The list of elements is shown in Table 7.13.

Table 7.13 Problem 48: table of elements

Elements	Set	QN	Data				
			Name	Parameter	Type	Belonging	Value
Positions	i, $i' = 1\ldots n$	I_U	Distance	$D_{ii'}$	C (Km)	S	–
Teachers with position	$k = 1\ldots r$	I_U	Occupied position	O_k	O	W	–
			Requested position	P_k	O	W	–
			Grade	GT_k	C	W	–
New teachers	$j = 1\ldots m$	I_U	Requested position in order	PO_{je}	O	S	–
			Grade	GN_j	C	W	–
System	–	I_U	Teacher satisfaction with a place	S_4	C	W	–
Election order	$e = 1\ldots 3$	I_U	Satisfaction of new teachers	S_p	C	W	–
				PO_{je}			

Decision Activities

Allocate positions (*i*) to teachers with position (*k*)

$$\forall k, i/i = O_k \ \vee \ i = P_k : \alpha_{ki} = 1 \text{ if teacher } k \text{ is allocated position } i; \ 0 \text{ otherwise.} \quad (48.1)$$

Allocate positions(*i*) to new teachers (*j*)

$$\forall j, i : \beta_{ji} = 1 \text{ if teacher } j \text{ is allocated position } i; 0 \text{ otherwise.} \quad (48.2)$$

Specifications

– Quantitative selection specifications:

Activities	Elements selecting	Selectable elements	Selection	Constraints
$\forall i, k : \alpha_{ki}$	Teachers with position ($k = 1\ldots r$)	Positions $\forall i/i = O_k \ \vee \ i = P_k$	=1	$\forall k : \sum_{i/i=O_k \ \vee \ i=P_k} \alpha_{ki} = 1$ (48.3)
$\forall j, i : \beta_{ji}$	New teachers ($j = 1\ldots m$)	Positions ($i = 1\ldots n$)	=1	$\forall j : \sum_{i=1}^{n} \beta_{ji} = 1$ (48.4)
$\forall i, k : \alpha_{ki}$ $\forall j, i : \beta_{ji}$	Positions ($i = 1\ldots n$)	Teachers with position $k/O_k = i \ \vee \ P_k = i$	=1	$\forall i : \sum_{k/O_k=i \ \vee \ P_k=i} \alpha_{ki} + \sum_{j=1}^{m} \beta_{ji} = 1$ (48.5)

(continued)

Activities	Elements selecting	Selectable elements	Selection	Constraints
		and New teachers ($j = 1...m$)		

– Allocation norms (propositional logic specifications):

We integrate the norms into four propositions according to the type of teacher involved and the grade:

- Teacher with position compared to a teacher with a position and a lesser grade
- Teacher with position with respect to a new teacher with a lesser grade
- New teacher compared to new teacher with a lower grade
- New teacher compared to a teacher with a position and a lower grade

Teacher with position compared to a teacher with a position and a lesser grade:

$$\begin{aligned}
&\forall k, k'/k \neq k' \& \mathrm{GT}_k > \mathrm{GT}_{k'}, i: \\
&\text{IF } P_k = i \text{ AND } P_{k'} = i \text{ AND } \alpha_{ki} = 0 \text{ THEN } \alpha_{k'i} = 0 \\
&\Rightarrow f_{45} \Rightarrow \forall k, k'/k \neq k' \& \mathrm{GT}_k > \mathrm{GT}_{k'}, i/P_k = i \ \& \ P_{k'} = i: \\
&\text{IF } \alpha_{ki} = 0 \text{ THEN } \alpha_{k'i} = 0 \\
&\Rightarrow f_7, f_{13} \Rightarrow \forall k, k'/k \neq k' \& \mathrm{GT}_k > \mathrm{GT}_{k'}, i/P_k = i \ \& \ P_{k'} = i: \\
&1 - \alpha_{ki} \leq 1 - \alpha_{k'i}
\end{aligned} \tag{48.6}$$

Teacher with position with respect to a new teacher with a lesser grade

$$\begin{aligned}
&\forall j, k/\mathrm{GN}_j < \mathrm{GT}_k, i: \text{IF } P_k = i \quad Y \ \alpha_{ki} = 0 \text{ THEN } \beta_{ji} = 0 \\
&\Rightarrow f_{45} \Rightarrow \forall j, k/\mathrm{GN}_j < \mathrm{GT}_k, i/P_k = i: \text{IF } \alpha_{ki} = 0 \text{ THEN } \beta_{ji} = 0 \\
&\Rightarrow f_7, f_{13} \Rightarrow \forall j, k/\mathrm{GN}_j < \mathrm{GT}_k, i/P_k = i: 1 - \alpha_{ki} \leq 1 - \beta_{ji}
\end{aligned} \tag{48.7}$$

New teacher compared to a new teacher with a lower grade

$$
\begin{aligned}
&\forall j, j'/j \neq j' \& \mathrm{GN}_j > \mathrm{GN}_{j'}, i, e:\\
&\text{IF } \mathrm{PO}_{je} = i \text{ AND } \mathrm{PO}_{j'e} = i \text{ AND } \sum_{e'=1}^{e-1} \beta_{j\mathrm{PO}_{je'}} = 0 \text{ THEN } \beta_{j'i} = 0\\
&\Rightarrow f_{45} \Rightarrow \forall j, j'/j \neq j' \& \mathrm{GN}_j > \mathrm{GN}_{j'}, i, e/\& \mathrm{PO}_{je} = i \ \& \mathrm{PO}_{j'e} = i:\\
&\text{IF } \sum_{e'=1}^{e-1} \beta_{j\mathrm{PO}_{je'}} = 0 \text{ THEN } \beta_{j'i} = 0\\
&\Rightarrow f_7, f_{13} \left(\sum_{e'=1}^{e-1} \beta_{j\mathrm{PO}_{je'}} \text{ binary expression} \right) \Rightarrow\\
&\forall j, j'/j \neq j' \& \mathrm{GN}_j > \mathrm{GN}_{j'}, i, e/\mathrm{PO}_{je} = i \ \& \mathrm{PO}_{j'e} = i:\\
&1 - \sum_{e'=1}^{e-1} \beta_{j\mathrm{PO}_{je'}} \leq 1 - \beta_{j'i}
\end{aligned}
\tag{48.8}
$$

New teacher compared to a teacher with a position and a lower grade

$$
\begin{aligned}
&\forall j, k/\mathrm{GN}_j > \mathrm{GT}_k, i, e:\\
&\text{IF } \mathrm{PO}_{je} = i \text{ AND } P_k = i \text{ AND } \sum_{e'=1}^{e-1} \beta_{j\mathrm{PO}_{je'}} = 0 \text{ THEN } \alpha_{ki} = 0\\
&\Rightarrow f_{45} \Rightarrow \forall j, k/\mathrm{GN}_j > \mathrm{GT}_k, i, e/\mathrm{PO}_{je} = i \ \& \ P_k = i:\\
&\text{IF } \sum_{e'=1}^{e-1} \beta_{j\mathrm{PO}_{je'}} = 0 \text{ THEN } \alpha_{ki} = 0\\
&\Rightarrow f_7, f_{13} \Rightarrow \forall j, k/\mathrm{GN}_j > \mathrm{GT}_k, \ i, e/\mathrm{PO}_{je} = i \ \& \ P_k = i:\\
&1 - \sum_{e'=1}^{e-1} \beta_{j\mathrm{PO}_{je'}} \leq 1 - \alpha_{ki}
\end{aligned}
\tag{48.9}
$$

Objective Criterion
Maximize allocation satisfaction.
Satisfaction of teachers with position:
Auxiliary Calculation

$$
\forall k: y_k = \text{Satisfaction of teacher } k
$$

$$
\forall k: y_k = S_4 \alpha_{kP_k} \tag{48.10}
$$

Satisfaction of new teachers:

First, we calculate the shortest distance from the position assigned to the requested positions. If the maximum distance had been measured in order to assess a low value, an upper bound calculation would have been enough since the objective function would have adjusted the value of the bound to the maximum. The modeling would have been as follows:

Upper Bound Calculation

$\forall j : x_j^{\text{MAX}}$ = Upper bound of the distance from the position assigned to the new teacher j

$$\forall i, j, e/\text{PO}_{je} > 0 : x_j^{\text{MAX}} \geq D_{\text{PO}_{je}i}\beta_{ji}$$

However, when measuring the shortest distance between the requests of the new teacher, that strategy is not valid, since the objective function would adjust it to 0. For this case, I propose a strategy based on a simple auxiliary calculation that is associated with a calculation of data:

Auxiliary Calculation

$$\forall j : x_j^{\text{MIN}} = \text{Minimum distance from teacher } j$$

$$\forall j : x_j^{\text{MIN}} = \sum_{i=1}^{n} \min_{e/\text{PO}_{je}>0} \left(D_{i\text{PO}_{je}}\right)\beta_{ji} \tag{48.11}$$

This strategy could also have been used in the case of working with the maximum distance.

We also need the requested position that has turned out to have the shortest distance with the allocated position, since the satisfaction value of the order of priority of this requested place must be subtracted x_j^{MIN}.

The difficulty arises because the minimum distance may coincide in more than one of the requested positions. In that case, we will use the one with the greatest satisfaction as the closest one. We can calculate the requested position with a binary logical calculation:

Binary Logical Calculation

$$\forall j, i, e : \delta_{jie} = 1 \text{ if the requested position in priority order } e \text{ has the shortest distance with the position } i \text{ assigned to teacher } j. \tag{48.12}$$

$$\forall j, i, e = 1 : \delta_{jie} = 1 \text{ IF AND ONLY IF } \beta_{ji} = 1 \text{ AND } D_{i\text{PO}_{je}} = \min_{e'/\text{PO}_{je'}>0}\left(D_{i\text{PO}_{je'}}\right)$$

$$\forall j,i,e>1:\delta_{jie}=1 \text{ IF AND ONLY IF}$$

$$\beta_{ji}=1 \text{ AND } D_{i\mathrm{PO}_{je}}=\min_{e'/\mathrm{PO}_{je'}>0}\left(D_{i\mathrm{PO}_{je'}}\right) \text{ AND } \left(D_{i\mathrm{PO}_{je''}}>D_{i\mathrm{PO}_{je}}\right)_{\forall e''<e}$$

Modeling:

$$\Rightarrow f_{S_V}\Rightarrow \forall j,i,e=1:\text{IF } \delta_{jie}=1 \text{ THEN } \beta_{ji}=1 \text{ AND}$$
$$D_{iPO_{je}}=\min_{e'/\mathrm{PO}_{je'}>0}\left(D_{i\mathrm{PO}_{je'}}\right)$$
$$\Rightarrow f_{S_V}\Rightarrow \forall j,i,e>1:$$
$$\text{IF } \delta_{jie}=1 \text{ THEN } \beta_{ji}=1 \text{ AND } D_{i\mathrm{PO}_{je}}=\min_{e'/\mathrm{PO}_{je'}>0}\left(D_{i\mathrm{PO}_{je'}}\right) \text{ AND}$$
$$\left(D_{i\mathrm{PO}_{je''}}>D_{i\mathrm{PO}_{je}}\right)_{\forall e''<e}$$

$$\Rightarrow f_3\Rightarrow \forall j,i,e=1:$$
$$\text{IF NOT}\left(\beta_{ji}=1 \text{ AND } D_{i\mathrm{PO}_{je}}=\min_{e'/\mathrm{PO}_{je'}>0}\left(D_{i\mathrm{PO}_{je'}}\right)\right) \text{ THEN } \delta_{jie}=0$$
$$\Rightarrow f_3\Rightarrow \forall j,i,e>1:$$
$$\text{IF NOT}\left(\beta_{ji}=1 \text{ AND } D_{i\mathrm{PO}_{je}}=\min_{e'/\mathrm{PO}_{je'}>0}\left(D_{i\mathrm{PO}_{je'}}\right) \text{ AND } \left(D_{i\mathrm{PO}_{je''}}>D_{i\mathrm{PO}_{je}}\right)_{\forall e''<e}\right)$$
$$\text{THEN } \delta_{jie}=0$$
$$\Rightarrow \forall j,i,e=1:\text{IF } \beta_{ji}=0 \text{ OR}$$
$$D_{i\mathrm{PO}_{je}}\neq\min_{e'/\mathrm{PO}_{je'}>0}\left(D_{i\mathrm{PO}_{je'}}\right) \text{ THEN } \delta_{jie}=0\Rightarrow f_{40}\Rightarrow \forall j,i,$$
$$e=1:\text{IF } \beta_{ji}=0 \text{ THEN } \delta_{jie}=0\Rightarrow f_3,f_{13}\Rightarrow \forall j,i,e=1:\delta_{jie}\leq\beta_{ji} \tag{48.13}$$

$$\Rightarrow f_{40}\Rightarrow:\text{IF } D_{i\mathrm{PO}_{je}}\neq\min_{e'/\mathrm{PO}_{je'}>0}\left(D_{i\mathrm{PO}_{je'}}\right) \text{ THEN } \delta_{jie}=0$$
$$\Rightarrow f_{45}\Rightarrow \forall j,i,e=1/D_{i\mathrm{PO}_{je}}\neq\min_{e'/\mathrm{PO}_{je'}>0}\left(D_{i\mathrm{PO}_{je'}}\right):\delta_{jie}=0 \tag{48.14}$$

$\Rightarrow \forall j, i, e > 1:$

IF $\beta_{ji} = 0$ OR $D_{iPO_{je}} \neq \min_{e'/PO_{je'}>0}\left(D_{iPO_{je'}}\right)$ OR NOT$\left(\left(D_{iPO_{je''}} > D_{iPO_{je}}\right)_{\forall e''<e}\right)$

THEN $\delta_{jie} = 0$

$\Rightarrow f_{40} \Rightarrow \forall j, i, e > 1:$ IF $\beta_{ji} = 0$ THEN $\delta_{jie} = 0$

$\Rightarrow f_3, f_{13} \Rightarrow \forall j, i, e > 1: \delta_{jie} \leq \beta_{ji}$

$$(48.15)$$

$$\Rightarrow f_{40} \Rightarrow: \text{IF } D_{iPO_{je}} \neq \min_{e'/PO_{je'}>0}\left(D_{iPO_{je'}}\right) \; THEN \;\; \delta_{jie} = 0$$

$$\Rightarrow f_{45} \Rightarrow \forall j, i, e > 1/D_{iPO_{je}} \neq \min_{e'/PO_{je'}>0}\left(D_{iPO_{je'}}\right) : \delta_{jie} = 0 \tag{48.16}$$

$$\Rightarrow f_{40} \Rightarrow: \text{IF NOT}\left(\left(D_{iPO_{je''}} > D_{iPO_{je}}\right)_{\forall e''<e}\right) \text{ THEN } \delta_{jie} = 0$$

$$\Rightarrow f_{45} \Rightarrow \forall j, i, e/e > 1\&\text{NOT}\left(\left(D_{iPO_{je''}} > D_{iPO_{je}}\right)_{\forall e''<e}\right) : \delta_{jie} = 0 \tag{48.17}$$

And finally, with an auxiliary calculation we obtain the satisfaction value.

Auxiliary Calculation

$$\forall j : z_j = \text{Satisfaction of the new teacher } j$$

$$\forall j : z_j = \sum_{e/PO_{je}>0} S_e \sum_{i=1}^{n} \delta_{jie} - x_j^{MIN} \tag{48.18}$$

Note that if the assigned post coincides with one of the requested posts, automatically the minimum distance is 0; therefore, x_j^{MIN} is canceled and δ_{jie} takes a value of 1 in the requested post that coincides with the assigned post.

Objective Function

$$\text{Max} \sum_{k=1}^{r} y_k + \sum_{j=1}^{m} z_j \tag{48.19}$$

Complete Model

$$\text{Max} \sum_{k=1}^{r} y_k + \sum_{j=1}^{m} z_j \tag{48.19}$$

s.t.

$$\forall k: \sum_{i/i=O_k \ \vee \ i=P_k} \alpha_{ki} = 1 \tag{48.3}$$

$$\forall j: \sum_{i=1}^{n} \beta_{ji} = 1 \tag{48.4}$$

$$\forall i: \sum_{k/O_k=i \ \vee \ P_k=i} \alpha_{ki} + \sum_{j=1}^{m} \beta_{ji} = 1 \tag{48.5}$$

$$\forall k, k'/k \neq k' \& \mathrm{GT}_k > \mathrm{GT}_{k'}, i/P_k = i \ \& \ P_{k'} = i : 1 - \alpha_{ki} \leq 1 - \alpha_{k'i} \tag{48.6}$$

$$\forall j, k/\mathrm{GN}_j < \mathrm{GT}_k, i/P_k = i : 1 - \alpha_{ki} \leq 1 - \beta_{ji} \tag{48.7}$$

$$\forall j, j'/j \neq j' \& \mathrm{GN}_j > \mathrm{GN}_{j'}, i, e/\mathrm{PO}_{je} = i \ \& \mathrm{PO}_{j'e} = i : 1 - \sum_{e'=1}^{e-1} \beta_{j\mathrm{PO}_{je'}} \leq 1 - \beta_{j'i} \tag{48.8}$$

$$\forall j, k/\mathrm{GN}_j > \mathrm{GT}_k, i, e/\mathrm{PO}_{je} = i \ \& \ P_k = i : 1 - \sum_{e'=1}^{e-1} \beta_{j\mathrm{PO}_{je'}} \leq 1 - \alpha_{ki} \tag{48.9}$$

$$\forall k: y_k = S_4 \alpha_{kA_k} \tag{48.10}$$

$$\forall j: x_j^{\mathrm{MIN}} = \sum_{i=1}^{n} \min_{e/\mathrm{PO}_{je}>0} \left(D_{i\mathrm{PO}_{je}}\right) \beta_{ji} \tag{48.11}$$

$$\forall j, i, e: \delta_{jie} \leq \beta_{ji} \tag{48.13, 48.15}$$

$$\forall j, i, e/D_{i\mathrm{PO}_{je}} \neq \min_{e'/\mathrm{PO}_{je'}>0} \left(D_{i\mathrm{PO}_{je'}}\right) : \delta_{jie} = 0 \tag{48.14, 48.16}$$

$$\forall j, i, e/e > 1 \& \mathrm{NOT}\left(\left(D_{i\mathrm{PO}_{je''}} > D_{i\mathrm{PO}_{je}}\right)_{\forall e''<e}\right) : \delta_{jie} = 0 \tag{48.17}$$

$$\forall j: z_j = \sum_{e/\mathrm{PO}_{je}>0} S_e \sum_{i=1}^{n} \delta_{jie} - x_j^{\mathrm{MIN}} \tag{48.18}$$

$$\forall k, i/i = O_k \ \vee \ i = P_k : \alpha_{ki} \ \text{binarias} \tag{48.1}$$

$$\forall j, i: \beta_{ji} \ \text{binaries} \tag{48.2}$$

$$\forall j, i, e: \delta_{jie} \ \text{binaries} \tag{48.12}$$

$$\forall k: y_k \geq 0$$

$$\forall j : x_j^{\text{MIN}} \geq 0$$

$$\forall j : z_j \geq 0$$

Implementation in LINGO

```
MODEL:
 DATA:
  r=@OLE('C:\Files_\48_Excel.xlsx','rr');
  m=@OLE('C:\Files_\48_Excel.xlsx','m');
 ENDDATA
 CALC:
  n=m+r;
 ENDCALC
SETS:
 Positions/1..n/;
 TeachersP/1..r/:O,P,GT,y;
 NewTeachers/1..m/:GN,z,xMIN;
 Order/1..3/:S;
 TPP(TeachersP,Positions):Alpha;
 NTP(NewTeachers,Positions):Beta;
 ONT(Order,NewTeachers):PO;
 PP(Positions,Positions):D;
 NTPO(NewTeachers,Positions,Order):Delta;
ENDSETS
DATA:
 O=@OLE('C:\Files_\48_Excel.xlsx','O');
 P=@OLE('C:\Files_\48_Excel.xlsx','P');
 GT=@OLE('C:\Files_\48_Excel.xlsx','GT');
 GN=@OLE('C:\Files_\48_Excel.xlsx','GN');
 PO=@OLE('C:\Files_\48_Excel.xlsx','PO');
 D=@OLE('C:\Files_\48_Excel.xlsx','Distance');
 S =@OLE('C:\Files_\48_Excel.xlsx','S');
 S4 =@OLE('C:\Files_\48_Excel.xlsx','SysS4');
 @OLE('C:\Files_\48_Excel.xlsx','Sol_T_P')=@WRITEFOR(TeachersP(k):
@SUM(Positions(i):i*Alpha(k,i)));
@OLE('C:\Files_\48_Excel.xlsx','Sol_NT')=@WRITEFOR(NewTeachers
(j):@SUM(Positions(i):i*Beta(j,i)));
ENDDATA
CALC:
@FOR(Positions(i):@FOR(Positions(ii)|i#LT#ii:D(ii,i)=D(i,ii)));
ENDCALC
!48.19;
MAX = @SUM(TeachersP(k):y(k))+@SUM(NewTeachers(j):z(j));
!48.3;
@FOR(TeachersP(k):@SUM(Positions(i)|i#EQ#O(k) #OR# i#EQ#P(k):Alpha
(k,i))=1);
!48.4;
@FOR(NewTeachers(j):@SUM(Positions(i):Beta(j,i))=1);
!48.5;
@FOR(Positions(i):@SUM(TeachersP(k)|i#EQ#O(k) #OR# i#EQ#P(k):Alpha
(k,i)) + @SUM(NewTeachers(j):Beta(j,i))=1);
!48.6;
@FOR(TeachersP(k):@FOR(TeachersP(kk)|k#NE#kk#AND#GT(k)#GT#GT(kk):
```

```
@FOR(Positions(i)|P(k)#EQ#i#AND#P(kk)#EQ#i: 1-Alpha(k,i)<=1-Alpha
(kk,i))));
!48.7;
@FOR(NewTeachers(j):@FOR(TeachersP(k)|GN(j)#LT#GT(k):@FOR
(Positions(i)|P(k)#EQ#i:1-Alpha(k,i)<=1-Beta(j,i))));
!48.8;
@FOR(NewTeachers(j):@FOR(NewTeachers(jj)|j#NE#jj#AND#GN(j)#GT#GN
(jj):@FOR(Positions(i):@FOR(Order(e)|e#EQ#1#AND#PO(e,j)
#EQ#i#AND#PO(e,jj)#EQ#i:1<=1-Beta(jj,i))))));
@FOR(NewTeachers(j):@FOR(NewTeachers(jj)|j#NE#jj#AND#GN(j)#GT#GN
(jj):@FOR(Positions(i):@FOR(Order(e)|e#GT#1#AND#PO(e,j)
#EQ#i#AND#PO(e,jj)#EQ#i: 1-@SUM(Order(ee)|ee#LE#e-1:Beta(j,PO(ee,
j)))<=1-Beta(jj,i))))));
!48.9;
@FOR(NewTeachers(j):@FOR(TeachersP(k)|GN(j)#GT#GT(k):@FOR
(Positions(i):@FOR(Order(e)|e#EQ#1#AND#PO(e,j)#EQ#i#AND#P(k)
#EQ#i:1<=1-Alpha(k,i))))));
@FOR(NewTeachers(j):@FOR(TeachersP(k)|GN(j)#GT#GT(k):@FOR
(Positions(i):@FOR(Order(e)|e#GT#1#AND#PO(e,j)#EQ#i#AND#P(k)
#EQ#i:1-@SUM(Order(ee)|ee#LE#e-1:Beta(j,PO(ee,j)))<=1-Alpha(k,
i))))));
!48.10;
@FOR(TeachersP(k):y(k)=S4*Alpha(k,P(k)));
!48.11;
@FOR(NewTeachers(j):xMIN(j)=@SUM(Positions(i):@MIN(Order(e)|PO(e,
j)#GT#0:D(i,PO(e,j)))*Beta(j,i)));
!48.13;!48.15;
@FOR(NewTeachers(j):@FOR(Positions(i):@FOR(Order(e): Delta(j,i,e)
<=Beta(j,i))));
!48.14;!48.16;
@FOR(NewTeachers(j):@FOR(Order(e)|PO(e,j)#GT#0:@FOR(Positions(i)|
D(i,PO(e,j))#NE#@MIN(Order(ee)|PO(ee,j)#GT#0:D(i,PO(ee,j)))):Delta
(j,i,e)=0)));
!48.17;
@FOR(NewTeachers(j):@FOR(Order(e)|e#GT#1 #AND# PO(e,j)#GT#0:@FOR
(Order(ee)|ee#LT#e #AND# PO(ee,j)#GT#0:@FOR(Positions(i)| D(i,PO
(ee,j))#LT#D(i,PO(e,j)):Delta(j,i,e)=0))));
!48.18;
@FOR(NewTeachers(j):z(j)=@SUM(Order(e)|PO(e,j)#GT#0:S(e)*@SUM
(Positions(i):Delta(j,i,e)))-xMIN(j));
!48.1;
@FOR(TeachersP(k):@FOR(Positions(i)|i#EQ#O(k) #OR# i#EQ#P(k):@BIN
(Alpha(k,i))));
!48.2;
@FOR(NewTeachers(j):@FOR(Positions(i):@BIN(Beta(j,i))));
!48.12;
@FOR(NewTeachers(j):@FOR(Positions(i):@FOR(Order(e): @BIN(Delta(j,
i,e)))));
END
```

Solution

It is shown in Fig. 7.18.

```
Solution Report - 48_LingoE
Global optimal solution found.
Objective value:                              85207.00
Objective bound:                              85207.00
Infeasibilities:                              0.000000
Extended solver steps:                               0
Total solver iterations:                           156
Elapsed runtime seconds:                          1.52


Export Summary Report
---------------------
Transfer Method:        OLE BASED
Workbook:               C:\Files_\48_Excel.xlsx
Ranges Specified:               1
    Sol_T_P
Ranges Found:                   1
Range Size Mismatches:          0
Values Transferred:            12
```

Fig. 7.18 Problem 48: LINGO solution report

7.13 Problem 49. Placement of Tables

We are going to create a model to distribute a set of rectangular tables in a room.

We have 15 tables, each one with a length and a width, and a number of seats known. We consider the width as the shortest side of the table.

We have a rectangular room with known length and width.

The tables can be placed in two positioning modes (Fig. 7.19): The width of the table parallel to the width of the room (Mode 1) or the width of the table parallel to the length of the room (Mode 2).

When placing the tables, one must take into account that:

- Between two tables placed there must be at least 2 m, so that the staff can move, on any side of the table.
- Between each table placed and the wall, at least one meter.

Maximize the number of seats that can be placed.

Modeling

To position an object in space is to measure its position with respect to each axis; therefore, the placement of a vertex of the table implies two decision variables, one per axis.

We can work only with measurable decisions, measuring the position of two vertices of the table and imposing in the specifications that the position of the table placed must be parallel to the width or length of the room. These would really be the primary decisions of the problem with the greatest flexibility.

However, we can use an evolution of those decisions, that of working with the position of a single vertex and introducing logical decisions about the positioning mode.

Fig. 7.19 Placement of tables in the room

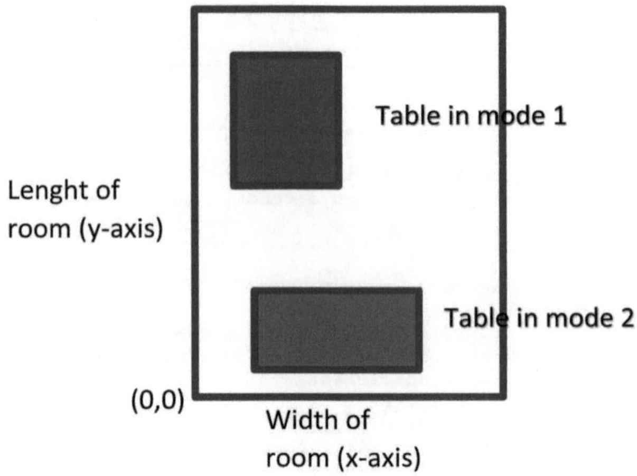

The list of elements is shown in Table 7.14.

Table 7.14 Problem 49: table of elements

			DATA				
Elements	Set	QN	Name	Parameter	Type	Belonging	Value
Tables	$i = 1 \ldots 15$	I_M	Length	L_i	C(m)	W	–
			Width	W_i	C(m)	W	–
			Seats	S_i	I	W	–
Room	–	I_U	Length	LR	C(m)	W	–
			Width	WR	C(m)	W	–
Position modes	$k = 1, 2$	I_U					

Decision Activities

Place tables (i)

$\forall i : x_i =$ position of Table i on the x-axis (with respect to the width of the room)

$\forall i : y_i =$ position of Table i on the y-axis (with respect to the length of the room)

Assign modes (k) to tables (i)

$$\forall i, k : \alpha_{ik} = 1 \text{ if we assign Table } i \text{ in mode } k; 0 \text{ otherwise.} \quad (49.1)$$

Specifications

– Quantitative selection specification:

Activities	Elements selecting	Selectable elements	Selection	Constraints
$\forall i, k : \alpha_{ik}$	Tables ($i = 1\ldots n$)	Modes ($k = 1, 2$)	≤ 1	$\forall i : \sum_{k=1}^{2} \alpha_{ik} \leq 1$ (49.2)

- Implicit relationship between variables: We could establish that a table without positioning mode $\left(\sum_{k=1}^{2} \alpha_{ik} = 0\right)$ cannot be placed in the room ($x_i = y_i = 0$), but it is not necessary to establish this specification. If the table does not have a mode assigned, it will not be valued in the objective function, and the positioning specifications will not be established on them. Therefore, even if their variables x_i and y_i took positive values, they would be irrelevant in the solution.

However, if you want to make sure that the values of x_i and y_i are null for the unplaced tables, you can model this specification.

Between two placed tables, there must be at least 2 m on either side of the table (propositional logic specification).

$$\forall i,j/i < j : \text{IF } \sum_{k=1}^{2} \alpha_{ik} = 1 \text{ AND } \sum_{k=1}^{2} \alpha_{jk} = 1 \text{ THEN}$$
$$\left(\left(x_i + W_i\alpha_{i1} + L_i\alpha_{i2} - x_j\right) \leq -2 \text{ OR } \left(x_j + W_j\alpha_{j1} + L_j\alpha_{j2} - x_i\right) \leq -2\right) \text{ OR}$$
$$\left(\left(y_i + L_i\alpha_{i1} + W_i\alpha_{i2} - y_j\right) \leq -2 \text{ OR } \left(y_j + L_j\alpha_{j1} + W_j\alpha_{j2} - y_i\right) \leq -2\right)$$

The input of the proposition expresses that Tables i and j have been placed, in either of the two possible modes.

The output expresses all the placement possibilities:

Table i in mode 1 and Table j in mode 1:

$$\left(\left(x_i + W_i - x_j\right) \leq -2 \text{ OR } \left(x_j + W_j - x_i\right) \leq -2\right)$$
$$\text{OR}\left(\left(y_i + L_i - y_j\right) \leq -2 \text{ OR } \left(y_j + L_j - y_i\right) \leq -2\right)$$

The proposition expresses that we can allow the overlap with respect to one axis, but not with respect to both. On the other hand, we need to consider the two possibilities in relation to which table goes first on each axis.

The rest of the cases would be extracted from the general proposition in an equivalent way:

Table i in mode 1 and Table j in mode 2
Table i in mode 2 and Table j in mode 1
Table i in mode 2 and Table j in mode 2

Obviously, each possible situation could have been proposed independently.

Modeling:

$$\Rightarrow f_{37}\left(\sum_{k=1}^{2}\alpha_{ik}\text{ binary expresion}\right) \Rightarrow \forall i,j/i<j:\text{IF }\sum_{k=1}^{2}\alpha_{ik}+\sum_{k=1}^{2}\alpha_{jk}\geq 2\text{ THEN}$$
$$\left(\left(x_i+W_i\alpha_{i1}+L_i\alpha_{i2}-x_j\right)\leq -2\ O\ \left(x_j+W_j\alpha_{j1}+L_j\alpha_{j2}-x_i\right)\leq -2\right)\ O$$
$$\left(\left(y_i+L_i\alpha_{i1}+W_i\alpha_{i2}-y_j\right)\leq -2\ O\ \left(y_j+L_j\alpha_{j1}+W_j\alpha_{j2}-y_i\right)\leq -2\right)$$
$$\Rightarrow \forall i,j/i<j:\omega_{ij}^{1}=1\text{ IF AND ONLY IF }\left(x_i+W_i\alpha_{i1}+L_i\alpha_{i2}-x_j\right)\leq -2 \tag{49.3}$$

$$\Rightarrow \forall i,j/i<j:\omega_{ij}^{2}=1\text{ IF AND ONLY IF }\left(x_j+W_j\alpha_{j1}+L_j\alpha_{j2}-x_i\right)\leq -2 \tag{49.4}$$

$$\Rightarrow \forall i,j/i<j:\omega_{ij}^{3}=1\text{ IF AND ONLY IF }\left(y_i+L_i\alpha_{i1}+W_i\alpha_{i2}-y_j\right)\leq -2 \tag{49.5}$$

$$\Rightarrow \forall i,j/i<j:\omega_{ij}^{4}=1\text{ IF AND ONLY IF }\left(y_j+L_j\alpha_{j1}+W_j\alpha_{j2}-y_i\right)\leq -2 \tag{49.6}$$

$$\Rightarrow f_{S_V}\Rightarrow \forall i,j/i<j:\text{IF }\omega_{ij}^{1}=1\text{ THEN }\left(x_i+W_i\alpha_{i1}+L_i\alpha_{i2}-x_j\right)\leq -2$$
$$\Rightarrow f_{15}\left(\text{UB}_{x_i+W_i\alpha_{i1}+L_i\alpha_{i2}-x_j}=\text{WR}-W_j-1\right)\Rightarrow$$
$$\forall i,j/i<j:x_i+W_i\alpha_{i1}+L_i\alpha_{i2}-x_j\leq -2+(WR-W_i-1)\left(1-\omega_{ij}^{1}\right) \tag{49.7}$$

$$\Rightarrow f_{S_V}\Rightarrow \forall i,j/i<j:\text{IF }\omega_{ij}^{2}=1\text{ THEN }\left(x_j+W_j\alpha_{j1}+L_j\alpha_{j2}-x_i\right)\leq -2$$
$$\Rightarrow f_{15}\left(\text{UB}_{x_j+W_j\alpha_{j1}+L_j\alpha_{j2}-x_i}=\text{WR}-W_j-1\right)\Rightarrow$$
$$\forall i,j/i<j:x_j+W_j\alpha_{j1}+L_j\alpha_{j2}-x_i\leq -2\omega_{ij}^{2}+\left(\text{WR}-W_j-1\right)\left(1-\omega_{ij}^{2}\right) \tag{49.8}$$

$$\Rightarrow f_{S_V}\Rightarrow \forall i,j/i<j:\text{IF }\omega_{ij}^{3}=1\text{ THEN }\left(y_i+L_i\alpha_{i1}+W_i\alpha_{i2}-y_j\right)\leq -2$$
$$\Rightarrow f_{15}\left(\text{UB}_{y_i+L_i\alpha_{i1}+W_i\alpha_{i2}-y_j}=\text{LR}-W_i-1\right)\Rightarrow$$
$$\forall i,j/i<j:y_i+L_i\alpha_{i1}+W_i\alpha_{i2}-y_j\leq -2\omega_{ij}^{3}+(\text{LR}-W_i-1)\left(1-\omega_{ij}^{3}\right) \tag{49.9}$$

$$\Rightarrow f_{S_V}\Rightarrow \forall i,j/i<j:IF\ \omega_{ij}^{4}=1\text{ THEN }\left(y_j+L_j\alpha_{j1}+W_j\alpha_{j2}-y_i\right)\leq -2$$
$$\Rightarrow f_{15}\left(\text{UB}_{y_j+L_j\alpha_{j1}+W_j\alpha_{j2}-y_i}=LR-W_j-1\right)\Rightarrow$$
$$\forall i,j/i<j:y_j+L_j\alpha_{j1}+W_j\alpha_{j2}-y_i\leq -2\omega_{ij}^{4}+\left(\text{LR}-W_j-1\right)\left(1-\omega_{ij}^{4}\right) \tag{49.10}$$

$$\Rightarrow \forall i,j/i<j : \text{IF } \sum_{k=1}^{2} \alpha_{ik} + \sum_{k=1}^{2} \alpha_{jk} \geq 2 \text{ THEN } \omega_{ij}^1 + \omega_{ij}^2 + \omega_{ij}^3 + \omega_{ij}^4 \geq 1$$

$$\Rightarrow f_{18}\left(\text{UB}_{\sum_{k=1}^{2}\alpha_{ik}+\sum_{k=1}^{2}\alpha_{jk}} = 2\right) \Rightarrow \forall i,j/i<j : \sum_{k=1}^{2} \alpha_{ik} + \sum_{k=1}^{2} \alpha_{jk} \leq 1 + \lambda_{ij} \tag{49.11}$$

$$\Rightarrow f_{21}\left(\text{LB}_{\omega_{ij}^1+\omega_{ij}^2+\omega_{ij}^3+\omega_{ij}^4} = 0\right) \Rightarrow \forall i,j/i<j : \omega_{ij}^1 + \omega_{ij}^2 + \omega_{ij}^3 + \omega_{ij}^4 \geq \lambda_{ij} \tag{49.12}$$

– Between each placed table and the wall, at least one meter.

We could dispense with this specification if we had removed the 1-m separation contour from the room. Anyway, we keep it so as not to lose modeling possibilities.

$$\begin{aligned} &\forall i : \text{IF } \sum_{k=1}^{2} \alpha_{ik} = 1 \text{ THEN } x_i \geq 1 \text{ AND } y_i \geq 1 \text{ AND} \\ &x_i + W_i\alpha_{i1} + L_i\alpha_{i2} \leq \text{WR} - 1 \\ &\text{AND } y_i + L_i\alpha_{i1} + W_i\alpha_{i2} \leq \text{LR} - 1 \end{aligned}$$

The last two conditions can be carried out directly without an input condition. The first two not, since values below 1 in x and y will mean that the table has not been placed:

$$\forall i : x_i + W_i\alpha_{i1} + L_i\alpha_{i2} \leq \text{WR} - 1 \tag{49.13}$$

$$\forall i : \ y_i + L_i\alpha_{i1} + W_i\alpha_{i2} \leq \text{LR} - 1 \tag{49.14}$$

$$\Rightarrow f_{41} \Rightarrow \forall i : \text{IF } \sum_{k=1}^{2} \alpha_{ik} = 1 \text{ THEN } x_i \geq 1$$

$$\Rightarrow f_{15}(\text{LB}_{x_i} = 0) \Rightarrow \forall i : x_i \geq \sum_{k=1}^{2} \alpha_{ik} \tag{49.15}$$

$$\Rightarrow f_{41} \Rightarrow \forall i : \text{IF } \sum_{k=1}^{2} \alpha_{ik} = 1 \text{ THEN } y_i \geq 1$$

$$\Rightarrow f_{15}(\text{LB}_{y_i} = 0) \Rightarrow \forall i : y_i \geq \sum_{k=1}^{2} \alpha_{ik} \tag{49.16}$$

Objective Function

$$\text{Max} \quad \sum_{i=1}^{15} S_i \sum_{k=1}^{2} \alpha_{ik} \tag{49.17}$$

Complete Model

$$\text{Max} \quad \sum_{i=1}^{15} S_i \sum_{k=1}^{2} \alpha_{ik} \tag{49.17}$$

s.t.

$$\forall i : \sum_{k=1}^{2} \alpha_{ik} \leq 1 \tag{49.2}$$

$$\forall i,j/i<j : x_i + W_i\alpha_{i1} + L_i\alpha_{i2} - x_j \leq -2 + (\text{WR} - W_i - 1)\left(1 - \omega_{ij}^1\right) \tag{49.7}$$

$$\forall i,j/i<j : x_j + W_j\alpha_{j1} + L_j\alpha_{j2} - x_i \leq -2\omega_{ij}^2 + \left(\text{WR} - W_j - 1\right)\left(1 - \omega_{ij}^2\right) \tag{49.8}$$

$$\forall i,j/i<j : y_i + L_i\alpha_{i1} + W_i\alpha_{i2} - y_j \leq -2\omega_{ij}^3 + (\text{LR} - W_i - 1)\left(1 - \omega_{ij}^3\right) \tag{49.9}$$

$$\forall i,j/i<j : y_j + L_j\alpha_{j1} + W_j\alpha_{j2} - y_i \leq -2\omega_{ij}^4 + \left(\text{LR} - W_j - 1\right)\left(1 - \omega_{ij}^4\right) \tag{49.10}$$

$$\forall i,j/i<j : \sum_{k=1}^{2} \alpha_{ik} + \sum_{k=1}^{2} \alpha_{jk} \leq 1 + \lambda_{ij} \tag{49.11}$$

$$\forall i,j/i<j : \omega_{ij}^1 + \omega_{ij}^2 + \omega_{ij}^3 + \omega_{ij}^4 \geq \lambda_{ij} \tag{49.12}$$

$$\forall i : x_i + W_i\alpha_{i1} + L_i\alpha_{i2} \leq \text{WR} - 1 \tag{49.13}$$

$$\forall i : \ y_i + L_i\alpha_{i1} + W_i\alpha_{i2} \leq \text{LR} - 1 \tag{49.14}$$

$$\forall i : x_i \geq \sum_{k=1}^{2} \alpha_{ik} \tag{49.15}$$

$$\forall i : y_i \geq \sum_{k=1}^{2} \alpha_{ik} \tag{49.16}$$

$$\forall i : x_i \geq 0$$

$$\forall i : y_i \geq 0$$

$$\forall i, k : \alpha_{ik} \text{ binaries} \qquad (49.1)$$

$$\forall i,j : \omega_{ij}^1, \omega_{ij}^2, \omega_{ij}^3, \omega_{ij}^4 \text{ binaries} \qquad (49.3,\ 49.4,\ 49.5,\ 49.6)$$

$$\forall i,j : \lambda_{ij} \text{ binaries} \qquad (49.18)$$

Implementation in LINGO

```
MODEL:
SETS:
 Tables/1..15/:L,W,S,x,y;
 Modes/1..2/;
 MM(Tables,Modes): Alpha;
 MeMe(Tables,Tables):Omega1,Omega2,Omega3,Omega4,Lamda;
ENDSETS
DATA:
 L= 2 3 3.5 3 3.5 3 3.5 2 3 3.5 3 3.5 3 3.5 2 3 3.5 3 3.5 3 3.5 3 3.5 3 3.5 ;
 W= 1 1 1.2 1 1.2 1 1.2 1 1 1.2 1 1.2 1 1.2 1 1 1.2 1 1.2 1 1.2 1 1.2 1 1.2;
 S= 4 6 8 6 6 8 4 6 8 4 6 8 4 6 8 4 6 8 4 6 8 4 6 8 8;
 LR=35;
 WR=10;
ENDDATA
!49.17;
MAX = @SUM(Tables(i):S(i)*@SUM(Modes(k):Alpha(i,k)));
!49.2;
@FOR(Tables(i):@SUM(Modes(k):Alpha(i,k))<=1);
!49.7;
@FOR(Tables(i):@FOR(Tables(j)|i#LT#j: x(i)+W(i)*Alpha(i,1)+L(i)
*Alpha(i,2)-x(j)<=-2+(WR-W(i)+-1)*(1-Omega1(i,j))));
!49.8;
@FOR(Tables(i):@FOR(Tables(j)|i#LT#j: x(j)+W(j)*Alpha(j,1)+L(j)
*Alpha(j,2)-x(i)<=-2+(WR-W(j)-1)*(1-Omega2(i,j))));
!49.9;
@FOR(Tables(i):@FOR(Tables(j)|i#LT#j: y(i)+L(i)*Alpha(i,1)+W(i)
*Alpha(i,2)-y(j)<=-2+(LR-W(i)-1)*(1-Omega3(i,j))));
!49.10;
@FOR(Tables(i):@FOR(Tables(j)|i#LT#j: y(j)+L(j)*Alpha(j,1)+W(j)
*Alpha(j,2)-y(i)<=-2+(LR-W(j)-1)*(1-Omega4(i,j))));
!49.11;
@FOR(Tables(i):@FOR(Tables(j)|i#LT#j:@SUM(Modes(k):Alpha(i,k))
+@SUM(Modes(k):Alpha(j,k))<=1+Lamda(i,j)));
!49.12;
@FOR(Tables(i):@FOR(Tables(j)|i#LT#j:Omega1(i,j)+Omega2(i,j)
+Omega3(i,j)+Omega4(i,j)>=Lamda(i,j)));
!49.13;
@FOR(Tables(i):x(i)+W(i)*Alpha(i,1)+L(i)*Alpha(i,2)<=WR-1);
!49.14;
@FOR(Tables(i):y(i)+L(i)*Alpha(i,1)+W(i)*Alpha(i,2)<=LR-1);
!49.15;
@FOR(Tables(i):x(i)>=@SUM(Modes(k):Alpha(i,k)));
!49.16;
@FOR(Tables(i):y(i)>=@SUM(Modes(k):Alpha(i,k)));
```

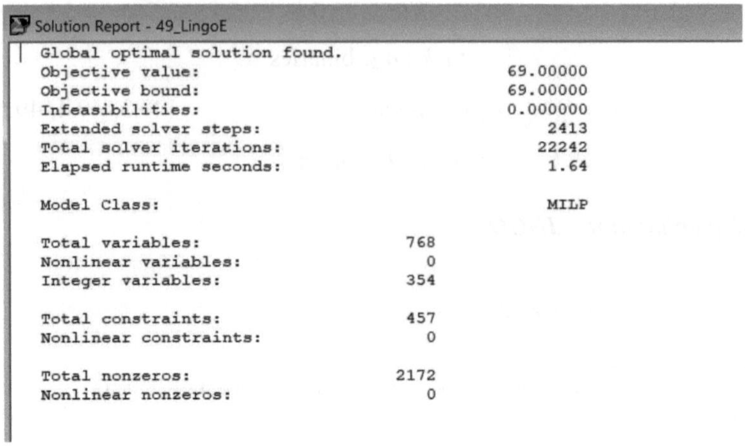

Solution Report - 49_LingoE

```
Global optimal solution found.
Objective value:                              69.00000
Objective bound:                              69.00000
Infeasibilities:                              0.000000
Extended solver steps:                            2413
Total solver iterations:                         22242
Elapsed runtime seconds:                          1.64

Model Class:                                      MILP

Total variables:                    768
Nonlinear variables:                  0
Integer variables:                  354

Total constraints:                  457
Nonlinear constraints:                0

Total nonzeros:                    2172
Nonlinear nonzeros:                   0
```

Fig. 7.20 Problem 49: LINGO solution report

```
!49.1;
@FOR(Tables(i):@FOR(Modes(k):@BIN(Alpha(i,k))));
!49.3;
@FOR(Tables(i):@FOR(Tables(j)|i#LT#j:@BIN(Omega1(i,j))));
!49.4;
@FOR(Tables(i):@FOR(Tables(j)|i#LT#j:@BIN(Omega2(i,j))));
!49.5;
@FOR(Tables(i):@FOR(Tables(j)|i#LT#j:@BIN(Omega3(i,j))));
!49.6;
@FOR(Tables(i):@FOR(Tables(j)|i#LT#j:@BIN(Omega4(i,j))));
!49.18;
@FOR(Tables(i):@FOR(Tables(j)|i#LT#j:@BIN(Lamda(i,j))));
END
```

Solution

The solution is shown in Fig. 7.20.

7.14 Problem 50. Cutting Stock Problem (Cheng et al. 1994)

Given a set of rectangular wooden boards existing in a warehouse and a set of rectangular output boards that we need to obtain from the cutting of the warehouse boards, it is about planning the cuts of the warehouse boards in such a way that we minimize the square meters leftovers from cut warehouse boards, that is, minimize waste. If a warehouse board is not processed, it is not considered.

It is necessary for the model to provide me with how each output board is going to be cut, that is, the cutting scheme of the output boards with respect to the warehouse board from which it is obtained.

To facilitate the actual cutting of the boards, we could impose that the output boards that are cut from a warehouse board are together on at least two sides to other

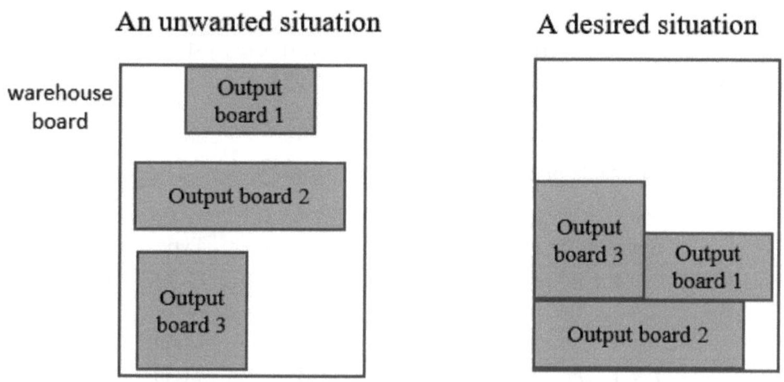

Fig. 7.21 Unwanted situation and desired situation

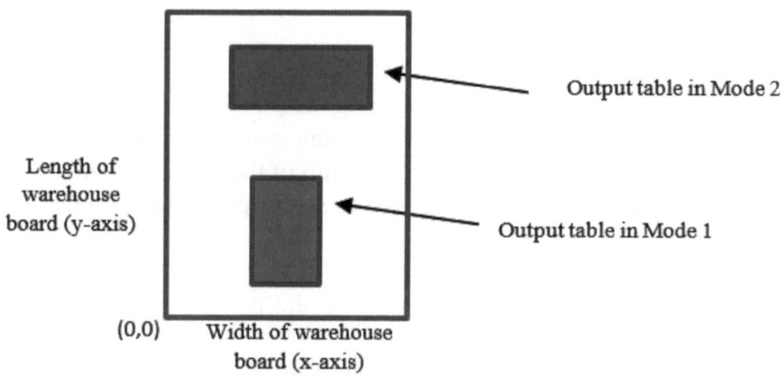

Fig. 7.22 Positioning modes

boards cut from that warehouse board or with two ends of the warehouse board (Fig. 7.21).

We will model this specification; however, we will not include it in the final model. In the model, we will promote this situation within the objective criteria, minimizing the value of the cutting coordinates in addition to the excess meters of the cut warehouse boards.

We know the width and length of each warehouse board. From the output boards, we know the width, length, and number of units.

We consider both for warehouse boards and for output boards, the width as the side with the shortest length.

It is assumed that there are enough warehouse boards to be able to get all the output boards.

Note that the cutting saw consumes 4 mm of the board in the cutting process.

Two possible cutting positions are considered (Fig. 7.22): the width of the output board parallel to the width of the warehouse board (mode 1) and the width of the output board parallel to the length of the warehouse board (mode 2).

Modeling

Each output board to be cut must be individualized in the system. Do not treat boards of the same size as a collective element (Table 7.15).

Table 7.15 Problem 50: table of elements

Elements	Set	QN	Data				
			Name	Parameter	Type	Belonging	Value
Output boards	$i = 1 \ldots n$	I_M	Length	L_i	C (cm)	W	–
			Width	W_i	C(cm)	W	–
Warehouse boards	$j = 1 \ldots m$	I_U	Length	LB_j	C(cm)	W	–
			Width	WB_j	C(cm)	W	–
Cutting modes	$k = 1, 2$	I_U					
System	–	I_U	Cutting width	C	C(cm)	W	0.4

Cutting a board is equivalent to establishing the position of a vertex of the board to be obtained on the warehouse board. This means considering the measurement on each axis. The x-axis will correspond to the width of the warehouse table and the y-axis with length. In addition to that, we need to set the positioning mode (mode 1 or mode 2).

Decision Activities

If we propose cutting of output boards with the participation of all warehouse boards, we shall:

Cut output boards (*i*) from warehouse boards (*j*)

$$\forall i,j : x_{ij} = \text{position of the output board } i \text{ on the } x\text{-axis}$$
$$\text{(relative to the width of the warehouse boards)}$$

$$\forall i,j : y_{ij} = \text{position of the output board } i \text{ on the } y\text{-axis}$$
$$\text{(relative to the length of the warehouse boards)}$$

We know that later in the specifications, it is necessary to impose that an output board is only cut from a warehouse board. Therefore, we can take advantage of this to divide the decision:

We eliminate the participation of the warehouse boards in the cutting process by only measuring the position of the output board, regardless of the warehouse board where it is cut.

We incorporate a logical decision to know on which warehouse board each output board is cut.

This second setup comes from the first. We use it due to the fact that the first setup will need logical calculations to know in which warehouse board each output board has been cut, so we will obtain an equivalence in the variables with respect to the second setup.

Cut output boards (i)

$\forall i : x_i =$ position of the output board i position on the x-axis when cutting

$\forall i : y_i =$ position of the output board i position on the y-axis when cutting

Assign warehouse boards (j) to output boards (i)

$$\forall i, j : \beta_{ij} = 1 \text{ we assign warehouse board } j \text{ to output board } i \quad (50.1)$$

For any of the settings, we need to add the decision about the output board cutting mode.

Assign cutting modes (k) to output boards (i)

$$\forall i, k : \alpha_{ik} = 1 \text{ if we assign cutting mode } k \text{ to output board } i; \; 0 \text{ otherwise.} \quad (50.2)$$

Specifications

- Quantitative selection specifications:

Activities	Elements selecting	Selectable elements	Selection	Constraints
$\forall i, k : \alpha_{ik}$	Output boards ($i = 1\ldots n$)	Cutting modes ($k = 1, 2$)	=1	$\forall i : \sum_{k=1}^{2} \alpha_{ik} = 1$ (50.3)
$\forall i, j : \beta_{ij}$	Output boards ($i = 1\ldots n$)	Warehouse boards ($j = 1\ldots m$)	=1	$\forall i : \sum_{j=1}^{m} \beta_{ij} = 1$ (50.4)

- Cut the output board within the dimensions of the assigned warehouse board (propositional logic specification).

If the output board is cut from one position with $x_i = 0$ or $y_i = 0$, the cutting width C on that axis must not be taken into account; it is only taken into account if the cut starts from the inside of the axis. We must collect that in a logical calculation.

Binary Logical Calculation

$$\forall i : \pi x_i = 1 \text{ if cut with } x_i = 0; 0 \text{ otherwise} \quad (50.5)$$

$$\begin{aligned}
&\forall i : \pi x_i = 1 \text{ IF AND ONLY IF } x_i = 0\\
&\Rightarrow f_{S_v} \Rightarrow \forall i : \text{IF } \pi x_i = 1 \text{ THEN } x_i = 0\\
&\Rightarrow f_{\text{LB}} \Rightarrow \forall i : \text{IF } \pi x_i = 1 \text{ THEN } x_i \leq 0\\
&\Rightarrow f_{\text{LB}}\left(\text{UB}_{x_i} = \max_j \{\text{WB}_j\} - W_i\right) \Rightarrow\\
&\Rightarrow \forall i : x_i \leq \left(\max_j \{\text{WB}_j\} - W_i\right)(1 - \pi x_i)
\end{aligned} \tag{50.6}$$

$$\forall i : \pi y_i = 1 \text{ if cut with } y_i = 0; 0 \text{ otherwise.} \tag{50.7}$$

$$\begin{aligned}
&\forall i : \pi y_i = 1 \text{ IF AND ONLY IF } y_i = 0\\
&\Rightarrow f_{S_v} \Rightarrow \forall i : \text{IF } \pi y_i = 1 \text{ THEN } y_i = 0\\
&\Rightarrow f_{\text{LB}} \Rightarrow \forall i : \text{IF } \pi y_i = 1 \text{ THEN } y_i \leq 0\\
&\Rightarrow f_{\text{LB}}\left(\text{UB}_{y_i} = \max_j \{\text{LB}_j\} - W_i\right) \Rightarrow\\
&\Rightarrow \forall i : y_i \leq \left(\max_j \{\text{LB}_j\} - W_i\right)(1 - \pi y_i)
\end{aligned} \tag{50.8}$$

Specification modeling:

$$\begin{aligned}
&\forall i,j : \text{IF } \beta_{ij} = 1 \text{ THEN } C(1 - \pi x_i) + x_i + W_i\alpha_{i1} + L_i\alpha_{i2} \leq \text{WB}_j\\
&\Rightarrow f_{14}\left(\text{UB}_{C(1-\pi x_i)+x_i+W_i\alpha_{i1}+L_i\alpha_{i2}} = \max_j \{\text{WB}_j - W_i\}\right)\\
&\Rightarrow \forall i,j : P(1 - \pi x_i) + x_i + W_i\alpha_{i1} + L_i\alpha_{i2} \leq\\
&\text{WB}_j + \left(\max_{j'} \{\text{WB}_{j'}\} - W_i - \text{WB}_j\right)(1 - \beta_{ij})
\end{aligned} \tag{50.9}$$

$$\begin{aligned}
&\forall i,j : \text{IF } \beta_{ij} = 1 \text{ THEN } C(1 - \pi y_i) + y_i + L_i\alpha_{i1} + W_i\alpha_{i2} \leq \text{WB}_j\\
&\Rightarrow f_{14}\left(\text{UB}_{C(1-\pi y_i)+y_i+L_i\alpha_{i1}+W_i\alpha_{i2}} = \max_j \{\text{LB}_j - W_i\}\right)\\
&\Rightarrow \forall i,j : C(1 - \pi y_i) + y_i + L_i\alpha_{i1} + W_i\alpha_{i2} \leq\\
&\text{LB}_j + \left(\max_{j'} \{\text{LB}_{j'}\} - W_i - \text{LB}_j\right)(1 - \beta_{ij})
\end{aligned} \tag{50.10}$$

– Two output boards from the same warehouse board cannot intersect (propositional logic specification)

To avoid overlaps between two Tables i and i', one of the following rules must be satisfied:

- i before i' on the x-axis
- i' before i on the x-axis

- i before i' on the y-axis
- i' before i on the y-axis

And only one at least has to be satisfied, since two pieces can overlap the position with respect to one axis but never with respect to both.

$$\forall i, i'/i < i', j : \text{IF } \beta_{ij} = 1 \text{ AND } \beta_{i'j} = 1 \text{ THEN}$$
$$\begin{pmatrix} x_i + W_i\alpha_{i1} + L_i\alpha_{i2} \leq x_{i'} \text{ OR } x_{i'} + W_{i'}\alpha_{i'1} + L_{i'}\alpha_{i'2} \leq x_i \\ \text{OR } y_i + L_i\alpha_{i1} + W_i\alpha_{i2} \leq y_{i'} \text{ OR } y_{i'} + L_{i'}\alpha_{i'1} + W_{i'}\alpha_{i'2} \leq y_i \end{pmatrix}$$

$$\forall i, i'/i < i' : \omega^1_{ii'} = 1 \text{ IF AND ONLY IF } x_i + W_i\alpha_{i1} + L_i\alpha_{i2} - x_{i'} \leq 0 \quad (50.11)$$

$$\Rightarrow f_{S_v} \Rightarrow \forall i, i'/i < i' :$$
$$\text{IF } \omega^1_{ii'} = 1 \text{ THEN } \quad x_i + W_i\alpha_{i1} + L_i\alpha_{i2} - x_{i'} \leq 0$$
$$\Rightarrow f_{14}\left(\text{UB}_{x_i + W_i\alpha_{i1} + L_i\alpha_{i2} - x_{i'}} = \max_j \{\text{WB}_j\} - W_i\right) \Rightarrow \quad (50.12)$$
$$\forall i, i'/i < i' : x_i + W_i\alpha_{i1} + L_i\alpha_{i2} - x_{i'} \leq \left(\max_j \{\text{WB}_j\} - W_i\right)(1 - \omega^1_{ii'})$$

$$\forall i, i'/i < i' : \omega^2_{ii'} = 1 \text{ IF AND ONLY IF } x_{i'} + W_{i'}\alpha_{i'1} + L_{i'}\alpha_{i'2} - x_i \leq 0 \quad (50.13)$$

$$\Rightarrow f_{S_v} \Rightarrow \forall i, i'/i < i' : \text{IF } \omega^2_{ii'} = 1 \text{ THEN } \quad x_{i'} + W_{i'}\alpha_{i'1} + L_{i'}\alpha_{i'2} - x_i \leq 0$$
$$\Rightarrow f_{14}\left(\text{UB}_{x_{i'} + A_{i'}\alpha_{i'1} + L_{i'}\alpha_{i'2} - x_i} = \max_j \{\text{WB}_j\} - W_{i'}\right) \Rightarrow$$
$$\forall i, i'/i < i' : x_{i'} + W_{i'}\alpha_{i'1} + L_{i'}\alpha_{i'2} - x_i \leq \left(\max_j \{\text{WB}_j\} - W_{i'}\right)(1 - \omega^2_{ii'}) \quad (50.14)$$

$$\forall i, i'/i < i' : \omega^3_{ii'} = 1 \quad \text{IF AND ONLY IF } y_i + L_i\alpha_{i1} + W_i\alpha_{i2} - y_{i'} \leq \ 0 \quad (50.15)$$

$$\Rightarrow f_{S_v} \Rightarrow \forall i, i'/i < i' : \text{IF } \omega^3_{ii'} = 1 \text{ THEN } \quad y_i + L_i\alpha_{i1} + W_i\alpha_{i2} - y_{i'} \leq 0$$
$$\Rightarrow f_{14}\left(\text{UB}_{y_i + L_i\alpha_{i1} + W_i\alpha_{i2} - y_{i'}} = \max_j \{\text{LB}_j\} - W_i\right) \Rightarrow$$
$$\forall i, i'/i < i' : y_i + L_i\alpha_{i1} + W_i\alpha_{i2} - y_{i'} \leq \left(\max_j \{\text{LB}_j\} - W_i\right)(1 - \omega^3_{ii'}) \quad (50.16)$$

$$\forall i, i'/i < i' : \omega^4_{ii'} = 1 \quad \text{IF AND ONLY IF } y_{i'} + L_{i'}\alpha_{i'1} + W_{i'}\alpha_{i'2} - y_i \leq 0 \quad (50.17)$$

$$\Rightarrow f_{S_v} \Rightarrow \forall i,i'/i<i' : IF\ \ \omega_{ii'}^4 = 1 \ \text{ THEN } \quad y_{i'} + L_{i'}\alpha_{i'1} + W_{i'}\alpha_{i'2} - y_i \le 0$$
$$\Rightarrow f_{14}\left(\text{UB}_{y_{i'}+L_{i'}\alpha_{i'1}+W_{i'}\alpha_{i'2}-y_i} = \max_j \{\text{LB}_j\} - W_{i'}\right) \Rightarrow$$
$$\forall i,i'/i<i' : y_{i'} + L_{i'}\alpha_{i'1} + W_{i'}\alpha_{i'2} - y_i \le \left(\max_j \{\text{LB}_j\} - W_{i'}\right)\left(1-\omega_{ii'}^4\right) \tag{50.18}$$

$$\begin{aligned}&\Rightarrow f_{34} \Rightarrow \forall i,i'/i<i',j : \text{IF } \beta_{ij} = 1 \text{ AND } \beta_{i'j} = 1\\ &\text{THEN } \omega_{ii'}^1 + \omega_{ii'}^2 + \omega_{ii'}^3 + \omega_{ii'}^4 \ge 1\\ &\Rightarrow f_{37} \Rightarrow \forall i,i'/i<i',j : \text{IF } \beta_{ij} + \beta_{i'j} \ge 2\\ &\text{THEN } \omega_{ii'}^1 + \omega_{ii'}^2 + \omega_{ii'}^3 + \omega_{ii'}^4 \ge 1\\ &\Rightarrow f_{18}\left(\text{UB}_{\beta_{ij}+\beta_{i'j}} = 2\right) \Rightarrow \forall i,i'/i<i',j : \beta_{ij} + \beta_{i'j} \le 1 + \lambda_{ii'j}\end{aligned} \tag{50.19}$$

$$\begin{aligned}&\Rightarrow f_{21}\left(LB_{\omega_{ij}^1+\omega_{ij}^2+\omega_{ij}^3+\omega_{ij}^4} = 0\right) \Rightarrow\\ &\Rightarrow \forall i,i'/i<i',j : \omega_{ii'}^1 + \omega_{ii'}^2 + \omega_{ii'}^3 + \omega_{ii'}^4 \ge \lambda_{ii'j}\end{aligned} \tag{50.20}$$

- In an output board, the two sides of the reference vertex must be in ends of the warehouse board or joined with other output board cut in the same warehouse board. (This specification will not be included in the final model.)

$$\begin{aligned}&\forall i,j : \text{IF } \beta_{ij} = 1 \quad \text{THEN}\\ &\left(x_i = 0 \quad (\text{OR } (x_{i'} + W_{i'}\alpha_{i'1} + L_{i'}\alpha_{i'2} + C = x_i \text{ AND } \beta_{i'j} = 1))_{\forall i' \ne i}\right)\\ &\text{AND } \left(y_i = 0 \ (\text{OR } (\ y_{i''} + L_{i''}\alpha_{i''1} + W_{i''}\alpha_{i''2} + C = y_i \text{ AND } \beta_{i''j} = 1))_{\forall i'' \ne i}\right)\end{aligned}$$

$$\begin{aligned}&\Rightarrow f_{\text{LB}} \Rightarrow \forall i,j : \text{IF } \beta_{ij} = 1 \quad \text{THEN}\\ &\left(x_i \le 0 \ (\text{OR } (x_{i'} + W_{i'}\alpha_{i'1} + L_{i'}\alpha_{i'2} + C = x_i \text{ AND } \beta_{i'j} = 1))_{\forall i' \ne i}\right)\\ &\text{AND } \left(y_i \le 0 \ (\text{OR } (\ y_{i''} + L_{i''}\alpha_{i''1} + W_{i''}\alpha_{i''2} + C = y_i \text{ AND } \beta_{i''j} = 1))_{\forall i'' \ne i}\right)\end{aligned}$$

$$\begin{aligned}
&\Rightarrow f_{41} \Rightarrow \forall i,j : \text{IF } \beta_{ij} = 1 \quad \text{THEN} \\
&x_i \leq 0 \ \left(\text{OR}\left(x_{i'} + W_{i'}\alpha_{i'1} + L_{i'}\alpha_{i'2} + C = x_i \ \text{ AND } \beta_{i'j} = 1\right)\right)_{\forall i' \neq i} \\
&\Rightarrow f_{41} \Rightarrow \forall i,j : \text{IF } \beta_{ij} = 1 \quad \text{THEN} \\
&y_i \leq 0 \ \left(\text{OR } \left(\ y_{i'} + L_{i'}\alpha_{i'1} + W_{i'}\alpha_{i'2} + C = y_i \ \text{ AND } \beta_{i'j} = 1\right)\right)_{\forall i' \neq i}
\end{aligned} \tag{50.21}$$

$$\begin{aligned}
&\forall i, i'/i \neq i' : \delta_{ii'} = 1 \quad \text{IF AND ONLY IF} \\
&x_{i'} + W_{i'}\alpha_{i'1} + L_{i'}\alpha_{i'2} - x_i = -C
\end{aligned}$$

$$\begin{aligned}
&\Rightarrow f_{S_V} \Rightarrow \forall i, i'/i \neq i' : \text{IF } \delta_{ii'} = 1 \text{ THEN} \\
&x_{i'} + W_{i'}\alpha_{i'1} + L_{i'}\alpha_{i'2} - x_i = -C \\
&\Rightarrow f_{16}\left(\text{UB}_{x_{i'}+W_{i'}\alpha_{i'1}+L_{i'}\alpha_{i'2}-x_i} = \max_j \left\{\text{WB}_j\right\} - W_{i'}\right) \\
&\Rightarrow \ \forall i, i'/i \neq i' : x_{i'} + W_{i'}\alpha_{i'1} + L_{i'}\alpha_{i'2} - x_i \leq \\
&-C + \left(\max_j \left\{\text{WB}_j\right\} - W_{i'} + C\right) \ (1 - \delta_{ii'})
\end{aligned} \tag{50.22}$$

$$\begin{aligned}
&\Rightarrow f_{16}\left(LB_{x_{i'}+W_{i'}\alpha_{i'1}+L_{i'}\alpha_{i'2}-x_i} = W_{i'} - \left(\max_j \left\{\text{WB}_j\right\} - W_i\right)\right) \\
&\Rightarrow \forall i, i'/i \neq i' : x_{i'} + W_{i'}\alpha_{i'1} + L_{i'}\alpha_{i'2} - x_i \geq \\
&-C\delta_{ii'} + \left(W_{i'} - \left(\max_j \left\{\text{WB}_j\right\} - W_i\right)\right)(1 - \delta_{ii'})
\end{aligned} \tag{50.23}$$

$$\begin{aligned}
&\Rightarrow \forall i, i'/i \neq i' : \mu_{ii'} = 1 \quad \text{IF AND ONLY IF} \\
&y_{i'} + L_{i'}\alpha_{i'1} + W_{i'}\alpha_{i'2} - y_i = -C
\end{aligned} \tag{50.24}$$

$$\begin{aligned}
&\Rightarrow f_{S_V} \Rightarrow \forall i, i'/i \neq i' : IF \ \mu_{ii'} = 1 \text{ THEN} \\
&y_{i'} + L_{i'}\alpha_{i'1} + W_{i'}\alpha_{i'2} - y_i = -C \\
&\Rightarrow f_{16}\left(\text{UB}_{y_{i'}+L_{i'}\alpha_{i'1}+A_{i'}\alpha_{i'2}-y_i} = \max_j \left\{\text{LB}_j\right\} - W_{i'}\right) \\
&\Rightarrow \ \forall i, i'/i \neq i' : y_{i'} + L_{i'}\alpha_{i'1} + W_{i'}\alpha_{i'2} - y_i \leq \\
&-C + \left(\max_j \left\{\text{LB}_j\right\} - W_{i'} + C\right) \ (1 - \mu_{ii'})
\end{aligned} \tag{50.25}$$

$$\begin{aligned}
&\Rightarrow f_{16}\left(LB_{y_{i'}+L_{i'}\alpha_{i'1}+W_{i'}\alpha_{i'2}-y_i} = W_{i'} - \left(\max_j \left\{\text{LB}_j\right\} - W_i\right)\right) \\
&\Rightarrow \ \forall i, i'/i \neq i' : y_{i'} + L_{i'}\alpha_{i'1} + W_{i'}\alpha_{i'2} - y_i \geq \\
&-C\mu_{ii'} + \left(W_{i'} - \max_j \left\{\text{LB}_j\right\} - W_i\right)(1 - \mu_{ii'})
\end{aligned} \tag{50.26}$$

$$\Rightarrow \forall i,j : \text{IF } \beta_{ij} = 1 \quad \text{THEN } x_i \leq 0 \ \left(\text{OR}\left(\delta_{ii'} = 1 \ \text{ AND } \beta_{i'j} = 1\right)\right)_{\forall i' \neq i}$$
$$\Rightarrow \forall i,j : \text{IF } \beta_{ij} = 1 \quad \text{THEN } y_i \leq 0 \ \left(\text{OR } \left(\mu_{ii'} = 1 \ \text{ AND } \beta_{i'j} = 1\right)\right)_{\forall i' \neq i}$$
$$\Rightarrow f_{37} \Rightarrow \forall i,j : \text{IF } \beta_{ij} = 1 \quad \text{THEN } x_i \leq 0 \ \left(\text{OR}\left(\delta_{ii'} + \ \beta_{i'j} \geq 2\right)\right)_{\forall i' \neq i}$$
$$\Rightarrow f_{37} \Rightarrow \forall i,j : \text{IF } \beta_{ij} = 1 \quad \text{THEN } y_i \leq 0 \ \left(\text{OR } \left(\mu_{ii'} + \ \beta_{i'j} \geq 2\right)\right)_{\forall i' \neq i}$$
$$\Rightarrow (50.5) \Rightarrow \forall i,j : \text{IF } \beta_{ij} = 1 \quad \text{THEN } \pi x_i = 1 \ \left(\text{OR}\left(\delta_{ii'} + \ \beta_{i'j} \geq 2\right)\right)_{\forall i' \neq i}$$
$$\Rightarrow (50.7) \Rightarrow \forall i,j : \text{IF } \beta_{ij} = 1 \quad \text{THEN } \pi y_i = 1 \ \left(\text{OR}\left(\mu_{ii'} + \ \beta_{i'j} \geq 2\right)\right)_{\forall i' \neq i}$$
$$\forall i,j,i'/i \neq i' : \psi^1_{ii'j} = 1 \quad \text{IF AND ONLY IF } \delta_{ii'} + \ \beta_{i'j} \geq 2 \tag{50.27}$$

$$\forall i,j,i'/i \neq i' : \psi^2_{ii'j} = 1 \quad \text{IF AND ONLY IF } \mu_{ii'} + \ \beta_{i'j} \geq 2 \tag{50.28}$$

$$\Rightarrow f_{S_V} \Rightarrow \forall i,j,i'/i \neq i' : \text{IF } \psi^1_{ii'j} = 1 \ \text{ THEN } \delta_{ii'} + \ \beta_{i'j} \geq 2$$
$$\Rightarrow f_{S_V} \Rightarrow \forall i,j,i'/i \neq i' : \text{IF } \psi^2_{ii'j} = 1 \ \text{ THEN } \mu_{ii'} + \ \beta_{i'j} \geq 2 \tag{50.29}$$
$$\Rightarrow f_{15}\left(\text{LB}_{\delta_{ii'} + \ \beta_{i'j}} = 0\right) \Rightarrow \forall i,j,i'/i \neq i' : \delta_{ii'} + \ \beta_{i'j} \geq 2\psi^1_{ii'j}$$

$$\Rightarrow f_{15}\left(\text{LB}_{\mu_{ii'} + \ \beta_{i'j}} = 0\right) \Rightarrow \forall i,j,i'/i \neq i' : \mu_{ii'} + \ \beta_{i'j} \geq 2\psi^2_{ii'j} \tag{50.30}$$

$$\Rightarrow \forall i,j : \text{IF } \beta_{ij} = 1 \quad \text{THEN } \pi x_i = 1 \ \left(\text{OR } \psi^1_{ii'j} = 1\right)_{\forall i' \neq i}$$
$$\Rightarrow \forall i,j : \text{IF } \beta_{ij} = 1 \quad \text{THEN } \pi y_i = 1 \ \left(\text{OR } \psi^2_{ii'j} = 1\right)_{\forall i' \neq i}$$
$$\Rightarrow f_{34} \Rightarrow \forall i,j : \text{IF } \beta_{ij} = 1 \quad \text{THEN } \pi x_i + \sum_{i'/i \neq i'} \psi^1_{ii'j} \geq 1$$
$$\Rightarrow f_{34} \Rightarrow \forall i,j : IF \ \beta_{ij} = 1 \quad \text{THEN } \pi y_i + \sum_{i'/i \neq i'} \psi^2_{ii'j} \geq 1 \tag{50.31}$$
$$\Rightarrow f_{15}\left(\text{LB}_{\pi x_i + \sum_{i'/i \neq i'} \psi^1_{ii'j}} = 0\right) \Rightarrow \forall i,j : \ \pi x_i + \sum_{i'/i \neq i'} \psi^1_{ii'j} \geq \beta_{ij}$$

$$\Rightarrow f_{15}\left(\mathrm{LB}_{\pi y_i+\sum\limits_{i'/i\neq i'}\psi^2_{ii'j}}=0\right)\Rightarrow\forall i,j:\ \ \pi y_i+\sum_{i'/i\neq i'}\psi^2_{ii'j}\geq\ \beta_{ij} \tag{50.32}$$

Objective Criterion

Minimize the excess m^2 of the cut boards + the distance from the cutting points to (0, 0).

We calculate which boards have been cut:

Binary Logical Calculation

$$\forall j:\theta_j=1 \text{ if the warehouse board } j \text{ is cut; 0 otherwise.} \tag{50.33}$$

$$\forall j:\theta_j=1\quad \text{IF AND ONLY IF }\sum_{i=1}^{n}\beta_{ij}>0$$

$$\Rightarrow f_{S_V}\Rightarrow\forall j:\text{IF }\sum_{i=1}^{n}\beta_{ij}>0\text{ THEN }\theta_j=1 \tag{50.34}$$

$$\Rightarrow f_3,f_7,f_{14}\left(\mathrm{UB}_{\sum\limits_{i=1}^{n}\beta_{ij}}=n\right)\Rightarrow\forall j:\sum_{i=1}^{n}\beta_{ij}\leq n\theta_j$$

Objective Function

$$\text{Min }\sum_{j=1}^{m}\left((\mathrm{WB}_j\times\mathrm{LB}_j)\theta_j-\sum_{i=1}^{n}(W_i\times L_i)\beta_{ij}\right)+\sum_{i=1}^{n}(x_i+y_i) \tag{50.35}$$

Complete Model

$$\text{Min }\sum_{j=1}^{m}\left((\mathrm{WB}_j\times\mathrm{LB}_j)\theta_j-\sum_{i=1}^{n}(W_i\times L_i)\beta_{ij}\right)+\sum_{i=1}^{n}(x_i+y_i) \tag{50.35}$$

s.t.

$$\forall i:\sum_{k=1}^{2}\alpha_{ik}=1 \tag{50.3}$$

$$\forall i:\sum_{j=1}^{m}\beta_{ij}=1 \tag{50.4}$$

$$\forall i:x_i\leq\left(\max_j\{\mathrm{WB}_j\}-W_i\right)(1-\pi x_i) \tag{50.6}$$

$$\forall i: y_i \leq \left(\max_j \{\mathrm{LB}_j\} - W_i\right)(1 - \pi y_i) \tag{50.8}$$

$$\begin{aligned}\forall i,j: C(1-\pi x_i) + x_i + W_i\alpha_{i1} + L_i\alpha_{i2} \\ \leq \mathrm{WB}_j + \left(\max_{j'} \{\mathrm{WB}_{j'}\} - W_i - \mathrm{WB}_j\right)(1-\beta_{ij})\end{aligned} \tag{50.9}$$

$$\begin{aligned}\forall i,j: C(1-\pi y_i) + y_i + L_i\alpha_{i1} + W_i\alpha_{i2} \\ \leq \mathrm{LB}_j + \left(\max_{j'} \{\mathrm{LB}_{j'}\} - W_i - \mathrm{LB}_j\right)(1-\beta_{ij})\end{aligned} \tag{50.10}$$

$$\forall i,i'/i<i': x_i + W_i\alpha_{i1} + L_i\alpha_{i2} - x_{i'} \leq \left(\max_j \{\mathrm{WB}_j\} - W_i\right)(1-\omega^1_{ii'}) \tag{50.12}$$

$$\begin{aligned}\forall i,i'/i<i': x_{i'} + W_{i'}\alpha_{i'1} + L_{i'}\alpha_{i'2} - x_i \\ \leq \left(\max_j \{\mathrm{WB}_j\} - W_{i'}\right)(1-\omega^2_{ii'})\end{aligned} \tag{50.14}$$

$$\forall i,i'/i<i': y_i + L_i\alpha_{i1} + W_i\alpha_{i2} - y_{i'} \leq \left(\max_j \{\mathrm{LB}_j\} - W_i\right)(1-\omega^3_{ii'}) \tag{50.16}$$

$$\begin{aligned}\forall i,i'/i<i': y_{i'} + L_{i'}\alpha_{i'1} + W_{i'}\alpha_{i'2} - y_i \\ \leq \left(\max_j \{\mathrm{LB}_j\} - W_{i'}\right)(1-\omega^4_{ii'})\end{aligned} \tag{50.18}$$

$$\forall i,i'/i<i',j: \beta_{ij} + \beta_{i'j} \leq 1 + \lambda_{ii'j} \tag{50.19}$$

$$\forall i,i'/i<i',j: \omega^1_{ii'} + \omega^2_{ii'} + \omega^3_{ii'} + \omega^4_{ii'} \geq \lambda_{ii'j} \tag{50.20}$$

$$\begin{aligned}\forall i,i'/i \neq i': x_{i'} + W_{i'}\alpha_{i'1} + L_{i'}\alpha_{i'2} - x_i \\ \leq -C + \left(\max_j \{\mathrm{WB}_j\} - W_{i'} + C\right)\ (1-\delta_{ii'})\end{aligned} \tag{50.22}$$

$$\begin{aligned}\forall i,i'/i \neq i': x_{i'} + W_{i'}\alpha_{i'1} + L_{i'}\alpha_{i'2} - x_i \\ \geq -C\delta_{ii'} + \left(W_{i'} - \left(\max_j \{\mathrm{WB}_j\} - W_i\right)\right)(1-\delta_{ii'})\end{aligned} \tag{50.23}$$

$$\begin{aligned}\forall i,i'/i \neq i': y_{i'} + L_{i'}\alpha_{i'1} + W_{i'}\alpha_{i'2} - y_i \\ \leq -C + \left(\max_j \{\mathrm{LB}_j\} - W_{i'} + C\right)\ (1-\mu_{ii'})\end{aligned} \tag{50.25}$$

$$\forall i, i'/i \neq i' : y_{i'} + L_{i'}\alpha_{i'1} + W_{i'}\alpha_{i'2} - y_i$$

$$\geq -C\mu_{ii'} + \left(W_{i'} - \max_j \{\text{LB}_j\} - W_i\right)(1 - \mu_{ii'}) \quad (50.26)$$

$$\forall i, i'/i \neq i', j : \delta_{ii'} + \beta_{i'j} \geq 2\psi^1_{ii'j} \quad (50.29)$$

$$\forall i, i'/i \neq i', j : \mu_{ii'} + \beta_{i'j} \geq 2\psi^2_{ii'j} \quad (50.30)$$

$$\forall i, j : \pi x_i + \sum_{i'/i \neq i'} \psi^1_{ii'j} \geq \beta_{ij} \quad (50.31)$$

$$\forall i, j : \pi y_i + \sum_{i'/i \neq i'} \psi^2_{ii'j} \geq \beta_{ij} \quad (50.32)$$

$$\forall j : \sum_{i=1}^{n} \beta_{ij} \leq n\theta_j \quad (50.34)$$

$$\forall i : x_i \geq 0, y_i \geq 0$$

$$\forall i, k : \alpha_{ik} \text{ binaries} \quad (50.1)$$

$$\forall i, j : \beta_{ij} \textit{ binaries} \quad (50.2)$$

$$\forall i : \pi x_i, \pi y_i \text{ binaries} \quad (50.5, 50.7)$$

$$\forall i, i'/i < i' : \omega^1_{ii'}, \omega^2_{ii'}, \omega^3_{ii'}, \omega^4_{ii'} \text{ binaries} \quad (50.11, 50.13, 50.15, 5.17)$$

$$\forall i, i'/i \neq i' : \delta_{ii'} \text{ binaries} \quad (50.21)$$

$$\forall i, i'/i \neq i' : \mu_{ii'} \text{ binaries} \quad (50.24)$$

$$\forall i, i'/i \neq i', j : \psi^1_{ii'j}, \psi^2_{iii'j} \text{ binaries} \quad (50.27, 50.28)$$

$$\forall j : \theta_j \text{ binaries} \quad (50.33)$$

$$\forall i, i'/i < i', j : \lambda_{ii'j} \text{ binaries} \quad (50.36)$$

Implementation in LINGO

```
MODEL:
 DATA:
  n=@OLE('C:\Files_\50_Excel.xlsx','n');
  m=@OLE('C:\Files_\50_Excel.xlsx','m');
 ENDDATA
SETS:
 WarehouseB/1..m/:LB,WB,Tita;
 OutPutB/1..n/:L,W,x,y,Pix,Piy;
 Modes/1..2/;
 SA(OutPutB,WarehouseB):Beta;
```

```
 SM(OutPutB,Modes):Alpha;
 SS(OutPutB,OutPutB):Omega1,Omega2,Omega3,Omega4,Delta,Gamma;
 SSA(OutPutB,OutPutB,WarehouseB):Lamda,UR1,UR2;
ENDSETS
DATA:
 LB=@OLE('C:\Files_\50_Excel.xlsx','LB');
 WB=@OLE('C:\Files_\50_Excel.xlsx','WB');
 L=@OLE('C:\Files_\50_Excel.xlsx','LO');
 W=@OLE('C:\Files_\50_Excel.xlsx','WO');
 C=0.4;
@OLE('C:\Files_\50_Excel.xlsx','WBoard')=@WRITEFOR(OutPutB(i):
@SUM(WarehouseB(j):j*Beta(i,j)));
 @OLE('C:\Files_\50_Excel.xlsx','x')=x;
 @OLE('C:\Files_\50_Excel.xlsx','y')=y;
@OLE('C:\Files_\50_Excel.xlsx','Mode')=@WRITEFOR(OutPutB(i):@SUM
(Modes(k):k*Alpha(i,k)));
ENDDATA
!50.35;
MIN = @SUM(WarehouseB(j):(WB(j)*LB(j))*Tita(j)- @SUM(OutPutB(i):(W
(i)*L(i))*Beta(i,j))) + @SUM(OutPutB(i):x(i)+y(i));
!50.3;
@FOR(OutPutB(i):@SUM(Modes(k):Alpha(i,k))=1);
!50.4;
@FOR(OutPutB(i):@SUM(WarehouseB(j):Beta(i,j))=1);
!50.6;
@FOR(OutPutB(i):x(i)<=(@MAX(WarehouseB(j):WB(j))-W(i))*(1-Pix
(i)));
!50.8;
@FOR(OutPutB(i):y(i)<=(@MAX(WarehouseB(j):LB(j))-W(i))*(1-Piy
(i)));
!50.9;
@FOR(OutPutB(i):@FOR(WarehouseB(j):C*(1-Pix(i))+x(i)+W(i)*Alpha
(i,1)+L(i)*Alpha(i,2)<=WB(j)+(@MAX(WarehouseB(jj):WB(jj))-W(i)-WB
(j))*(1-Beta(i,j))));
!50.10;
@FOR(OutPutB(i):@FOR(WarehouseB(j):C*(1-Piy(i))+y(i)+L(i)*Alpha
(i,1)+W(i)*Alpha(i,2)<=LB(j)+(@MAX(WarehouseB(jj):LB(jj))-W(i)-LB
(j))*(1-Beta(i,j))));
!50.12;
@FOR(OutPutB(i):@FOR(OutPutB(ii)|i#LT#ii:x(i)+W(i)*Alpha(i,1)+L
(i)*Alpha(i,2)-x(ii)<=(@MAX(WarehouseB(j):WB(j))-W(i))*(1-Omega1
(i,ii))));
!50.14;
@FOR(OutPutB(i):@FOR(OutPutB(ii)|i#LT#ii:x(ii)+W(ii)*Alpha(ii,1)
+L(ii)*Alpha(ii,2)-x(i)<=(@MAX(WarehouseB(j):WB(j))-W(ii))*
(1-Omega2(i,ii))));
!50.16;
@FOR(OutPutB(i):@FOR(OutPutB(ii)|i#LT#ii:y(i)+L(i)*Alpha(i,1)+W
(i)*Alpha(i,2)-y(ii)<=(@MAX(WarehouseB(j):LB(j))-W(i))*(1-Omega3
```

```
(i,ii))));
!50.18;
@FOR(OutPutB(i):@FOR(OutPutB(ii)|i#LT#ii:y(ii)+L(ii)*Alpha(ii,1)
+W(ii)*Alpha(ii,2)-y(i)<=(@MAX(WarehouseB(j):LB(j))-W(ii))*
(1-Omega4(i,ii))));
!50.19;
@FOR(OutPutB(i):@FOR(OutPutB(ii)|i#LT#ii:@FOR(WarehouseB(j):Beta
(i,j)+Beta(ii,j)<=1 + Lamda(i,ii,j))));
!50.20;
@FOR(OutPutB(i):@FOR(OutPutB(ii)|i#LT#ii:@FOR(WarehouseB(j):
Omega1(i,ii)+Omega2(i,ii)+Omega3(i,ii)+Omega4(i,ii)>=Lamda(i,ii,
j))));
!50.34;
@FOR(WarehouseB(j):@SUM(OutPutB(i):Beta(i,j))<=n*Tita(j));
!50.1;
@FOR(SM(i,k):@BIN(Alpha(i,k)));
!50.2;
@FOR(SA(i,j):@BIN(Beta(i,j)));
!50.5;
@FOR(OutPutB(i):@BIN(Pix(i)));
!50.7;
@FOR(OutPutB(i):@BIN(Piy(i)));
!50.11;
@FOR(OutPutB(i):@FOR(OutPutB(ii)|i#LT#ii:@BIN(Omega1)));
!50.13;
@FOR(OutPutB(i):@FOR(OutPutB(ii)|i#LT#ii:@BIN(Omega2)));
!50.15;
@FOR(OutPutB(i):@FOR(OutPutB(ii)|i#LT#ii:@BIN(Omega3)));
!50.17;
@FOR(OutPutB(i):@FOR(OutPutB(ii)|i#LT#ii:@BIN(Omega4)));
!50.21;
@FOR(OutPutB(i):@FOR(OutPutB(ii)|i#NE#ii:@BIN(Delta)));
!50.24;
@FOR(OutPutB(i):@FOR(OutPutB(ii)|i#NE#ii:@BIN(Gamma)));
!50.27;
@FOR(OutPutB(i):@FOR(OutPutB(ii)|i#NE#ii:@FOR(WarehouseB(j):@BIN
(UR1))));
!50.28;
@FOR(OutPutB(i):@FOR(OutPutB(ii)|i#NE#ii:@FOR(WarehouseB(j):@BIN
(UR2))));
!50.33;
@FOR(WarehouseB(j):@BIN(Tita(j)));
!50.36;
@FOR(OutPutB(i):@FOR(OutPutB(ii)|i#LT#ii:@FOR(WarehouseB(j):@BIN
(Lamda))));
END
```

```
Solution Report - 50_LingoE
Global optimal solution found.
Objective value:                                 36080.00
Objective bound:                                 36080.00
Infeasibilities:                                 0.000000
Extended solver steps:                            4950674
Total solver iterations:                         50800893
Elapsed runtime seconds:                          2188.01



  Export Summary Report
---------------------
Transfer Method:       OLE BASED
Workbook:              C:\Files_\50_Excel.xlsx
Ranges Specified:              1
    Mode
Ranges Found:                  1
Range Size Mismatches:         0
Values Transferred:            8

Model Class:                                         MILP

Total variables:                    1236
Nonlinear variables:                   0
Integer variables:                   852

Total constraints:                   437
Nonlinear constraints:                 0

Total nonzeros:                     1944
Nonlinear nonzeros:                    0
```

Fig. 7.23 Problem 50: LINGO solution report

Solution

It is shown in Figs. 7.23 and 7.24.

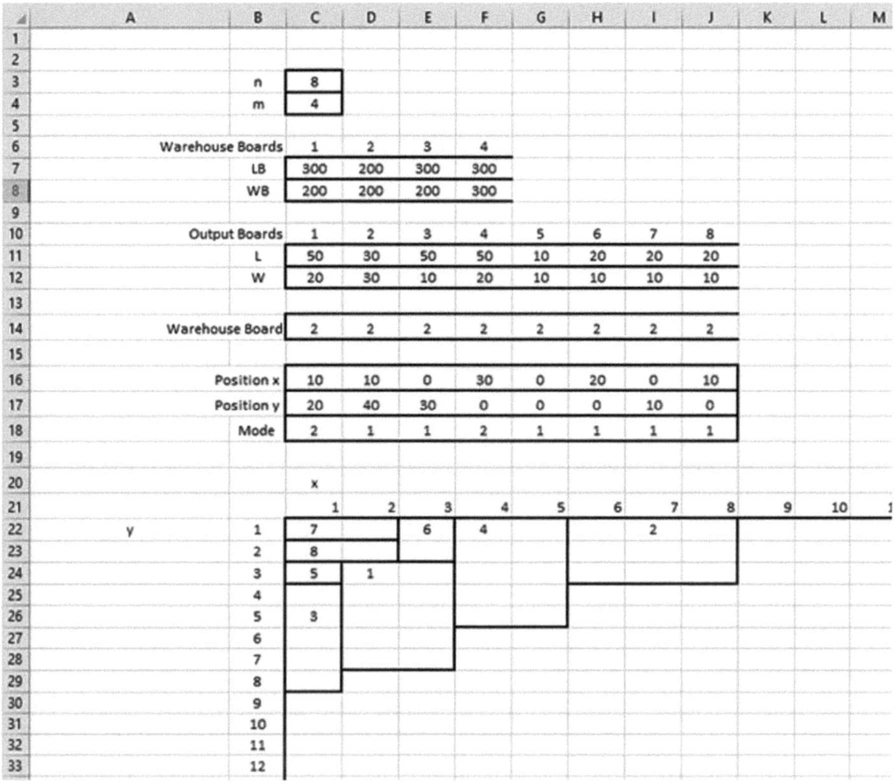

Fig. 7.24 "50_Excel.xlsx"

References

Albayrak G, Albayrak U (2015) In: Engineering Optimization IV – Rodrigues et al (eds) Optimization of the ready-mixed concrete delivery system using transportation algorithm. Taylor & Francis Group, London

Cheng CH, Feiring BR, Cheng TCE (1994) The cutting stock problem, a survey. Int J Product Econ 36(3):291–305. https://doi.org/10.1016/0925-5273(94)00045-X

Delahaye J-P (2006) The Science behind Sudoku. Sci Am 294:80–87

García JM (2021) Modelling in mathematical programming: methodology and technics. Springer, Cham. https://doi.org/10.1007/978-3-030-57250-1

Kira Games (2021) Unblock me. http://www.kiragames.com/faq.html. Accessed Dec 2021

MIX
Papier aus verantwortungsvollen Quellen
Paper from responsible sources
FSC® C105338

If you have any concerns about our products,
you can contact us on
ProductSafety@springernature.com

In case Publisher is established outside the EU,
the EU authorized representative is:
Springer Nature Customer Service Center GmbH
Europaplatz 3, 69115 Heidelberg, Germany

Printed by Libri Plureos GmbH
in Hamburg, Germany